AF389442

Clinical Pharmacology and Pharmacy of Antimicrobial Agents

Clinical Pharmacology and Pharmacy of Antimicrobial Agents

Editor

Dóra Kovács

Basel • Beijing • Wuhan • Barcelona • Belgrade • Novi Sad • Cluj • Manchester

Editor
Dóra Kovács
Department of Pharmacology
and Toxicology
University of Veterinary
Medicine Budapest
Budapest
Hungary

Editorial Office
MDPI
St. Alban-Anlage 66
4052 Basel, Switzerland

This is a reprint of articles from the Special Issue published online in the open access journal *Antibiotics* (ISSN 2079-6382) (available at: www.mdpi.com/journal/antibiotics/special_issues/Phar_Anti).

For citation purposes, cite each article independently as indicated on the article page online and as indicated below:

Lastname, A.A.; Lastname, B.B. Article Title. *Journal Name* **Year**, *Volume Number*, Page Range.

ISBN 978-3-0365-9109-4 (Hbk)
ISBN 978-3-0365-9108-7 (PDF)
doi.org/10.3390/books978-3-0365-9108-7

Contents

Review

Should Airway Interstitial Fluid Be Used to Evaluate the Pharmacokinetics of Macrolide Antibiotics for Dose Regimen Determination in Respiratory Infection?

Jianzhong Wang [1,†] , Xueying Zhou [2,†] , Sara T. Elazab [3] , Seung-Chun Park [4] and Walter H. Hsu [5,*]

1 Shanxi Key Laboratory for Modernization of TCVM, College of Veterinary Medicine, Shanxi Agricultural University, Taigu, Jinzhong 030810, China
2 Department of Veterinary Clinical Science, College of Veterinary Medicine, China Agricultural University, Beijing 100107, China
3 Department of Pharmacology, Faculty of Veterinary Medicine, Mansoura University, El-Mansoura 35516, Egypt
4 Laboratory of Veterinary Pharmacokinetics and Pharmacodynamics, College of Veterinary Medicine, Kyungpook National University, Daegu 41566, Republic of Korea
5 Department of Biomedical Sciences, College of Veterinary Medicine, Iowa State University, Ames, IA 50011-2042, USA
* Correspondence: whsu@iastate.edu
† These authors contributed equally to this work.

Abstract: Macrolide antibiotics are important drugs to combat infections. The pharmacokinetics (PK) of these drugs are essential for the determination of their optimal dose regimens, which affect antimicrobial pharmacodynamics and treatment success. For most drugs, the measurement of their concentrations in plasma/serum is the surrogate for drug concentrations in target tissues for therapy. However, for macrolides, simple reliance on total or free drug concentrations in serum/plasma might be misleading. The macrolide antibiotic concentrations of serum/plasma, interstitial fluid (ISF), and target tissue itself usually yield very different PK results. In fact, the PK of a macrolide antibiotic based on serum/plasma concentrations alone is not an ideal predictor for the in vivo efficacy against respiratory pathogens. Instead, the PK based on drug concentrations at the site of infection or ISF provide much more clinically relevant information than serum/plasma concentrations. This review aims to summarize and compare/discuss the use of drug concentrations of serum/plasma, airway ISF, and tissues for computing the PK of macrolides. A better understanding of the PK of macrolide antibiotics based on airway ISF concentrations will help optimize the antibacterial dose regimen as well as minimizing toxicity and the emergence of drug resistance in clinical practice.

Keywords: serum/plasma concentrations; interstitial concentrations; tissue concentrations; pharmacokinetics; macrolide antibiotics

---

Citation: Wang, J.; Zhou, X.; Elazab, S.T.; Park, S.-C.; Hsu, W.H. Should Airway Interstitial Fluid Be Used to Evaluate the Pharmacokinetics of Macrolide Antibiotics for Dose Regimen Determination in Respiratory Infection? *Antibiotics* **2023**, *12*, 700. https://doi.org/10.3390/antibiotics12040700

Academic Editor: Dóra Kovács

Received: 3 February 2023
Revised: 7 March 2023
Accepted: 8 March 2023
Published: 3 April 2023

1. Introduction

Macrolide antibiotics are a family of compounds featured by the existence of a macrocyclic lactone ring of ≥ 12 members [1,2]. The macrolide molecule is hydrophobic and is distributed in the extracellular fluid. In consideration of their satisfactory bioavailability via oral administration, superior tissue penetration and broad efficacy against many pulmonary pathogens, macrolides are extensively used as first-line antibiotics for the treatment of respiratory bacterial infections [3].

Pharmacokinetics (PK) describe the chronological movement of drugs within the body, i.e., the time course of the drug concentrations of serum/plasma or tissue fluid. Understanding the PK plays an important role in monitoring the antibiotic exposure in a patient. PK are most frequently evaluated by measuring the drug concentrations of serum/plasma. In addition to PK, the optimal dosing of an antibiotic also relies on the pharmacodynamics (PD) of the drug. The PD of a drug describe the relationship between drug concentration and pharmacological activity. Usually, the measurement of drug exposure is based on serum/plasma concentration–time course data [4]. Traditional PD are based on serum/plasma concentrations of a drug which achieve equilibrium with tissues. The selection of antibiotic concentrations aims to obtain an ideal exposure that should maximize the antibacterial activity and minimize antibiotic resistance [4–7]. Thus, most investigators have focused on the drug concentrations of serum/plasma for the PK/PD studies. However, the interstitial tissue is a site invaded by most bacteria [8]. The concentration of an antibiotic in the interstitial fluid (ISF) of the target tissue is essential for assessing the antibacterial effects [8]. Therefore, researchers realize that only the free (unbound) antibiotic concentration in the ISF of the target site is in charge of the antibacterial activity and is more applicable in determining clinical efficacy than serum/plasma concentration [9–12]. However, for most antibiotics, serum/plasma concentrations have been used to calculate PK/PD for the determination of the optimal dose regimens [11]. Meanwhile, the traditional PD parameters based on the serum/plasma concentration of a macrolide antibiotic are not suitable for the management of respiratory infections due to their much higher concentrations in the respiratory tract than serum/plasma [4]. Plasma/serum concentrations of macrolide antibiotics (e.g., clarithromycin, tildipirosin, gamithromycin, tilmicosin, telithromycin, and tulathromycin) in animals following the administration of recommended doses are, overall, substantially lower than their minimum inhibitory concentrations (MICs) [13]. The aforementioned studies indicated that the length of time when the drug concentration remains higher than MIC of the pathogen at the infection site, which provides more therapeutically relevant data than depending on serum/plasma concentrations [14].

This review aims to address the use of serum/plasma concentrations and pulmonary tissue and ISF concentrations for the PK/PD of macrolide antibiotics.

2. ISF Concentrations of Macrolide Antibiotics in the Lower Respiratory Tract

For most drugs in general, the measurement of their concentrations in the serum/plasma is the surrogate for drug concentrations in target tissues for therapy [4]. However, most infections appear in tissues instead of the blood [15]. There is increasing interest in the relationship between the PK and PD of antimicrobial agents [11,16,17], such as MIC, C_{max}/MIC, and AUC_{24}/MIC, which rely on the serum/plasma concentration as the PK input value and MIC as the PD input value [9]. Further understanding of PK and PD is attainable via meticulous PK investigations, the use of the free drug concentration of serum/plasma in confirming PK values, and the analysis of drug concentrations of tissues and ISF [18]. The precise assessment of the drug concentration at the infection site is necessary for the optimal therapy in patients. Measurements in special compartments (such as epithelial lining fluid) or confirming concentrations in ISF contributes to our understanding of macrolide concentrations at the infection site, thereby leading to the high therapeutic efficacy of macrolides. Following a macrolide administration, different tissues may contain different macrolide concentrations. In a mouse model, it was found that after clarithromycin administration to *Streptococcus pneumoniae*-inoculated mice, the incubation of lung and thigh tissues yielded very different bacteria counts: a drastic reduction in the lung and an increase in the thigh [19]. Thus, consistent bacterial killing was observed in the lung model of infection whereas no drug effect was seen in the thigh model. These results implied that following the macrolide administration, its concentration is tissue-dependent. The lungs may have a much higher macrolide concentration than the thigh. Others also reported that lung tissues contain higher macrolide concentrations than those of muscle [20], fat [21], and skin [22]. In addition, higher clarithromycin concentrations were found in pulmonary epithelial lining fluid (PELF) than the serum of mice [19]. In another mouse study, ~19-fold higher clarithromycin concentrations were found in lungs than plasma following a 100 mg/kg clarithromycin dose [23]. The unique characteristic of macrolide antibiotics with a higher concentration in lungs and PELF than serum/plasma is an exciting phenomenon.

A series of physicochemical factors might facilitate the pulmonary preference of macrolides, including low molecular weight and an extensive level of dissociation/ionization at the plasma pH of 7.4 [24,25]. According to the supposable concept of ion trapping, basic/ionized chemicals are easily trapped in the acidic environment of organelles after being transported into cells [26,27]. Macrolides are chemically lipophilic and basic (PK$_a$ > 7), with a low molecular weight of <1 KDa [28]. Normal airway surface liquid and alveolar subphase fluid (pH of 6.92 ± 0.01) are both more acidic than plasma [29]. The presence of pneumonia in bronchi was associated with a lower pH than that in noninfected bronchi: 6.48 ± 0.12 vs. 6.69 ± 0.13 (p < 0.05) [30]. The low degree of ionization under the plasma pH is likely to promote a higher drug concentration in acidic sites, e.g., the PELF of patients with pneumonia [29,31]. Therefore, the higher lung concentrations of macrolide antibiotics may yield higher clinical efficacy. However, a higher local drug concentration is not always consistent with the clinical efficacy, as local situations might be an influential factor of drug activity. Erythromycin may be accumulated markedly in an acidic site, but it is unstable at a low pH and may turn into a dysfunctional molecule in an acidic environment [32]. In addition, the capability of drug penetration into the extravascular space is influenced by a number of factors, including the extent of plasma protein and tissue binding, the drug's molecular charge and size as well as lipid solubility, and blood flow at the site of infection [15,33]. Macrolide antibiotics are basic compounds, poorly soluble in water, which are mostly absorbed in the alkaline intestinal environment [34]. Macrolides are highly liposoluble and consequently penetrate well into the lower respiratory tract and its secretions [34]. Plasma protein binding is variable from one macrolide compound to another. At therapeutic concentrations, protein-bound erythromycin accounts for 80–90% of the total drug present in the blood, and the protein-bound fraction is 95% for roxithromycin [34]. Protein binding appears to affect PK parameters, e.g., decreasing distribution volume and renal elimination. It is likely that the lack of plasma protein in PELF plays a role in the higher PELF concentration of a macrolide than serum/plasma. Macrolides reversibly penetrate cells and are highly partitioned into many cells [33]. Particularly in the case of azithromycin, the long serum half-life is partially a result of this intracellular partitioning [33]. Moreover, one of the extravascular delivery mechanisms of macrolides is mediated via phagocytes, which is considered to be an essential distribution pathway during respiratory infections [22, 26,27].

As a primary component of the immune defense system, PELF is distributed continuously along the lower respiratory tract and is a potential target of bacterial infection [28]. Compared to plasma, the higher PELF concentration of tulathromycin helps improve antibacterial efficacy during respiratory infections, which is similar to other macrolides [28,35,36]. Macrolides are also inclined to concentrate in bronchoalveolar lavage (BAL) cells and PELF much more than plasma [24,28].

The PK/PD studies based on PELF will aid in understanding the interpretative value of drug concentrations of PELF. Additionally, understanding the relationship between serum/plasma concentrations and concentrations at the infection/target site may be difficult. Can serum/plasma concentrations of a macrolide antibiotic reflect those in the ISF of the target tissue, particularly the respiratory tract? How much difference is there among the serum/plasma concentration, PELF concentration, and pulmonary tissue concentration of a macrolide antibiotic? The following section will address this question.

3. Concentrations of Macrolide Antibiotics in Plasma/Serum, Airway Fluid, and Tissues

Since the respiratory tract tissue and its epithelial lining fluid contain higher macrolide concentrations than plasma/serum, tremendous research interest has been focused on the area of macrolide concentrations in the airway tissues and associated ISF [12,37]. However, the PK of macrolide antibiotics (e.g., erythromycin, tylosin, azithromycin, clarithromycin, tildipirosin, gamithromycin, tilmicosin, and tulathromycin) in plasma/serum and target tissues, particularly those of the respiratory tract, have not been fully elucidated. Macrolide concentrations are ~10-fold higher in the airway epithelial fluid than in plasma [38]. In a human study, the researchers examined the pharmacokinetics of azithromycin in plasma, lung tissue and PELF in patients after oral administration of 500 mg azithromycin for 3 d. This study examined the pharmacokinetics of azithromycin in plasma, lung tissue, and PELF in patients after the oral administration of 500 mg azithromycin for 3 d. It was found that the plasma azithromycin concentrations were only ~10% and ~1% those of bronchial fluid (BF) and lung, respectively (Figure 1) [39]. The results showed that after azithromycin administration, the drug concentrations were: lung > PELF > plasma. In another two human studies, upon the administration of a single 500 mg oral dose of azithromycin, the tissue concentrations exceeded the minimum inhibitory concentration that inhibited 90% of likely pathogens (MIC_{90}), and phagocytic concentrations reached >200 times serum concentrations [40,41]. In another human study, concentrations of clarithromycin and azithromycin in PELF exceeded serum concentrations by 20-fold 24 h after the last dose of drug administration [42]. Telithromycin also has excellent penetration into bronchopulmonary tissues. In humans, concentrations of telithromycin in PEFL exceeded serum concentrations by 12-fold 24 h after the last dose of drug administration [42,43].

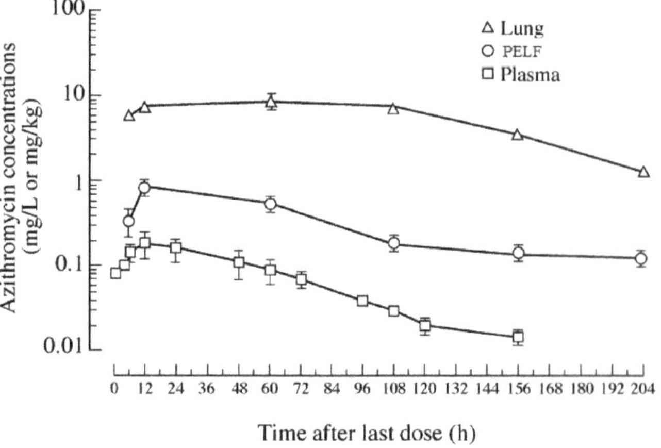

Figure 1. Concentrations vs. time semilogarithmic plot of azithromycin 500 mg per human patient once daily for 3 days in lung tissue, PELF and plasma. (Modified from Danesi et al. [39]). Copyright © 2003, Oxford University Press.

This phenomenon has been documented in cattle after the administration of tildipirosin as well [13,44,45]. The findings in cattle by Menge et al. [44] showed that the tildipirosin concentrations in lungs collected postmortem (4–240 h after dosing) exceeded those in BF (the ratio of lung/BF concentrations was ~2.5:1) and the ratio of lung/plasma was 27.6:1–214.5:1 (Figure 2). The results showed that after tildipirosin administration, the drug concentrations were: lung > PELF > plasma. The ratio of tildipirosin concentrations of BF/plasma determined from the same animals increased from 5.2:1 (4 h) to 72.3:1 (240 h or 10 days after administration) and then declined to 56.0:1 at the last collection time point of BF and plasma (504 h or 21 days) [44]. The plasma concentrations of tildipirosin were below the MIC_{90} of all three pathogens throughout the study. In contrast, the lung had drug concentrations higher than the MIC_{90} of *H. somni* for 18 d and the MIC_{90} of 2 other pathogens for >28 d. PELF had drug concentrations higher than the MIC_{90} of *H. somni* for 3 d and the MIC_{90} of 2 other pathogens for 21 d [44].

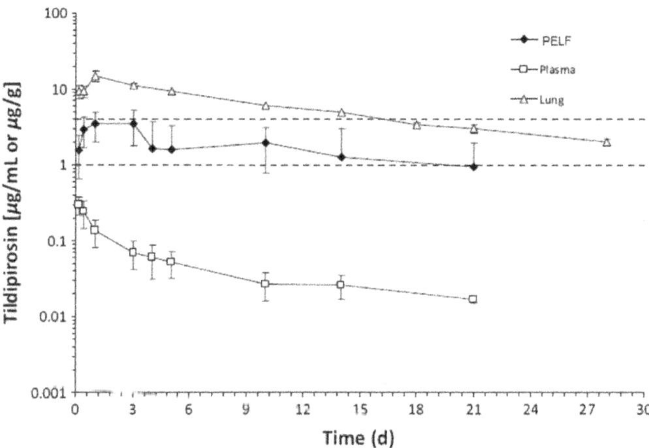

Figure 2. Tildipirosin concentration vs. time (d) in bovine plasma (µg/mL), lung tissue, and PELF (µg/g) from cattle following a single SC administration at 4 mg/kg body weight. The dotted lines represent the MIC_{90} for *Mannheimia haemolytica* and *Pasteurella multocida* (1 µg/mL) and for *Histophilus somni* (4 µg/mL) (Modified from [44]). Copyright © 2012 John Wiley & Sons, Inc.

Additionally, the concentrations of gamithromycin (Figures 3 and 4) [46,47] and tulathromycin (Figure 5) in bovine lung tissues were also much higher than those of plasma [48]. The results from Giguere, S. et al. (2011) showed that after gamithromycin administration, the drug concentrations were: BAL cells > lung > PELF > plasma [46]. Gamithromycin concentrations in BAL cells and blood neutrophils were 26–732 and 33–563 times higher than concurrent plasma concentrations, respectively. The ratios ranged 4.7:1–127:1 for PELF/plasma, 16:1–650:1 for lung tissue/plasma, and 3.2:1–2135:1 for BAL/plasma [46]. Both BAL cells and lung had drug concentrations above MIC_{90} for >14 d; PELF had drug concentrations higher than MIC_{90} for >6 d; and plasma had drug concentrations below MIC_{90} throughout the study [46]. The results from Berghaus, L.J. et al. (2012) showed that after gamithromycin administration, the drug concentrations were: BAL cells = neutrophils > PELF > plasma [47]. Both BAL cells and neutrophils had drug concentrations above the MIC_{90} of both pathogens for >336 h (14 d); PELF had drug concentrations higher than the MIC_{90} of *R. equi* for 48 h (2 d) and higher than the MIC_{90} of *S. zooepidemicus* for >144 h (6 d). Plasma had drug concentrations below the MIC_{90} of both pathogens throughout the study [47]. The plasma concentrations of tulathromycin in cattle were substantially lower than the lung tissue concentrations, with AUC_{0-360} (area under the plasma concentration–time curve) values for PELF, PELF cells, and lung homogenate of cattle (Figure 5) [48]. The results showed that after tulathromycin administration, the drug concentrations were: PELF cells > lungs > PELF > plasma. Likewise, tulathromycin concentrations in PELF were ~10 times higher than plasma concentrations with the IV/IM route in horses [49]. The AUC_{0-17} days in BF and PELF were 223 and 90 times, respectively, the corresponding values of plasma. The PELF and BF concentration profiles of tulathromycin revealed that there was disparity in the local PK of tulathromycin at various anatomical structures of the lung. Therefore, drug concentrations in PELF may not be a precise alternative of drug concentrations for different pulmonary compartments [50]. The equilibrium between plasma–BF and plasma–PELF does not occur instantaneously, and the concentration–time profiles in both compartments are different [51,52]. The PELF and BF drug concentrations are mainly affected by the alveolar area and bronchial section, respectively. The larger blood flow and surface area of the airway compartment compared with the alveolar and bronchial areas may account for, at least partially, the differences in tulathromycin profiles at those two lung compartments [50].

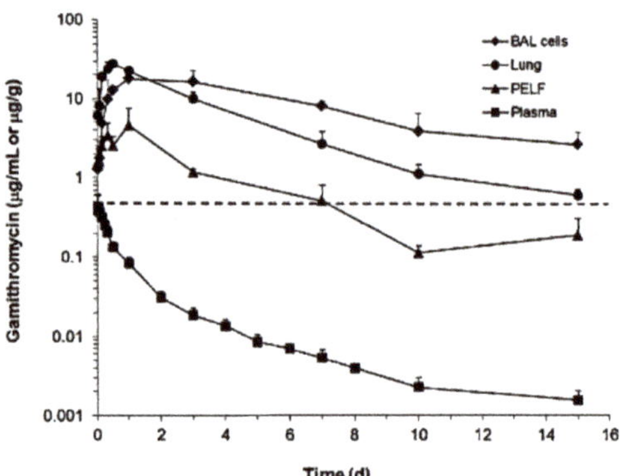

Figure 3. Gamithromycin concentrations in plasma, bronchoalveolar lavage (BAL) cells and pulmonary epithelial lining fluid (PELF) (µg/mL) and lung tissue (µg/g) of healthy Angus calves following gamithromycin administration (6 mg/kg, SC). The dotted line represents the MIC_{90} for *Mannheimia haemolytica* (0.5 µg/mL) [46].

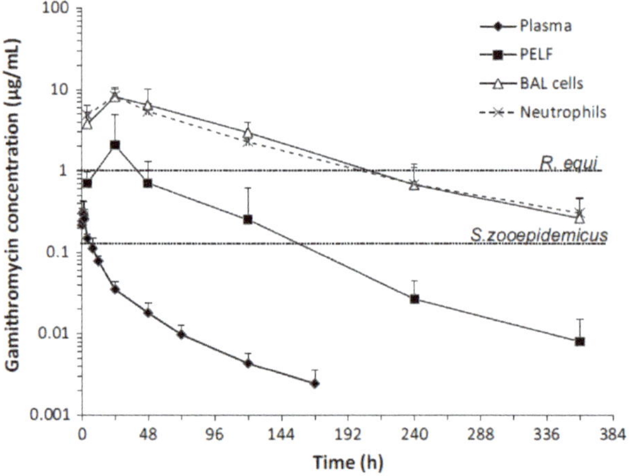

Figure 4. Gamithromycin concentrations in plasma, BAL cells, PELF, and neutrophils of 6 foals following a single IM dose of gamithromycin (6 mg/kg). The dotted horizontal lines represent the MIC_{90} of *Rhodococcus equi* (1 µg/mL) and *Streptococcus zooepidemicus* (0.125 µg/mL) isolates [47]. Copyright © 2012 John Wiley & Sons, Inc.

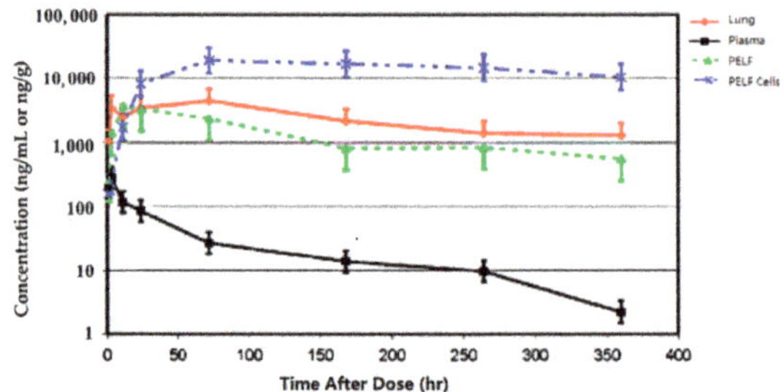

Figure 5. Tulathromycin concentrations for 360 h following a single 2.5 mg/kg IM dose to cattle (Modified from Cox et al. [48]).

Similarly, the C_{max} and AUC of tilmicosin in foals were higher for PELF than for serum (Figure 6) [53]. The results showed that after tilmicosin administration, the drug concentrations were: lung > BAL cells > PELF = neutrophils > serum. Following intragastric administration, the ratios of PELF/plasma telithromycin concentrations in foals were 5.89:1 and 5.64:1 at 4 and 24 h, respectively [54].

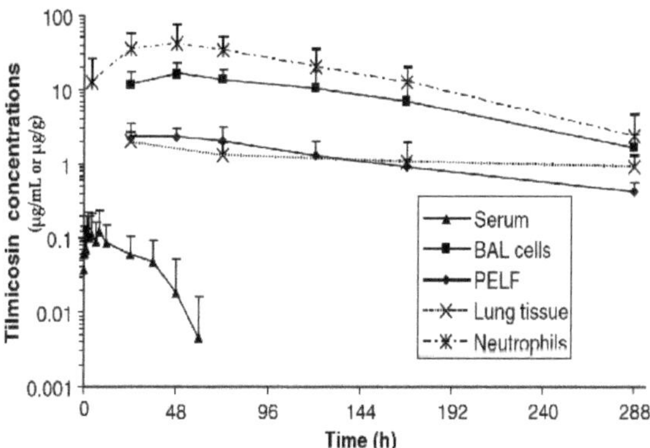

Figure 6. Tilmicosin concentrations in serum, bronchoalveolar lavage (BAL) cells, pulmonary epithelial lining fluid (PELF) (µg/mL), and lung tissue (µg/g) of 7 foals following a single IM dose of tilmicosin (10 mg/kg) [53]. Copyright © 2006 John Wiley & Sons, Inc.

The summary of the ratio thereof and the selected PK parameters based on the mean lung/PELF and plasma concentrations of several macrolide antimicrobials are shown in Table 1.

Table 1. Summary of selected PK parameters determined based on mean macrolide antibiotic concentrations in plasma, lung, and pulmonary epithelial lining fluid (PELF).

Drug	Species	Regimen	PK Parameter	Lung	Plasma	PELF	Ratio of Lung/Plasma	Ratio of PELF/Plasma	References
Tidipirosin	Cattle	SC 4 mg/kg BW	**PK Parameter (units)**						[44]
			AUC_{last} (µg·h/mL)	3846.735	24.208	882.45	159	37	
			AUC_{inf} (µg·h/mL)	4497.563	28.973	1233.75	155	43	
			$T_{1/2}$ (h)	242	216	267	1	1	
Gamithromycin	Cattle	SC 6 mg/kg BW	**PK Parameter (units)**						[46]
			AUC_{last} (µg·h/mL)	2154	7.82	334	275.4v	42.7	
			AUC_{inf} (µg·h/mL)	2235	7.95	348	281.1	43.7	
			C_{max} (µg·h/mL)	27.8	0.433	4.61	64.6	3.2	
			T_{max} (h)	12.01	1.00	24.0	12.01	24	
			$T_{1/2}$ (h)	93.0	62.0	-	1.5	NR	
Tulathromycin	Cattle	SC 2.5 mg/kg BW	**PK Parameter (units)**						[48]
			AUC0-360 (µg·h/mL)	867	9.26	492	93.6	53.13	
			C_{max} (ng/mL)	4510	277	3730	16.28	35.13	
			T_{max} (h)	72	3	11	24	3.67	
			$T_{1/2}$ (h)	279	64	330	4.35	5.15	
Tulathromycin	Pig	IM 2.5 mg/kg BW	**PK Parameter (units)**						[50–52]
			AUC_{last} (ng·h/mL)	NR	5670	435000	NR	76.7	
			AUC_{inf} (ng·h/mL)	NR	5650	473500	NR	83.8	
			C_{max} (ng/mL)	NR	458	5235	NR	11.4	
			$T_{1/2}$ (h)	NR	55.82	97.65	NR	1.75	

Table 1. *Cont.*

Drug	Species	Regimen	PK Parameter	Lung	Plasma	PELF	Ratio of Lung/Plasma	Ratio of PELF/Plasma	References
Azithromycin	human	PO 500 mg/once daily for 3 days	**PK Parameter (units)**						[39]
			AUClast (mg·h/L)	1245.4	11.62	70.29	107.2	6.04	
			Cmax (mg/L)	8.93	0.18	0.83	5.55	4.6	
Azithromycin	Foals	IG 10 mg/kg BW	**PK Parameter (units)**						[55]
			AUCinf (μg·h/mL)	NR	7.70	247	NR	35.2	
			C_{max} (μg/mL)	NR	0.83	10.00	NR	5.4	
			$T_{1/2}$ (h)	NR	25.7	34.8	NR	1.35	
Clarithromycin	Foals	IG 10 mg/kg BW	**PK Parameter (units)**						[55]
			AUC_{last} (mg·h/L)	NR	4.76	629	NR	132.14	
			C_{max} (mg/L)	NR	0.94	48.96	NR	52	
Tilmicosin	Foals	IM 10 mg/kg BW	**PK Parameter (units)**						[53]
			AUC_{inf} (μg·h/mL)	711	5.76	461	123.4	80	
			C_{max} (μg/mL)	1.90	0.19	2.91	10	15.3	
			$T_{1/2}$ (h)	193.3	18.4	73.1	10.5	3.97	

PO = Oral route; SC = subcutaneous route; IV = intravenous route6; IG = intragastric route; BW = body weight; NR = not reported; Conc = concentration AUC_{last}, area under the plasma concentration–time curve from time 0 to the last quantifiable timepoint (t_{last}); AUC_{inf}, area under the plasma concentration–time curve from time 0 to infinity; C_{max}, maximum plasma concentration; T_{max}, time of occurrence of C_{max}; $t_{1/2}$, elimination half-life.4. Which Matrix (Lung Tissue, ISF, or Plasma/Serum) Plays a Vital Role in Determining Macrolide Antibiotic Dose Regime?

Because macrolide antibiotics achieve concentrations in lung tissue and ISF that are many times the plasma/serum concentrations, the plasma-free drug concentrations may never reach the lung ISF level. In addition, the vast majority of bacterial pathogens are found in the ISF of the respiratory tract. These phenomena suggest that the macrolide concentration of the airway ISF is a more appropriate predictor of antibacterial activity than either lung tissue or plasma/serum concentrations. Furthermore, pathophysiological responses to infection, including increased phagocyte trafficking to the ISF and vascular permeability, may facilitate local drug delivery [56]. Therefore, the plasma/serum concentration of macrolide antibiotics is not an ideal indicator of in vivo efficacy against pulmonary bacterial infections [57]. It has been stated that the concentrations of lung tissues and ISF are more relevant than those of plasma/serum in determining the outcome of the treatment with a macrolide antibiotic [58]. On the other hand, Toutain et al. had a different opinion in an article published in 2017 [59]. In this study, they found that the ISF collected from the subcutaneous tissue after tulathromycin administration in calves had lower antibiotic concentrations than plasma. Based on these results, they recommended "standard PK/PD concepts can be applied to determine a regimen for a macrolide". However, the ISF of different tissues has different macrolide concentrations; lungs have higher concentrations than muscle, fat, and skin [20–22]. In addition, pulmonary ISF concentrations of a macrolide are much higher than those of plasma [38]. It is possible that some of the reported high PELF concentrations of macrolides are due to the contamination by lysed airway lining cells during bronchoalveolar lavage [60]. However, much higher macrolide concentrations were found in PELF than those of serum/plasma using microdialysis, a minimally invasive ISF collection technique [51,58]. Nevertheless, higher PELF concentrations of macrolides can be collected without significant contamination by airway lining cells. Therefore, we suggest the PK/PD paradigm using PELF/BF concentrations as input values be applied to determine the optimal dose regimen for a macrolide antibiotic in the treatment of pulmonary bacterial infections.

How do pharmaceutical companies determine the optimal dose regimen of macrolide antibiotics? The determination of the optimal dose regimen for antibiotics has a clinical cure as its aim. It is widely acceptable that the optimal dosage for clinical use can only be established in a clinical setting using a dose titration approach [59]. The results of drug development studies have demonstrated that the pulmonary ISF concentrations of a number of macrolide antibiotics (e.g., tildipirosin (NADA141-334), gamithromycin (NADA 141-328), tilmicosin (NADA141-361), and tulathromycin (NADA 141-244)) in animals are consistently higher than the plasma concentrations of these antibiotics. In addition, following the designated dose regimen by the pharmaceutical company, the tissue ISF levels of macrolide antibiotics are maintained above the MICs of major pathogens for an extended period of time (see the aforementioned NADAs); this would justify the clinical use of the designated dose regimen for the macrolide antibiotic. However, to the best of our knowledge, the ISF-based PK paradigm has not been used for the determination of PD as well as optimal dose regimen for macrolide antibiotics in clinical practice. In light of the challenges to the applicability of the PK/PD paradigm to macrolides, the use of lung ISF-derived PK data should improve the optimal dose regimen determination of a macrolide antibiotic for the treatment of lower respiratory tract bacterial infection.

4. Approaches for Airway ISF Collection

Since the ISF of the respiratory tract is the target site for most pathogenic bacteria and both BF and PELF contain much higher macrolide concentrations than plasma/serum, many more studies are warranted to evaluate the relevance of macrolide concentrations in the ISF of the respiratory tract. BF and PELF have been considered the principal target sites of the bacterial infections of the respiratory tract. The methods employed for collecting BF and PELF include the direct collection of BF, the pulmonary tissue homogenization technique, the BAL technique, and the microdialysis technique [60]. The tissue homogenization technique is no longer considered a valid technique for collecting ISF because the supernatant of the homogenate harvested after centrifugation produces a sample containing both intracellular and extracellular fluid. This approach overestimates drug concentrations, particularly those of macrolides in the extracellular space, because the intracellular macrolide concentration is much higher than that of ISF [60]. In contrast, the use of BAL to collect BF and PELF will avoid the contamination associated with the tissue homogenization technique. However, the BAL technique including the use of bronchoscopy is associated with technical complexity [60]. Different investigators may have different results with the main variations being the dwell time, the aspiration pressure (high pressure may damage epithelial linings), the volume of fluid injected, and the number of BAL aspirates collected [61]. To circumvent the problem associated with the BAL technique, a number of novel techniques including microdialysis can be used to confirm drug concentrations in ISF [62]. Free drug concentrations in ISF can be quantified over time using microdialysis; this technique is the gold standard in both human medicine [63] and veterinary medicine [64]. The technique of microdialysis enables the monitoring of a variety of molecules in ISF [65,66]. Microdialysis is a minimally invasive sampling technique employing membrane probes implanted in the tissue of interest in awake and freely moving subjects [67]. Microdialysis is a technique that uses a probe with a semipermeable membrane to insert into a tissue of interest. The probe is constantly flushed with a physiological buffer solution that allows the drug under study to leak out of the membrane and to be analyzed in the solution. In vivo microdialysis is a common in vivo method that measures the unbound (free) drug in the extracellular fluid of various tissues. The method uses a porous fiber membrane that allows the free drug molecules to cross the membrane along the concentration gradient. Microdialysis is performed by placing the probe made of a hollow fiber membrane into the extracellular fluid of the tissue where the free drug is to be measured. The probe has two tubes attached: one for the inlet and the other for the outlet where a physiological buffer flows through. As the buffer flows, the free drug molecules leak out of the pores into the outlet tube, and thus they can be detected [68]. The details of this technique are described by others [68].

5. Conclusions

Although the change in plasma/serum concentration–time and the change in airway ISF–concentration–time of macrolide antibiotics are proportional, macrolide antibiotics in plasma/serum do not reflect the antibacterial activity of the airway ISF. Thus, if the ISF concentrations of a macrolide antibiotic are efficiently and precisely collected using the least invasive method, e.g., microdialysis, practical PK/PD parameters can be obtained for such a study. Nevertheless, the PK data of airway ISF, the site of bacterial infection, are more important in setting the optimal dose regimen of a macrolide than the PK data of plasma/serum.

Author Contributions: Substantial contributions to the research and preparation of the manuscript and figures, J.W., S.T.E. and X.Z.; substantial contributions to the conception, revision, and critical evaluation of the content, S.-C.P. and W.H.H. All authors have read and agreed to the published version of the manuscript.

Funding: This work was supported by the Fund Program for the Scientific Activities of Selected Returned Overseas Professionals in Shanxi Province (20210012); Central Funds Guiding the Local Science and Technology Development in Shanxi Province (YDZJSX2021A034); the Project of Scientific Research for Excellent Doctors, Shanxi Province, China (SXBYKY2021047); the Research Fund (Clinical Diagnosis and Treatment of Pet) for Young College Teachers in Ruipeng Commonwealth Foundation (RPJJ2020021); and the Project of Science and Technology Innovation Fund of Shanxi Agricultural University (2021BQ06).

Institutional Review Board Statement: Not applicable.

Informed Consent Statement: Not applicable.

Data Availability Statement: Not applicable.

Conflicts of Interest: The authors declare no conflict of interest.

Abbreviations

BF	bronchial fluid
BAL	bronchoalveolar lavage cells
ISF	interstitial fluid
MIC	minimum inhibitory concentrations
PELF	pulmonary epithelial lining fluid
PK	pharmacokinetics
PD	pharmacodynamics

References

1. Spagnolo, P.; Fabbri, L.M.; Bush, A. Long-term macrolide treatment for chronic respiratory disease. *Eur. Respir. J.* **2013**, *42*, 239–251. [CrossRef]
2. Mazzei, T.; Mini, E.; Novelli, A.; Periti, P. Chemistry and mode of action of macrolides. *J. Antimicrob. Chemother.* **1993**, *31* (Suppl. C), 1–9. [CrossRef]
3. Bearden, D.T.; Rodvold, K.A. Penetration of macrolides into pulmonary sites of infection. *Infect. Med.* **1999**, *16*, 480A–484A.
4. Drusano, G.L. Infection site concentrations: Their therapeutic importance and the macrolide and macrolide-like lass of antibiotics. *Pharmacotherapy* **2005**, *25*, 150S–158S. [CrossRef]
5. Barza, M. Anatomical barriers for antimicrobial agents. *Eur. J. Clin. Microbiol. Infect. Dis.* **1993**, *12* (Suppl. S1), S31–S35. [CrossRef]
6. Barza, M. Pharmacokinetics of antibiotics in shallow and deep compartments. *J. Antimicrob. Chemother.* **1993**, *31* (Suppl. D), 17–27. [CrossRef] [PubMed]
7. Landersdorfer, C.B.; Nation, R.L. Limitations of antibiotic MIC-based PK-PD metrics: Looking back to move forward. *Front. Pharmacol.* **2021**, *12*, 3024. [CrossRef] [PubMed]
8. Matzneller, P.; Krasniqi, S.; Kinzig, M.; Sörgel, F.; Hüttner, S.; Lackner, E.; Müller, M.; Zeitlinger, M. Blood, Tissue, and intracellular concentrations of azithromycin during and after end of therapy. *Int. J. Antimicrob. Agents.* **2013**, *57*, 1736–1742. [CrossRef] [PubMed]
9. Liu, P.; Müller, M.; Derendorf, H. Rational dosing of antibiotics: The use of plasma concentrations versus tissue concentrations. *Int. J. Antimicrob. Agents* **2002**, *19*, 285–290. [CrossRef]
10. Toutain, P.L.; Bousquet-Melou, A. Free drug fraction vs. free drug concentration: A matter of frequent confusion. *J. Vet. Pharmacol. Ther.* **2002**, *25*, 460–463. [CrossRef]
11. Toutain, P.L.; del Castillo, J.R.E.; Bousquet-Mélou, A. The pharmacokinetic–pharmacodynamic approach to a rational regimen for antibiotics. *Res. Vet. Sci.* **2002**, *73*, 105–114. [CrossRef] [PubMed]

12. Gonzalez, D.; Schmidt, S.; Derendorf, H. Importance of relating efficacy measures to unbound drug concentrations for anti-infective agents. *Clin. Microbiol. Rev.* **2013**, *26*, 274–288. [CrossRef]
13. Rose, M.; Menge, M.; Bohland, C.; Zschiesche, E.; Wilhelm, C.; Kilp, S.; Metz, W.; Allan, M.; Ropke, R.; Nurnberger, M. Pharmacokinetics of tildipirosin in porcine plasma, lung tissue, and bronchial fluid and effects of test conditions on in vitro activity against reference strains and field isolates of Actinobacillus pleuropneumoniae. *J. Vet. Pharmacol. Ther.* **2013**, *36*, 140–153. [CrossRef] [PubMed]
14. Huang, R.A.; Letendre, L.T.; Banav, N.; Fischer, J.; Somerville, B. Pharmacokinetics of gamithromycin in cattle with comparison of plasma and lung tissue concentrations and plasma antibacterial activity. *J. Vet. Pharmacol. Ther.* **2010**, *33*, 227–237. [CrossRef]
15. Giguere, S.; Tessman, R.K. Rational dosing of antimicrobial agents for bovine respiratory disease: The use of plasma versus tissue concentrations in predicting efficacy. *Int. J. Appl. Res. Vet. M* **2009**, *9*, 342–355.
16. Ball, P.; Baquero, F.; Cars, O.; File, T.; Garau, J.; Klugman, K.; Low, D.E.; Rubinstein, E.; Wise, R. Antibiotic therapy of community respiratory tract infections: Strategies for optimal outcomes and minimized resistance emergence. *J. Antimicrob. Chemother.* **2002**, *49*, 31–40. [CrossRef] [PubMed]
17. Togami, K.; Chono, S.; Morimoto, K. Distribution characteristics of clarithromycin and azithromycin, macrolide antimicrobial agents used for treatment of respiratory infections, in lung epithelial lining fluid and alveolar macrophages. *Biopharm. Drug Dispos.* **2011**, *32*, 389–397. [CrossRef]
18. Di, L.; Kerns, E.H. Chapter 37—Pharmacokinetic Methods. In *Drug-Like Properties*, 2nd ed.; Di, L., Kerns, E.H., Eds.; Academic Press: Boston, MA, USA, 2016; pp. 455–461.
19. Maglio, D.; Capitano, B.; Banevicius, M.A.; Geng, Q.; Nightingale, C.H.; Nicolau, D.P. Differential efficacy of clarithromycin in lung versus thigh infection models. *Chemotherapy* **2004**, *50*, 63–66. [CrossRef]
20. Bachtold, K.A.; Alcorn, J.M.; Boison, J.O.; Matus, J.L.; Woodbury, M.R. Pharmacokinetics and lung and muscle concentrations of tulathromycin following subcutaneous administration in white-tailed deer (*Odocoileus virginianus*). *J. Vet. Pharmacol. Ther.* **2016**, *39*, 292–298. [CrossRef]
21. Romanet, J.; Smith, G.W.; Leavens, T.L.; Baynes, R.E.; Wetzlich, S.E.; Riviere, J.E.; Tell, L.A. Pharmacokinetics and tissue elimination of tulathromycin following subcutaneous administration in meat goats. *Am. J. Vet. Res.* **2012**, *73*, 1634–1640. [CrossRef]
22. Kobuchi, S.; Kabata, T.; Maeda, K.; Ito, Y.; Sakaeda, T. Pharmacokinetics of macrolide antibiotics and transport into the interstitial fluid: Comparison among erythromycin, clarithromycin, and azithromycin. *Antibiotics* **2020**, *9*, 199. [CrossRef]
23. Okamoto, H.; Miyazaki, S.; Tateda, K.; Ishii, Y.; Yamaguchi, K. In vivo efficacy of telithromycin (HMR3647) against *Streptococcus pneumoniae* and *Haemophilus influenzae*. *Antimicrob. Agents Chemother.* **2001**, *45*, 3250–3252. [CrossRef] [PubMed]
24. Tulkens, P.M. Intracellular distribution and activity of antibiotics. *Eur. J. Clin. Microbiol. Infect. Dis.* **1991**, *10*, 100–106. [CrossRef] [PubMed]
25. Nyberg, K.; Johansson, U.; Johansson, A.; Camner, P. Phagolysosomal pH in alveolar macrophages. *Environ. Health Perspect.* **1992**, *97*, 149–152. [CrossRef] [PubMed]
26. McDonald, P.J.; Pruul, H. Phagocyte uptake and transport of azithromycin. *Eur. J. Clin. Microbiol. Infect. Dis.* **1991**, *10*, 828–833. [CrossRef] [PubMed]
27. Togami, K.; Chono, S.; Morimoto, K. Subcellular distribution of azithromycin and clarithromycin in rat alveolar macrophages (NR8383) in vitro. *Biol. Pharm. Bull.* **2013**, *36*, 1494–1499. [CrossRef] [PubMed]
28. Villarino, N.; Martín-Jiménez, T. Pharmacokinetics of macrolides in foals. *J. Vet. Pharmacol. Ther.* **2013**, *36*, 1–13. [CrossRef]
29. Ng, A.W.; Bidani, A.; Heming, T.A. Innate Host Defense of the lung: Effects of lung-lining fluid pH. *Lung* **2004**, *182*, 297–317. [CrossRef]
30. Bodem, C.R.; Lampton, L.M.; Miller, D.P.; Tarka, E.F.; Everett, E.D. Endobronchial, pH Relevance of aminoglycoside activity in gram-negative bacillary pneumonia. *Am. Rev. Respir. Dis.* **1983**, *127*, 39–41. [CrossRef]
31. Van Bambeke, F.; Barcia-Macay, M.; Lemaire, S.; Tulkens, P.M. Cellular pharmacodynamics and pharmacokinetics of antibiotics: Current views and perspectives. *Curr. Opin. Drug Discov. Devel.* **2006**, *9*, 218–230.
32. Kim, Y.H.; Heinze, T.M.; Beger, R.; Pothuluri, J.V.; Cerniglia, C.E. A kinetic study on the degradation of erythromycin A in aqueous solution. *Int. J. Pharm.* **2004**, *271*, 63–76. [CrossRef] [PubMed]
33. Nix, D.E.; Goodwin, S.D.; Peloquin, C.A.; Rotella, D.L.; Schentag, J.J. Antibiotic tissue penetration and its relevance: Impact of tissue penetration on infection response. *Antimicrob. Agents Chemother.* **1991**, *35*, 1953–1959. [CrossRef] [PubMed]
34. Periti, P.; Mazzei, T.; Mini, E.; Novelli, A. Clinical pharmacokinetic properties of the macrolide antibiotics. *Clin. Pharmacokinet.* **1989**, *16*, 193–214. [CrossRef]
35. Jacks, S.; Giguère, S.; Gronwall, P.R.; Brown, M.P.; Merritt, K.A. Pharmacokinetics of azithromycin and concentration in body fluids and bronchoalveolar cells in foals. *Am. J. Vet. Res.* **2001**, *62*, 1870–1875. [CrossRef] [PubMed]
36. Womble, A.Y.; Giguère, S.; Lee, E.A.; Vickroy, T.W. Pharmacokinetics of clarithromycin and concentrations in body fluids and bronchoalveolar cells of foals. *Am. J. Vet. Res.* **2006**, *67*, 1681–1686. [CrossRef]
37. Ambrose, P.G.; Bhavnani, S.M.; Ellis-Grosse, E.J.; Drusano, G.L. Pharmacokinetic-pharmacodynamic considerations in the design of hospital-acquired or ventilator-associated bacterial pneumonia studies: Look before you leap! *Clin. Infect. Dis.* **2010**, *51*, S103–S110. [CrossRef]
38. Kanoh, S.; Rubin, B.K. Mechanisms of Action and Clinical Application of Macrolides as Immunomodulatory Medications. *Clin. Microbiol. Rev.* **2010**, *23*, 590–615. [CrossRef]

39. Danesi, R.; Lupetti, A.; Barbara, C.; Ghelardi, E.; Chella, A.; Malizia, T.; Senesi, S.; Alberto Angeletti, C.; Del Tacca, M.; Campa, M. Comparative distribution of azithromycin in lung tissue of patients given oral daily doses of 500 and 1000 mg. *J. Antimicrob. Chemother.* **2003**, *51*, 939–945. [CrossRef]
40. Firth, A.; Prathapan, P. Azithromycin: The First Broad-spectrum Therapeutic. *Eur. J. Med. Chem.* **2020**, *207*, 112739. [CrossRef]
41. Kong, F.Y.; Rupasinghe, T.W.; Simpson, J.A.; Vodstrcil, L.A.; Fairley, C.K.; McConville, M.J.; Hocking, J.S. Pharmacokinetics of a single 1g dose of azithromycin in rectal tissue in men. *PLoS ONE* **2017**, *12*, e0174372. [CrossRef]
42. Zuckerman, J.M. Macrolides and ketolides: Azithromycin, clarithromycin, telithromycin. *Infect. Dis. Clin. N. Am.* **2004**, *18*, 621–649. [CrossRef] [PubMed]
43. Muller-Serieys, C.; Soler, P.; Cantalloube, C.; Lemaitre, F.; Gia, H.P.; Brunner, F.; Andremont, A. Bronchopulmonary disposition of the ketolide telithromycin (HMR 3647). *Antimicrob. Agents Chemother.* **2001**, *45*, 3104–3108. [CrossRef] [PubMed]
44. Menge, M.; Rose, M.; Bohland, C.; Zschiesche, E.; Kilp, S.; Metz, W.; Allan, M.; Ropke, R.; Nurnberger, M. Pharmacokinetics of tildipirosin in bovine plasma, lung tissue, and bronchial fluid (from live, nonanesthetized cattle). *J. Vet. Pharmacol. Ther.* **2012**, *35*, 550–559. [CrossRef] [PubMed]
45. Torres, F.; Santamaria, R.; Jimenez, M.; Menjón, R.; Ibanez, A.; Collell, M.; Azlor, O.; Fraile, L. Pharmacokinetics of tildipirosin in pig tonsils. *J. Vet. Pharmacol. Ther.* **2016**, *39*, 199–201. [CrossRef]
46. Giguere, S.; Huang, R.; Malinski, T.J.; Dorr, P.M.; Tessman, R.K.; Somerville, B.A. Disposition of gamithromycin in plasma, pulmonary epithelial lining fluid, bronchoalveolar cells, and lung tissue in cattle. *Am. J. Vet. Res.* **2011**, *72*, 326–330. [CrossRef]
47. Berghaus, L.J.; Giguère, S.; Sturgill, T.L.; Bade, D.; Malinski, T.J.; Huang, R. Plasma pharmacokinetics, pulmonary distribution, and in vitro activity of gamithromycin in foals. *J. Vet. Pharmacol. Ther.* **2012**, *35*, 59–66. [CrossRef]
48. Cox, S.R.; McLaughlin, C.; Fielder, A.E.; Yancey, M.; Bowersock, T.; Garcia-Tapia, D.; Bryson, L.; Lucas, M.J.; Robinson, J.A.; Nanjiani, I.; et al. Rapid and prolonged distribution of tulathromycin into lung homogenate and pulmonary epithelial lining fluid of holstein calves following a single subcutaneous administration of 2.5 mg/kg body weight. *Int. J. Appl. Res. Vet. Med.* **2010**, *8*, 129–137. [CrossRef]
49. Leventhal, H.R.; McKenzie, H.C.; Estell, K.; Council-Troche, M.; Davis, J.L. Pharmacokinetics and pulmonary distribution of Draxxin® (tulathromycin) in healthy adult horses. *J. Vet. Pharmacol. Ther.* **2021**, *44*, 714–723. [CrossRef]
50. Villarino, N.; Brown, S.A.; Martín-Jiménez, T. Understanding the pharmacokinetics of tulathromycin: A pulmonary perspective. *J. Vet. Pharmacol. Ther.* **2014**, *37*, 211–221. [CrossRef]
51. Villarino, N.; Lesman, S.; Fielder, A.; García-Tapia, D.; Cox, S.; Lucas, M.; Robinson, J.; Brown, S.A.; Martín-Jiménez, T. Pulmonary pharmacokinetics of tulathromycin in swine. Part 2: Intra-airways compartments. *J. Vet. Pharmacol. Ther.* **2013**, *36*, 340–349. [CrossRef]
52. Villarino, N.; Lesman, S.; Fielder, A.; García-Tapia, D.; Cox, S.; Lucas, M.; Robinson, J.; Brown, S.A.; Martín-Jiménez, T. Pulmonary pharmacokinetics of tulathromycin in swine. *Part I: Lung homogenate in healthy pigs and pigs challenged intratracheally with lipopolysaccharide of Escherichia coli.* *J. Vet. Pharmacol. Ther.* **2013**, *36*, 329–339. [CrossRef]
53. Womble, A.; Giguère, S.; Murthy, Y.V.S.N.; Cox, C.; Obare, E. Pulmonary disposition of tilmicosin in foals and in vitro activity against *Rhodococcus equi* and other common equine bacterial pathogens. *J. Vet. Pharmacol. Ther.* **2006**, *29*, 561–568. [CrossRef]
54. Javsicas, L.; Giguère, S.; Womble, A.Y. Disposition of oral telithromycin in foals and in vitro activity of the drug against macrolide-susceptible and macrolide-resistant *Rhodococcus equi* isolates. *J. Vet. Pharmacol. Ther.* **2010**, *33*, 383–388. [CrossRef] [PubMed]
55. Suarez-Mier, G.; Giguère, S.; Lee, E.A. Pulmonary disposition of erythromycin, azithromycin, and clarithromycin in foals. *J. Vet. Pharmacol. Ther.* **2007**, *30*, 109–115. [CrossRef] [PubMed]
56. Kobuchi, S.; Aoki, M.; Inoue, C.; Murakami, H.; Kuwahara, A.; Nakamura, T.; Yasui, H.; Ito, Y.; Takada, K.; Sakaeda, T. Transport of Azithromycin into extravascular space in rats. *Antimicrob. Agents Chemother.* **2016**, *60*, 6823–6827. [CrossRef] [PubMed]
57. Hunter, R.P.; Burnett, T.J.; Simjee, S.A. Letter to the editor. *Brit. Poultry Sci.* **2009**, *50*, 544–545. [CrossRef]
58. Winther, L. Antimicrobial drug concentrations and sampling techniques in the equine lung. *Vet. J.* **2012**, *193*, 326–335. [CrossRef]
59. Toutain, P.L.; Potter, T.; Pelligand, L.; Lacroix, M.; Illambas, J.; Lees, P. Standard PK/PD concepts can be applied to determine a regimen for a macrolide: The case of tulathromycin in the calf. *J. Vet. Pharmacol. Ther.* **2017**, *40*, 16–27. [CrossRef]
60. Dhanani, J.; Roberts, J.A.; Chew, M.; Lipman, J.; Boots, R.J.; Paterson, D.L.; Fraser, J.F. Antimicrobial chemotherapy and lung microdialysis: A review. *Int. J. Antimicrob. Agents* **2010**, *36*, 491–500. [CrossRef]
61. Mimoz, O.; Dahyot-Fizelier, C. Mini-broncho-alveolar lavage: A simple and promising method for assessment of antibiotic concentration in epithelial lining fluid. *Intensive Care Med.* **2007**, *33*, 1495–1497. [CrossRef]
62. Müller, M.; dela Peña, A.; Derendorf, H. Issues in pharmacokinetics and pharmacodynamics of anti-infective agents: Distribution in tissue. *Antimicrob. Agents Chemother.* **2004**, *48*, 1441–1453. [CrossRef]
63. Maselyne, J.; Adriaens, I.; Huybrechts, T.; De Ketelaere, B.; Millet, S.; Vangeyte, J.; Van Nuffel, A.; Saeys, W. Measuring the drinking behaviour of individual pigs housed in group using radio frequency identification (RFID). *Animal* **2016**, *10*, 1557–1566. [CrossRef]
64. Rottbøll, L.A.; Friis, C. Microdialysis as a tool for drug quantification in the bronchioles of anaesthetized pigs. *Basic Clin. Pharmacol. Toxicol.* **2014**, *114*, 226–232. [CrossRef] [PubMed]
65. Chefer, V.I.; Thompson, A.C.; Zapata, A.; Shippenberg, T.S. Overview of brain microdialysis. *Curr. Protoc. Neurosci.* **2009**, *47*, 7.1.1–7.1.28. [CrossRef] [PubMed]

66. Toutain, P.-L.; Pelligand, L.; Lees, P.; Bousquet-Melou, A.; Ferran, A.A.; Turnidge, J. The pharmacokinetic/pharmacodynamic paradigm for antimicrobial drugs in veterinary medicine: Recent advances and critical appraisal. *J. Vet. Pharmacol. Ther.* **2021**, *44*, 172–200. [CrossRef] [PubMed]
67. Hopper, D.L. Automated microsampling technologies and enhancements in the 3Rs. *ILAR J.* **2016**, *57*, 166–177. [CrossRef]
68. Janle, E.M.; Kissinger, P.T. Microdialysis and ultrafiltration. *Adv. Food Nutr. Res.* **1996**, *40*, 183–196. [CrossRef]

antibiotics

Review

Phenolic Compounds in Bacterial Inactivation: A Perspective from Brazil

Angélica Correa Kauffmann [1] and Vinicius Silva Castro [2,*]

[1] Chemistry Department, Federal University of Mato Grosso, Cuiaba 78060-900, MT, Brazil; angelica.kauffmann@ufmt.br
[2] Department of Biological Sciences, University of Lethbridge, Lethbridge, AB T1K 3M4, Canada
* Correspondence: v.castro@uleth.ca; Tel.: +1-825-438-1645

Abstract: Phenolic compounds are natural substances that are produced through the secondary metabolism of plants, fungi, and bacteria, in addition to being produced by chemical synthesis. These compounds have anti-inflammatory, antioxidant, and antimicrobial properties, among others. In this way, Brazil represents one of the most promising countries regarding phenolic compounds since it has a heterogeneous flora, with the presence of six distinct biomes (Cerrado, Amazon, Atlantic Forest, Caatinga, Pantanal, and Pampa). Recently, several studies have pointed to an era of antimicrobial resistance due to the unrestricted and large-scale use of antibiotics, which led to the emergence of some survival mechanisms of bacteria to these compounds. Therefore, the use of natural substances with antimicrobial action can help combat these resistant pathogens and represent a natural alternative that may be useful in animal nutrition for direct application in food and can be used in human nutrition to promote health. Therefore, this study aimed to (i) evaluate the phenolic compounds with antimicrobial properties isolated from plants present in Brazil, (ii) discuss the compounds across different classes (flavonoids, xanthones, coumarins, phenolic acids, and others), and (iii) address the structure–activity relationship of phenolic compounds that lead to antimicrobial action.

Keywords: heterocyclic compounds; phenolic compounds; pyran; food microbiology; microbial pathogen

Citation: Kauffmann, A.C.; Castro, V.S. Phenolic Compounds in Bacterial Inactivation: A Perspective from Brazil. *Antibiotics* **2023**, *12*, 645. https://doi.org/10.3390/antibiotics12040645

Academic Editor: Dóra Kovács

Received: 22 February 2023
Revised: 17 March 2023
Accepted: 22 March 2023
Published: 24 March 2023

1. Introduction

Few organisms have had such an influence on human history as microorganisms. This influence can be seen in the production of food such as beers, bread, cheeses, and fermented products; in drug production using bacteria as clonal vectors; and in the onset of diseases linked to these microorganisms. Although several microorganisms are a great ally to human beings in obtaining these foods and drugs, it is unquestionable that pathogenic microorganisms represent a strong opponent, with an impact on the lives of thousands of people over the centuries. As an example, we can mention that the adoption of antiseptic hygiene principles proposed by the then-physician Joseph Lister in 1867 [1] made him known as the father of modern surgery [2] due to the impact of the reduction in death rates after the adoption of these principles. After that, the discovery of penicillin by Alexander Fleming in 1928 brought a very powerful weapon to human beings in the fight against these pathogens [3]. During the following years, the investigation of new antibacterial compounds was exhaustively studied, and several new molecules were obtained for different classes of pathogens. However, as in any dispute, bacteria have developed different mechanisms to attenuate the power of these drugs, and we are currently living in the era of multiresistant microorganisms [4]. In this sense, the academic world works in different directions to solve these resistance mechanisms (or just to stop the spread of resistance genes), and in this way, the use of natural compounds can play a prominent role in the fight against these resistant microorganisms.

Brazil is an important country considering natural compounds since it has six biomes (Cerrado, Amazon, Atlantic Forest, Caatinga, Pantanal, and Pampa) [5], in addition to a vast number of plants, fruits, and native vegetables. Although there is still an immense number of substances to be discovered, several studies conducted recently in the country have elucidated compounds with antimicrobial potential, which could be used as one of the strategies to combat pathogenic bacteria. Among these compounds, we highlight the presence of phenolic groups. Phenolic compounds encompass many substances that are produced through the secondary metabolism of plants, fungi, and bacteria, in addition to being produced by chemical synthesis [6,7]. Studies indicate that there are several pharmacological activities associated with these compounds, including antibacterial activity [8,9]. Additionally, one of the main factors that suggest antibacterial action is related to the substituent groups contained in the phenolic compounds, which influences the increase in the lipophilicity of the molecule in the antibiofilm property and in the modulating action of antibiotics [10–15]. Therefore, our study addressed the different classes involving phenolic compounds from substances found in studies performed in Brazil.

2. Search Strategy

In this study, a literature search was performed using online databases: Google Scholar, Scifinder, PubMed, Wiley Online, Science Direct, and Federated Academic Community (CAFe—Brazil). The following terms were used: "Flavonoids AND Antibacterial activity AND Brazil"; "Coumarin AND Antibacterial Activity AND Brazil"; Phenolic Acids AND Antibacterial Activity AND Brazil"; "Lignans AND Antibacterial Activity AND Brazil"; "Anthraquinones AND Antibacterial activity AND Brazil"; "Xanthones AND Antibacterial Activity AND Brazil"; "Acetophenone AND Antibacterial Activity AND Brazil"; "Benzophenone AND Antibacterial Activity AND Brazil"; "Tannins AND Antibacterial Activity AND Brazil"; and "Phenylpropanoids AND Antibacterial Activity AND Brazil". Only articles containing phenolic compounds isolated from plants analyzed in the Brazilian territory were considered in this study. Information about the name of the substance, the plant from which the phenolic compound originated, the microorganisms tested for antimicrobial action, the inhibitory concentration, and the location where the plant was collected (municipality and state of Brazil) are described in Table 1.

Table 1. Phenolic Compounds isolated from plants in Brazil and their respective antibacterial activities.

Structure	Phenolic Compound	Plant	Bacteria	Inhibition	Location	Reference
			Flavonoids			
1	Ampelopsin	*Euphorbia tirucalli*	*S. aureus* (ATCC 6538) *E. coli* (ATCC 8739)	** MIC = 8 μg/mL MIC = 16 μg/mL	Araruna—PB *	[16]
2	Myricetin	*Euphorbia tirucalli*	*S. aureus* (ATCC 6538) *E. coli* (ATCC 8739)	MIC = 16 μg/mL MIC = 8 μg/mL	Araruna—PB	[16]
3	Dihydrokaempferol	*Mauritia flexuosa*	*S. aureus* (ATCC 29213) *S. aureus* (clinical isolate 155)	MIC = 250 μg/mL MIC = 250 μg/mL	Chapada Gaúcha—MG *	[17]
4	(+)-(2R)-naringenin	*Mauritia flexuosa*	*S. aureus* (ATCC 29213) *S. aureus* (clinical isolate 155)	MIC = 62.5 μg/mL MIC = 62.5 μg/mL	Chapada Gaúcha—MG	[17]
5	3,7,3′-trihydroxy-4′-methoxy-8-prenylisoflavone	*Vatairea guianensis*	*S. aureus* (MRSA) *E. faecium*	*** IC$_{50}$ = 6.8 μM IC$_{50}$ = 12.8 μM	Santana—AP *	[18]
6	8-(3-hydroxy-3-methylbutyl)-5,7,3′,4′-tetrahydroxyisoflavone	*Vatairea guianensis*	*S. aureus* (MRSA)	IC$_{50}$ = 29.6 μM	Santana—AP	[18]
7	8-(3-hydroxy-3-methylbutyl)-5,7,4′-trihydroxy-3′-methoxyisoflavone	*Vatairea guianensis*	*S. aureus* (MRSA) *E. faecium*	IC$_{50}$ = 37 μM IC$_{50}$ = 80.6 μM	Santana—AP	[18]
8	8-(3-hydroxy-3-methylbutyl)-5,7,3′-trihydroxy-4′-methoxyisoflavone	*Vatairea guianensis*	*S. aureus* (MRSA)	IC$_{50}$ = 49.0 μM	Santana—AP	[18]
9	3-O-α-L-rhamnopyranosylquercetin	*Clusia burlemarxii*	*B. subtilis* (ATCC 6633) *S. aureus* (ATCC 6538)	MIC = 50 μg/mL MIC = 100 μg/mL	Mucugê—BA *	[19]
10	3-O-α-L-rhamnopyranosylkaempferol	*Clusia burlemarxii*	*S. aureus* (ATCC 6538)	MIC = 25 μg/mL	Mucugê—BA	[19]

Table 1. *Cont.*

Structure	Phenolic Compound	Plant	Bacteria	Inhibition	Location	Reference
11	(*E*)-3′-*O*-β-D-glucopyranosyl-4,5,6,4′-tetrahydroxy-7,2′-dimethoxyaurone	*Gomphrena agrestis*	*S. epidermidis* (6epi) *S. epidermidis* (epiC) *P. aeruginosa* (ATCC 27853) *P. aeruginosa* (290D)	MIC = 0.1 mg/mL MIC = 0.5 mg/mL MIC = 0.5 mg/mL MIC = 0.5 mg/mL	Alto Paraíso—GO *	[20]
12	Tiliroside	*Gomphrena agrestis* *Herissantia tiubae*	*S. aureus* (ATCC 25923) *S. aureus* (SA-1199B)	MIC = 0.5 mg/mL MIC = 256 µg/mL	Alto Paraíso—GO Juazeirinho—PB *	[20,21]
13	2′-hydroxy-4,4′,6′-trimethoxychalcone	*Piper hispidum*	*S. aureus* (ATCC 25923)	MIC = 125 µg/mL	Maringá—PR *	[22]
14	2′-hydroxy-3,4,4′,6′-tetramethoxychalcone	*Piper hispidum*	*S. aureus* (ATCC 25923)	MIC = 250 µg/mL	Maringá—PR	[22]
15	3,2′-dihydroxy-4,4′,6′-trimethoxychalcone	*Piper hispidum*	*S. aureus* (ATCC 25923)	MIC = 125 µg/mL	Maringá—PR	[22]
16	Genkwanin	*Praxelis clematidea*	*S. aureus* (AS-1199B)	MIC = 64 µg/mL **** (Substance with Norfloxacin) MIC = 128 µg/mL (Nofloxacin) MIC = 16 µg/mL (Substance with Ethidium bromide) MIC = 32 µg/mL (Ethidium bromide)	Santa Rita—PB	[12]
17	7,4′-dimethylapigenin	*Praxelis clematidea*	*S. aureus* (AS-1199B)	MIC = 64 µg/mL (Substance with Norfloxacin) MIC = 128 µg/mL (Norfloxacin) MIC = 16 µg/mL (Substance with Ethidium bromide) MIC = 32 µg/mL (Ethidium bromide)	Santa Rita—PB	[12]
18	trimethylapigenin	*Praxelis clematidea*	*S. aureus* (AS-1199B)	MIC = 16 µg/mL (Substance with Norfloxacin) MIC = 128 µg/mL (Norfloxacin) MIC = 8 µg/mL (Substance with Ethidium bromide) MIC = 32 µg/mL (Ethidium bromide)	Santa Rita—PB	[12]
19	cirsimaritin	*Praxelis clematidea*	*S. aureus* (AS-1199B)	MIC = 32 µg/mL (Substance with Norfloxacin) MIC = 128 µg/mL (Norfloxacin) MIC = 8 µg/mL (Substance with Ethidium bromide) MIC = 32 µg/mL (Ethidium bromide)	Santa Rita—PB	[12]
20	tetramethylscutellarein	*Praxelis clematidea*	*S. aureus* (AS-1199B)	MIC = 8 µg/mL (Substance with Norfloxacin) MIC = 128 µg/mL (Norfloxacin) MIC = 2 µg/mL (Substance with Ethidium bromide) MIC = 32 µg/mL (Ethidium bromide)	Santa Rita—PB	[12]
			Xanthones			
21	3,4-dihydroxy-2-methoxyxanthone	*Kielmeyera variabilis*	*S. aureus* (SA-1199B) *S. aureus* (XU212) *S. aureus* (ATCC 25923) *S. aureus* (RN4220) *S. aureus* (EMRSA-15) *S. aureus* (EMRSA-16)	MIC = 32 mg/L MIC = 32–16 mg/L MIC = 64 mg/L MIC = 32 mg/L MIC = 64 mg/L MIC = 16 mg/L	Mogi Guaçu—SP *	[23]
22	5-hydroxy-1,3-dimethoxyxanthone	*Kielmeyera variabilis*	*S. aureus* (SA-1199B) *S. aureus* (ATCC 25923) *S. aureus* (EMRSA-16)	MIC = 128–64 mg/L MIC = 128 mg/L MIC = 64 mg/L	Mogi Guaçu—SP	[23]

Table 1. *Cont.*

Structure	Phenolic Compound	Plant	Bacteria	Inhibition	Location	Reference
23	4-hydroxy-2,3-dimethoxyxanthone	*Kielmeyera variabilis*	*S. aureus* (SA-1199B) *S. aureus* (XU212) *S. aureus* (EMRSA-16)	MIC = 128–64 mg/L MIC = MIC = 128 mg/L MIC = 64 mg/L	Mogi Guaçu—SP	[23]
24	3-hydroxy-2-methoxyxanthone	*Kielmeyera variabilis*	*S. aureus* (SA-1199B) *S. aureus* (XU212) *S. aureus* (ATCC 25923) *S. aureus* (RN4220) *S. aureus* (EMRSA-15) *S. aureus* (EMRSA-16)	MIC = 64 mg/L MIC = 64 mg/L MIC = 64 mg/L MIC = 64 mg/L MIC = 64 mg/L MIC = 32 mg/L	Mogi Guaçu—SP	[23]
25	2-hydroxy-1-methoxyxanthone	*Kielmeyera variabilis*	*S. aureus* (SA-1199B) *S. aureus* (XU212) *S. aureus* (ATCC 25923) *S. aureus* (RN4220) *S. aureus* (EMRSA-15) *S. aureus* (EMRSA-16)	MIC = 64 mg/L MIC = 128 mg/L MIC = 64 mg/L MIC = 64 mg/L MIC = 64 mg/L MIC = 32 mg/L	Mogi Guaçu—SP	[23]
26	Assiguxanthone B	*Kielmeyera variabilis*	*S. aureus* (ATCC 25923) *B. subtilis* (ATCC 6623)	MIC = 100 µg/mL MIC = 25 µg/mL	Mogi Guaçu—SP	[24]
27	1,3,7′trihydroxy-2-(3-methylbut-2-enyl)-xanthone	*Kielmeyera coriacea*	*S. aureus* (ATCC 25922) *E. coli* (ATCC 25922) *B. subtilis* (ATCC 6623) *P. aeruginosa* (ATCC 15442)	MIC = 12.5 µg/mL MIC > 100 µg/mL MIC = 12.5 µg/mL MIC > 100 µg/mL	Mogi Guaçu—SP	[25]
28	8-carboxymethyl-1,3,5,6-tetrahydroxyxanthone	*Leiothrix spiralis*	*S. aureus* (ATCC 25923) *B. subtilis* (ATCC 19659) *P. aeruginosa* (ATCC 27853)	MIC = 125 µg/mL MIC = 125 µg/mL MIC = 125 µg/mL	Diamantina—MG	[26]
29	1,5-dihydroxyxanthone	*Calophyllum brasiliense*	*B. cereus* (ATCC 14579) *S. aureus* (ATCC 6538P) *S. saprophyticus* (ATCC 35552) *S. agalactiae* (ATCC 13813)	MIC = 700 µg/mL MIC = 200 µg/mL MIC = 200 µg/mL MIC = 500 µg/mL	Florianópolis—SC	[27]
Coumarins						
30	Tanizin	*Helietta apiculato*	*B. cereus* (ATCC 33019) *Enterococcus ssp* (ATCC 6589) *E. aerogenes* (ATCC 13048) *P. aeruginosa* (ATCC 9027) *E. coli* (ATCC 25922) *B. cepacia* (ATCC 17759) *S. sonnei* (ATCC 25931) *S. Typhimurium* (ATCC 14028) *M. morganii* (ATCC 25829)	MIC = 12.5 µg/mL MIC = 50 µg/mL MIC = 25 µg/mL MIC = 25 µg/mL MIC = 50 µg/mL MIC = 25 µg/mL MIC = 25 µg/mL MIC = 25 µg/mL MIC = 12.55 µg/mL	Mata—RS	[28]
31	Gravellifenore	*Helietta apiculato*	*B. subtilis* (ATCC 6633) *B. cereus* (ATCC 33019) *Enterococcus ssp* (ATCC 6589) *E. aeruginosa* (ATCC 9027) *E. coli* (ATCC 25922) *B. cepacia* (ATCC 17759) *S. sonnei* (ATCC 25931) *S. Typhimurium* (ATCC 14028) *M. morganii* (ATCC 25829)	MIC = 50 µg/ml MIC = 12.5 µg/mL MIC = 50 µg/mL MIC = 25 µg/mL MIC = 50 µg/mL MIC = 3.12 µg/mL MIC = 12.55 µg/mL MIC = 50 µg/mL MIC = 6.25 µg/mL	Mata—RS	[28]
Phenolic Acids						
32	Gallic acid	*Himatanthus sucuuba*	*S. aureus* (MRSA) *S. epidermidis* (ATCC 12228) *P. mirabilis* (MRSA) *S. haemolyticus* (ATCC 2737) *E. coli* (ATCC 25922)	MIC = 31 µg/mL MIC = 31 µg/mL MIC = 62 µg/mL MIC = 62 µg/mL MIC = 125 µg/mL	Santarém—PA	[29]
33	Protocatechuic acid	*Calophyllum brasiliense*	*B. cereus* (ATCC 14579) *S. aureus* (ATCC 6538P) *S. saprophyticus* (ATCC 35552) *S. agalactiae* (ATCC 13813) *E. cloacae* (ATCC 35030) *E. coli* (ATCC 11775) *P. aeruginosa* (ATCC 35032) *P. mirabilis* (ATCC 14273) *S. Typhimurium* (ATCC 14028)	MIC = 500 µg/mL MIC = 200 µg/mL MIC = 200 µg/mL MIC = 200 µg/mL MIC = 400 µg/mL MIC = 400 µg/mL MIC = 800 µg/mL MIC = 500 µg/mL MIC = 700 µg/mL	Florianópolis—SC	[27]
Other Phenolic Compounds						
34	Dihydrodehydrodiconiferyl alcohol	*Styrax ferrugineus*	*S. aureus* (ATCC 12228)	MIC = 20 µg/mL	Mogi-Guaçu—SP	[30]
35	Lyoniresinol	*Clusia burlemarxii*	*S. aureus* (ATCC 6538)	MIC = 25 µg/mL	Mucugê—BA	[19]
36	Acid 3,3′-dimethoxyellagic acid-4-*O*-α-rhamnopyranoside	*Euphorbia tirucalli*	*S. aureus* (ATCC 6538) *E. coli* (ATCC 8739)	MIC = 64 µg/mL MIC = 128 µg/mL	Araruna—PB	[16]
37	(E)-methyl-4-hydroxy-3,5-dimethoxycinnamate	*Helietta apiculato*	*B. cereus* (ATCC 33019) *E. aerogenes* (ATCC 13048) *B. Ceparia* (ATCC 17759) *M. morganii* (ATCC 25829)	MIC = 12.5 µg/mL MIC = 25 µg/mL MIC = 50 µg/mL MIC = 50 µg/mL	Mata—RS	[28]

Table 1. *Cont.*

Structure	Phenolic Compound	Plant	Bacteria	Inhibition	Location	Reference
38	(E)-ethyl-4-hydroxy-3,5-dimethoxycinnamate	*Helietta apiculato*	*B. cereus* (ATCC 33019) *E. aerogenes* (ATCC 13048)	MIC = 50 μg/mL MIC = 50 μg/mL	Mata—RS	[28]
39	2,2-dimethyl-3,5-dihydroxy-7-(4-hydroxyphenyl)chromane	*Clusia burlemarxii*	*M. luteus* (ATCC 10240) *S. aureus* (ATCC 6538) *B. subtilis* (ATCC 6633) *S. mutans* (ATCC 5175)	MIC = 25 μg/mL MIC = 50 μg/mL MIC = 100 μg/mL MIC = 100 μg/mL	Mucugê—BA	[19]
40	Aucuparin	*Kielmeyera coriacea*	*S. aureus* (ATCC 25922) *E. coli* (ATCC 25922) *B. subtilis* (ATCC 6623) *P. aeruginosa* (ATCC 15442)	MIC = 12.5 μg/mL MIC = 100 μg/mL MIC = 3.12 μg/mL MIC = 100 μg/mL	Mogiguaçu—SP	[25]

Caption: * Brazilian States (PB = Paraíba; MG = Minas Gerais; AP = Amapá; BA = Bahia; GO = Goiás; PR = Paraná; SC = Santa Catarina; SP = São Paulo; RS = Rio Grande do Sul; PA = Pará). ** MIC = Minimum inhibitory concentration. *** IC50 = Inhibitory Concentration 50%. **** Substances combined with other antibiotics to measure the synergistic effect are included in parentheses.

3. Phenolic Compounds

Phenolic compounds are substances widely produced by plants, with more than two hundred thousand compounds currently known. In their basic structure, phenolic compounds have a ring with all sp2-hybridized carbons, a planar structure with angles of 120° and electrons π delocalized, with one or more hydroxyl group bonds (O–H) [6,31–33]. This class of substances is divided into phenolic acids and polyphenols and can be found combined with other phenolic groups, mono- or polysaccharide structures, or occur as derivatives of esters and methyl esters [8,34].

Phenolic compounds are also aromatic molecules and play important roles in plant growth and reproduction, acting as allopathic agents. When induced by biotic and abiotic stress, they synthesize phytoalexin substances as a plant defense mechanism. Many phenolic compounds are attractive to pollinators in addition to being responsible for the organoleptic characteristics of vegetable foods [32,34–36].

In general, there are two metabolic pathways by which phenolic compounds can be synthesized, the *shikimate pathway* (1), which occurs through the combination of phosphoenolpyruvate (an intermediate of the glycolytic pathway) and erythrose 4-phosphate (from the pentose phosphate pathway), which generates shikimic acid, responsible for the formation of the amino acids, phenylalanine and tyrosine. After the formation of these amino acids, they undergo deamination and generate cinnamic acid and, thus, enter the phenylpropanoid pathway, producing compounds such as flavonoids, isoflavonoids, coumarins, lignans, lignins, and stilbenes. Another known pathway is the *acetate pathway* (2), responsible for the formation of several substances, including aromatic polyketides (anthraquinones, xanthones, benzophenones, and acetophenones) (Figure 1). The formation of phenolic compounds by this route occurs through the reaction of an acetyl-CoA unit, originating from glycolysis, and malonyl-CoA units following aldol reactions, Claisen condensation reactions, and enolization [37,38].

Thus, phenolic compounds are found in abundance in vegetables, fruits, and cereals [39]. The high consumption of foods rich in phenolic compounds can prevent a series of diseases [40]. Furthermore, studies indicate great biological potential associated with these compounds [7], including antiviral, antioxidant, antitumor, antiallergic, anti-inflammatory, fungal, and antibacterial activities [41–47]. In the next topics, we will address the compounds formed from these pathways.

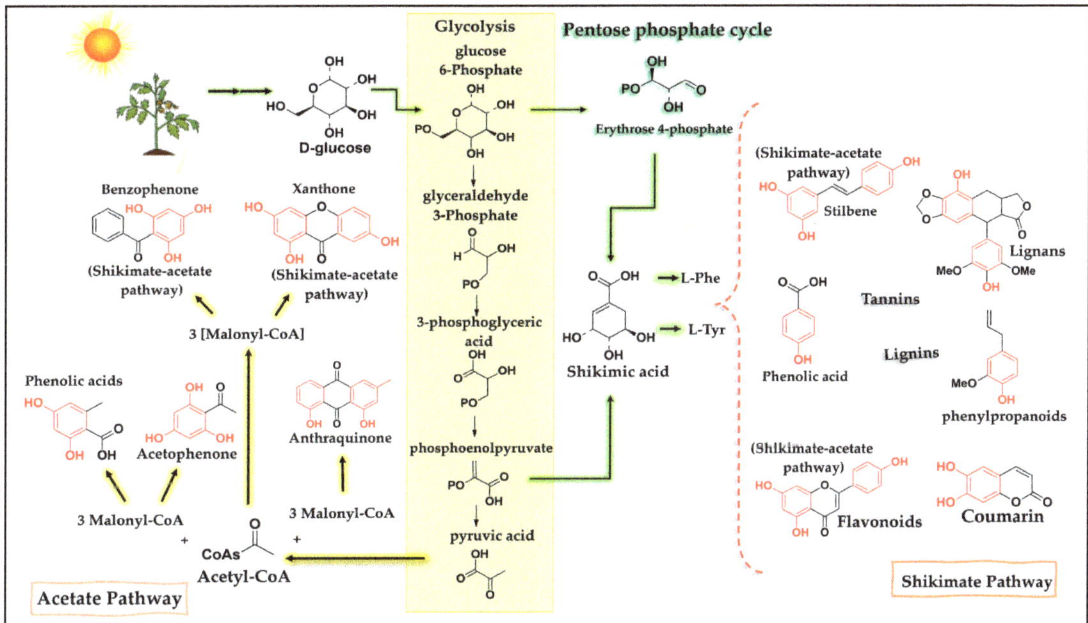

Figure 1. Simplified diagram of the shikimate and acetate pathways, which are responsible for the formation of phenolic compounds. Caption: The shikimate pathway occurs through the reaction between phosphoenolpyruvate and erythrose 4-phosphate, forming shikimic acid and leading to the formation of a variety of phenolic compounds. The acetate pathway occurs through the decarboxylation of pyruvic acid to form acetyl-CoA, which combines with malonyl-CoA units to form phenolic compounds.

4. Flavonoids

Flavonoids are polyphenolic compounds biosynthesized from the shikimate and acetate pathways, and they are the class of phenolic compounds with the highest number of reported substances, with more than 4000 types of naturally occurring flavonoids [37,48]. Structurally, flavonoids contain two aromatic rings (A and B), with fifteen carbon atoms in their basic skeleton, connected by a bridge of three carbon atoms, arranged in C6-C3-C6, providing a third ring [49]. They are structures that differ in the saturation of the C ring, in the position of aromatic ring B on carbons C2 or C3, and in hydroxylation patterns [50], causing subclasses of flavonoids: flavonols, flavones, flavanones, flavanol, flavanonols, isoflavones, aurones, anthocyanins, and chalcones (Figure 2). The pattern of substitution can occur in flavonoids of natural origin, such as hydroxyl, methyl, phenyl, glycosides, aliphatic, isoprenyl, aromatic acids, and methoxyl [51,52].

In plants, these compounds are mainly responsible for the red, blue, and purple pigments that color them [50], have UV protection functions, modulate enzymatic activity, attract or repel insects, are attractive to pollinators, or act as antiviral, fungal, and bacterial protectors [53,54]. Several natural flavonoids show good bacterial inhibition, acting as bactericides and bacteriostatic agents; however, the biological activity depends on the substituent groups of the flavonoid structure, which vary between structures [55,56]. The inhibitory potential of flavonoids may vary according to the molecule analyzed and, although the factors that lead flavonoids to bacterial inhibition are not fully elucidated in the literature, several studies point to the participation of flavonoids in the disruption of cell membranes [57]. This disruption would be derived from the partitioning of nonpolar compounds in the hydrophobic interior of the membrane and due to the interaction between the bacterial membrane and the formation of hydrogen bonds between polar groups of cell lipids with the more hydrophilic flavonoids [57]. In addition, other studies have pointed to an increase in cell permeability through a decrease in bilayer lipids [58]. Additionally, other studies have demonstrated the ability to generate reactive oxygen species (ROS) that can cause alterations in membrane permeability and lead to membrane damage [59].

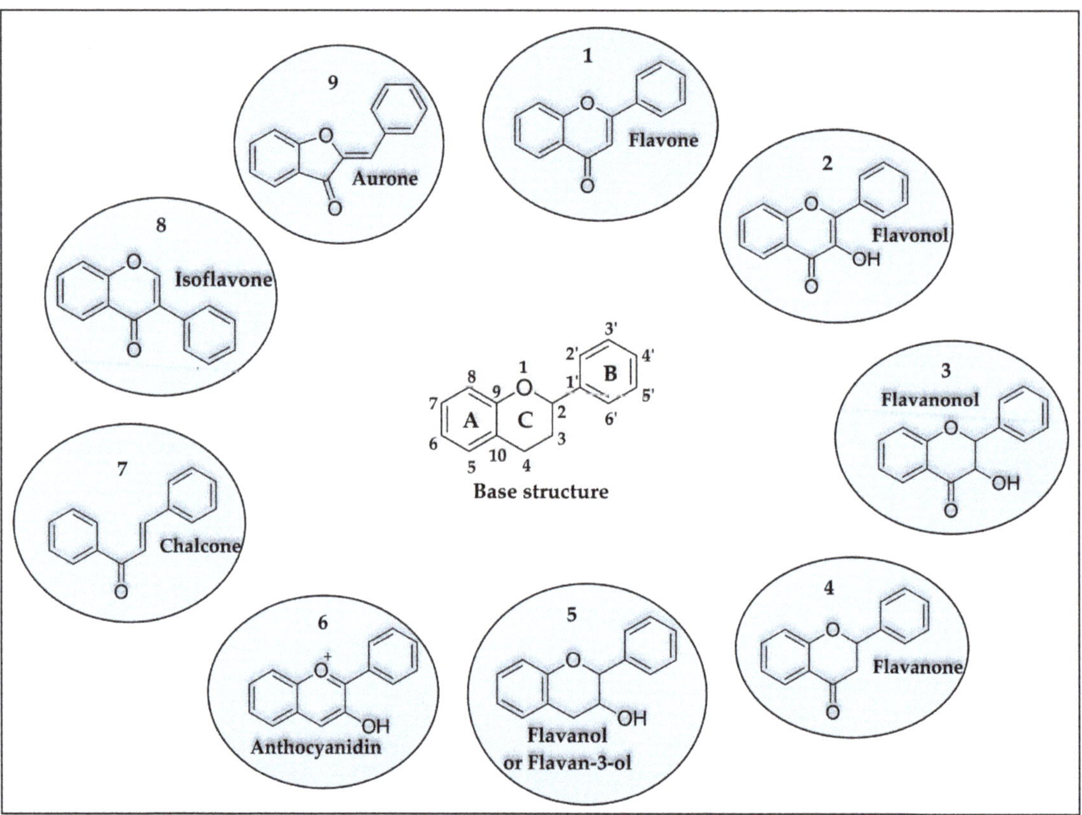

Figure 2. Base structure and classes of flavonoids. (**1**) flavone, (**2**) flavonol, (**3**) flavanonol, (**4**) flavanone, (**5**) flavanol, (**6**) anthocyanidin, (**7**) chalcone, (**8**) isoflavone, and (**9**) aurone.

In the Brazilian territory, there are several reports of isolated substances from the class of flavonoids as potential pathogen inhibitors. We can mention, for example, that, from the roots of *Euphorbia tirucalli* species, two flavonoids were isolated that proved to be potent inhibitors against the bacteria *Escherichia coli* (*E. coli*; ATCC 8739) and *Staphylococcus aureus* (*S. aureus*; ATCC 6538) [16]. This fact is interesting due to the antimicrobial potential of flavonoids being more related to Gram-positive bacteria than Gram-negative bacteria, and the lesser action on Gram-negative bacteria would be due to the presence of negatively charged LPS of the outer bacterial membrane [57]. When we analyzed the molecules tested in Brazil, the flavononol ampelopsin (**1**) and the flavonol myricetin (**2**) (Figure 3) showed higher inhibition values against *S. aureus* (Gram-positive) of 8 µg/mL and 16 µg/mL, respectively [16]. For *E. coli* (Gram-negative), both flavonoids showed antibacterial activity superior to that of the control antibiotic (tetracycline) with a minimum inhibitory concentration of 32 µg/mL, while ampelopsin presented a MIC of 16 µg/mL, and myricetin presented a MIC of 8 µg/mL (Table 1) [16]. An important point is that flavonol myricetin has also been described as a potent multitarget antivirulence agent against the pathogen *S. aureus*, with antibiofilm, antihemolytic, and antistaphyloxanthin properties [60].

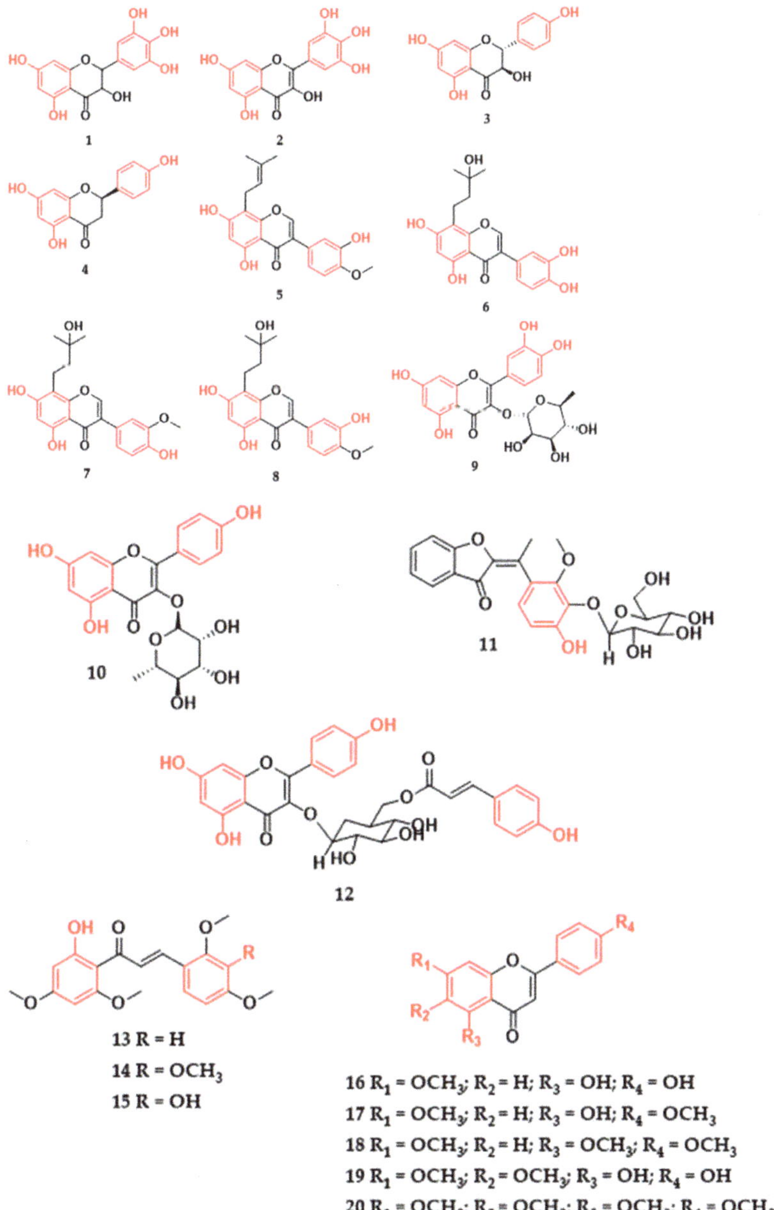

Figure 3. Flavonoids with antibacterial activity isolated from plants in Brazil. Ampelopsin (**1**), myricetin (**2**), dihydrokaempferol (**3**), (+)-(2R)-naringenin (**4**), 3,7,3′-trihydroxy-4′-methoxy-8-prenylisoflavone (**5**), 8-(3-hydroxy-3-methylbutyl)-5,7,3′,4′-tetrahydroxyisoflavone (**6**), 8-(3-hydroxy-3-methylbutyl)-5,7,4′-trihydroxy-3′-methoxyisoflavone (**7**), 8-(3-hydroxy-3-methylbutyl)-5,7,3′-trihydroxy-4′-methoxyisoflavone (**8**), 3-*O*-α-L-rhamnopyranosylquercetin (**9**), 3-*O*-α-L-rhamnopyranosylkaempferol (**10**), (*E*)-3′-*O*-β-D-glucopyranosyl-4,5,6,4′-tetrahydroxy-7,2′-dimethoxyaurone (**11**), tiliroside (**12**), 2′-hydroxy-4,4′,6′-trimethoxychalcone (**13**), 2′-hydroxy-3,4,4′,6′-tetramethoxychalcone (**14**), 3,2′-dihydroxy-4,4′,6′-trimethoxychalcone (**15**), genkwanin (**16**), 7,4′-dimethylapigenin (**17**), trimethylapigenin (**18**), cirsimaritin (**19**), tetramethylscutellarein (**20**).

In addition, it is suggested that the increase in antibacterial activity against *S. aureus* strains in flavonoids is related to the hydroxylation of carbons at positions C-5, C-7, C-3', and C4' [61], which may explain the high inhibitory value for the molecules. The structures of flavonoids (**1**) and (**2**) show the same substitution patterns, with hydroxyl groups on carbons C-3, C-5, C-7, C-3', C-4', and C-5', differing only in the unsaturation of the C ring. In work carried out relating the structure and activity of several flavonoids to the inhibition of *Escherichia coli*, the results showed more efficient inhibitory values in the flavonoids of the flavonol class [10]. Furthermore, some studies indicate that hydroxyl groups at the C-3 position are important for antibacterial activity and contribute to decreasing the cell membrane fluidity of *E. coli* bacteria, which may be one of the direct inhibitory mechanisms of flavonoids. In addition, it was observed that the hydrophobicity and electronic properties of flavonoids can determine their antibacterial activity against *E. coli* [10,52]. Therefore, although there may be a greater propensity for flavonoids to inactivate Gram-positive bacteria, some molecules may be equally effective in inactivating Gram-negative bacteria, ensuring a characteristic of flavonoids as multitarget compounds. Regarding compounds with action on Gram-positive bacteria, the flavonoids dihydroxykaempferol (**3**) and naringenin (**4**) (Figure 3), isolated from the stem of *Maurutia flexuosa* species, were shown to be active against methicillin-susceptible *Staphylococcus aureus* (MSSA) and methicillin-resistant *S. aureus* (MRSA) [17]. Dihydroxykaempferol presented a MIC value of 250 μg/mL for both strains, while the flavanone naringenin showed a more efficient MIC of 62.5 μg/mL (Table 1) [17]. It is important to note that there are works related to the flavonoid naringenin that describe its antibacterial potential against the bacteria *S. aureus* [61]. In a broth dilution assay, naringenin, when combined with an antibiotic, had considerably increased antibacterial activity against multidrug-resistant *S. aureus* strains [62]. In addition, this compound presented antibiofilm properties, and the probable mechanism of action that leads to inhibition of *S. aureus* occurs through the rupture of the bacterium's cytoplasmic membrane and binding to its genomic DNA [63,64].

Isoflavonoids are another subclass that has been reported to have antibacterial activities in Gram-positive bacteria and are one of the subclasses of flavonoids that are most frequently reported in plants of the *Fabaceae*/*Leguminosae* family [65]. They are structures that differ from flavones by the rearrangement of aromatic ring B in the C-2 to C-3 position through the action of the enzyme dependent on cytochrome P-450 [37]. The greatest structural variation of these compounds occurs by the substituent groups, hydroxyl, methoxyl, methylenedioxy, glycoside, and prenyl groups, which influence the antibacterial activity of these compounds [66]. Studies suggest that prenyl units on carbons at positions C-6, C-8, and C-3' and hydroxyl groups at positions C-5 and C-7 may contribute to the inhibition of methicillin-resistant *Staphylococcus aureus* strains [67,68]. It is hypothesized that prenyl groups enhance bacterial cell membrane penetration by the attachment of a strongly lipophilic arm to the molecule. Furthermore, diprenylation of isoflavones may be associated with lower minimum inhibitory concentration values against *E. coli* [69]. Isoflavonoids linked to phenyl groups have already been shown to be effective against Gram-positive bacteria such as *S. aureus* and *B. subtilis*, with greater antibacterial activity when the phenyl group is in positions C-6, C-8, C-3', and C-5' of the isoflavonoid [70]. Of the isoflavonoids isolated from the leaves of *Vaitera guianensis*, the substances that showed antibacterial activity against methicillin-resistant *S. aureus* were the prenylated isoflavones 5,7,3'-trihydroxy-4'-methoxy-8-prenylisoflavone (**5**), 8-(3-hydroxy-3-methylbutyl)-5,7,3',4'-tetrahydroxyisoflavone (**6**), 8-(3-hydroxy-3-methylbutyl)-5,7,4'-trihydroxy-3'-methoxyisoflavone (**7**), and 8-(3-hydroxy-3-methylbutyl)-5,7,3'-trihydroxy-4'-methoxyisoflavone (**8**) (Figure 3; Table 1). Compound (**5**) exhibited the highest *S. aureus* inhibition value at the half-maximal inhibitory concentration (IC$_{50}$) of 6.8 μM and was also active against *E. faecium* (IC50 of 12.8 μM) [18]. In work carried out to investigate the effect of soybean flavone on the DNA and RNA of *S. aureus* (ATCC 26112), the authors indicated that isoflavones were effective in inhibiting the activity of topoisomerase I and II enzymes, responsible for the dynamic control of changes in nucleic acids, and their inhibition affected nucleic acid synthesis and bacterial growth [71].

The flavonoid glycosides found in the leaves of the *Clusia burlemarxii* species showed promising antibacterial values. The flavonoid 3-*O*-α-L-rhamnopyranosylquercetin (**9**) (Figure 3) exhibited strong activity against *S. aureus* (ATCC 6538), while the flavonoid 3-*O*-α-L-rhamnopyranosylquercetin (**10**) showed moderate and low activity against *B. subtilis* (ATCC 6633) and *S. aureus* (ATCC 6538), respectively (Table 1) [19]. Both (E)-3′-*O*-β-D-glucopyranosyl-4,5,6,4′-tetrahydroxy-7,2′-dimethoxyaurone (**11**) and tiliroside (**12**) glycosides from *Gomphrena agrestis* species showed low inhibitory activity against bacterial strains (Table 1) [20]. However, when flavonoid (**12**) was tested in the presence of antibiotics at various concentrations, this flavonoid compound modulated the activity of the antibiotics and reduced the concentration required to inhibit *S. aureus* (AS-1199B). This activity may be related to the lipophilicity attributed to the flavonoid skeleton [21]. Similar results were observed with flavonoid glycosides in inhibiting the pathogen *E. coli* and methicillin-resistant *S. aureus*, and it was indicated that the glycosidic bond influences the low antibacterial activity of the flavonoid [10,72], which may be related to the lack of affinity for the phospholipid layer or to specific receptors in cell membranes [73]. In a comparative study of the flavonoids glucones and aglycones, it was observed that, although the compounds with *O*-glycosides presented no satisfactory values of inhibitory effect on the growth of *S. aureus* strains, this compound had a positive result when the authors evaluated the ability to reduce biofilm complex [74].

Through the phytochemical study of the leaves of *Piper hispidum* species collected in the southern region of Brazil, three flavonoids of the chalcone class were found that expressed antibacterial activity. Chalcones (**13–15**) (Figure 3) exhibited MIC values between 125–250 µg/mL for *S. aureus* (ATCC 25923) (Table 1) [22]. Chalcones (1,3-diaryl-2-propen-1-ones) present an open chain α,β-unsaturated flavonoid and are compounds of natural or synthetic origin, occur as cis and trans isomers, and act as precursors for several other natural flavonoid derivatives [37,75,76]. The substituent groups and their respective locations in the molecule interfere with the biological activity of chalcones. The introduction of hydroxyl substituents in the C-2 or C-4 positions of the B ring or C-2′ of the A ring of chalcone can influence the anti-MRSA activity; on the other hand, methoxyl groups linked to the structure prevent staphylococcal activity. It is very common for chalcones of natural origin to have a hydroxyl group at the C-2′ position, which helps stabilize the molecule. However, this presence may not be fundamental in its antibacterial activity [13,77,78]. Furthermore, studies suggest that only one hydroxyl group on the B ring of chalcones may not be sufficient to exhibit significant activity in inhibiting *S. aureus* [79]. In contrast, antibacterial activity against Gram-positive bacteria *S. aureus* and *B. cereus* may be related to the presence of an oxygenated substituent group on the C-4′ carbon, hydroxyl on the C-4 carbon, and an isoprenoid side chain, located on the C-3′ carbon [13].

In the last decade, the emergence of antibiotic-resistant strains has increased alarmingly [80]. As a result, the emergence of multiresistant bacteria occurs, leading to drug ineffectiveness, prolonging the duration of the disease, and leading to death [81]. A strategy to combat bacterial resistance is the use of compounds that can suppress cellular resistance mechanisms, increasing the effectiveness of existing antibiotics [82,83]. There are reports of compounds from the class of flavonoids that, when combined with drugs, potentiate the inhibitory effect of the drug [84–87]. As an example, the natural compounds genkwanin (**16**), 7,4′-dimethylapigenin (**17**), trimethylapigenin (**18**), cirsimaritin (**19**), and tetramethylscutellarein (**20**) (Figure 3), isolated from the aerial parts of *Praxelis clematidea* species, did not show significant results against *S. aureus* (SA-1199B0) when tested alone. However, these compounds proved to be good modulators of the activity of the antibiotics norfloxacin and ethidium bromide. In particular, compound (**20**) was able to reduce the MIC of drugs against pathogens by sixteen-fold (Table 1). This property may be related to the position of the methoxy groups in the 4′, 5, 6, and 7 positions of the flavone, which increases its lipophilicity [12]. In general, lipophilicity is an important factor in flavonoids for the inhibition of Gram-positive bacteria [88]. It is influenced by the pH and structural characteristics of the compounds (substituent groups and their respective locations in the backbone of the structure). It has been seen that hydroxyl groups and isopentenyl substituents on the A, B, and C rings of the flavonoid influence the lipophilicity of flavonoids and may be more effective in antibacterial activity [88].

5. Xanthones

Xanthones (derived from the Greek "yellow") are symmetric oxygenated heterocyclic derivatives with a basic skeleton of dibenzo-γ-pyrone [89,90]. They are polyphenolic secondary metabolites found in fungi, lichens, bacteria, and mainly in plants from the families *Gentianaceae*, *Polygalaceae*, *Clusiaceae*, and *Moraceae* [91,92]. They are divided into five groups: simple xanthones, glycosides, xanthonolignoids (frequent in the *Gentianaceae* family), prenylated xanthones (more frequent in the *Clusiaceae* family), and the xanthones of the miscellaneous group, which have unusual substitutions and are extracted from different plant and vegetable sources [93].

When produced by higher plants, xanthones are formed by a mixed biosynthetic pathway (shikimate-acetate) [93]. The phenylalanine produced in the shikimate pathway is oxidized to m-hydroxybenzoic acid, which reacts with three acetate units, forming the intermediate benzophenone, and through intramolecular reactions, xanthones are formed [92,94]. Molecular numbering occurs according to a biosynthetic convention. The carbons of the A ring (acetate pathway) of xanthones are numbered 1–4, while the B ring (shikimate pathway) is numbered 5–8 (Figure 4) [95].

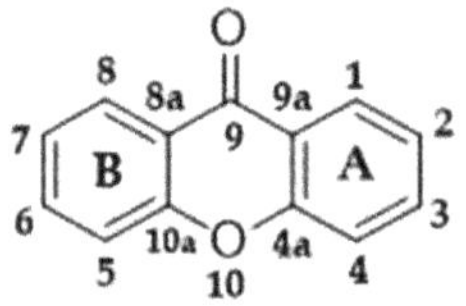

Figure 4. Basic structure of xanthones.

There are many reports on numerous pharmacological activities attributed to xanthones of natural origin, including anti-inflammatory [96,97], anti-Leptospira [98], anti-collagenase, anti-elastase, anti-hyaluronidase, anti-tyrosinase [99], anticancer [100,101], antidiabetic [102,103], and antifungal [104] activities, and many xanthones have been described as good bacterial inhibitors [105–108].

In a phytochemical investigation with *Kielmeyera variabilis* species collected in southeastern Brazil, oxygenated xanthones with a potential inhibitor of multidrug-resistant strains of *S. aureus* were isolated. The xanthones (**21–25**) (Figure 5) found in the leaves of this species exhibited MIC values ranging between 16–128 mg/L (Table 1). Among the compounds, xanthone 3,4-dihydroxy-2-methoxyxanthone (**21**), structurally differentiated by having a catechol group, exhibited greater antibacterial activity [23]. From the stems of the same plant species, the isolation of prenylated xanthone assiguxanthone B (**26**) (Figure 5) was reported, which proved efficient against *S. aureus* (ATCC 25923) and *B. subtilis* (ATCC 6623), with MICs of 100 μg/mL and 25 μg/mL, respectively (Table 1) [24].

The prenylated xanthone 1,3,7'trihydroxy-2-(3-methylbut-2-enyl)-xanthone (**27**) (Figure 5) isolated from *Kielmeyera coriacea* species proved to be a potent inhibitor against the Gram-positive bacteria *S. aureus* (ATCC 25923) and *B. subtilis* (ATCC 6623), exhibiting a MIC value of 12.5 μg/mL (Table 1) [25]. From the leaves of *Leithrix spiralis* species, the compound 8-carboxymethyl-1,3,5,6-tetrahydroxyxanthone (**28**) (Figure 5) was reported, showing inhibition values of 125 μg/mL against *S. aureus* (ATCC 25923), *B. subtilis* (ATCC 19659), and *P. aeruginosa* (ATCC 27853) (Table 1) [26]. In *Calophyllum brasiliense*, oxygenated xanthone 1,5-dihydroxyxanthone (**29**) (Figure 5) showed moderate activity against the strains *S. aureus*, *S. saprophyticus*, and *S. agalactiae* and weak activity against *B. cereus* (Table 1) [27].

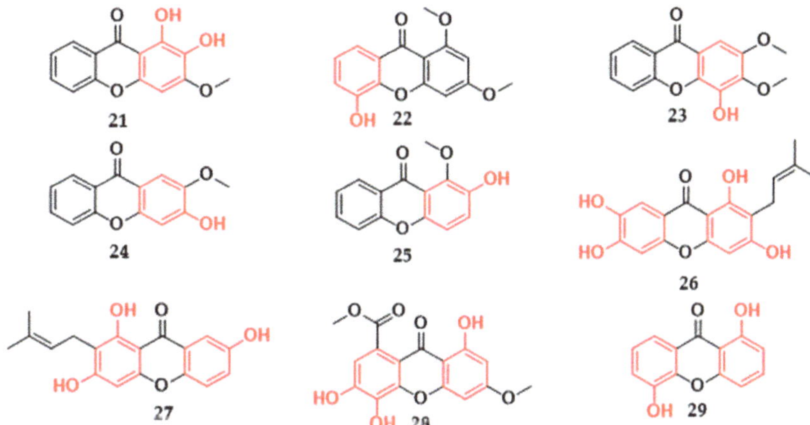

Figure 5. Xanthones with antibacterial activity isolated from plants in Brazil. 3,4-dihydroxy-2-methoxyxanthone (**21**), 5-hydroxy-1,3-dimethoxyxanthone (**22**), 4-hydroxy-2,3-dimethoxyxanthone (**23**), 3-hydroxy-2-methoxyxanthone (**24**), 2-hydroxy-1-methoxyxanthone (**25**), Assiguxanthone B (**26**), 1,3,7′trihydroxy-2-(3-methylbut-2-enyl)-xanthone (**27**), 8-carboxymethyl-1,3,5,6-tetrahydroxyxanthone (**28**), 1,5-dihydroxyxantona (**29**).

The substitution pattern in xanthone molecules is of high importance to understand their antimicrobial activity. Studies that relate structure and activity indicate that hydroxyl groups at the C-3 and C-6 positions and side chain prenyl groups at the C-2 carbon play a prominent antibacterial role against MRSA [11,109], as seen in compound (**26**). Prenylated xanthones have great antimicrobial potential, as the existence of apolar groups increases membrane permeability and can act as modulators of lipid affinity and cellular bioavailability [23]. However, a decrease in the anti-MRSA activity of synthetic xanthones modified with diethylamine groups was verified when long nonpolar alkyl chains were introduced in the structure, which confers hydrophobic characteristics to the structure and would have made it difficult to penetrate the peptidoglycan layer of the bacteria [110].

Methoxyl substituents at positions C-3 and C-7 and isoprenyl at positions C-2 and C-8 seem to contribute to better activity against Gram-positive bacteria, while geraniol and isoprenyl groups at positions C-4 and C-2 may contribute to a higher antibacterial activity against *P. aeruginosa*. However, bulky groups at the C-1 position may decrease the antibacterial activity [111]. In methicillin-resistant *S. aureus* strains, xanthones with hydroxyls at positions C-6, C-5, and C7, prenyl groups at C-4 and C-7, and dimethyl chromene rings at C-2 and C-3 present efficient activity [112]. Furthermore, Yan et al. [113], in a study evaluating xanthones with various substituents at the C-1, C-3, and C-6 positions, found that an acetyl substituent group at the C-1 position showed membrane selectivity against *S. aureus*, a higher inhibition of biofilm formation, and better antibacterial activity in vivo when compared to other xanthones [113].

Durães et al. [114] evaluated the potential of xanthones as efflux pump inhibitors in *Staphylococcus aureus* (272123) and *Salmonella enterica* serovar Typhimurium (SL1344) strains. It was observed that the introduction of hydroxyl groups in each aromatic ring at positions C-1 and C-7, or in the same plane as the ketone functional group of xanthone, as well as the presence of bulky groups at position C-1, was demonstrated to be efficient in inhibiting *Salmonella enterica*. For *Staphylococcus aureus*, methoxyl groups at the C-6 position can exert an important influence on the inhibition of the efflux pump; however, bulky groups at C-1 seem to impair the activity [114].

Similar to flavonoids, the mechanism of action of xanthones is also related to the ability to partition the bacterial cell membrane and has a greater relationship with Gram-positive bacteria. An important point is that xanthones have recently been widely explored due to the wide range of different molecular substitutions that can modulate several biological responses, such as those mentioned above [115]. This is because, in addition to the multiple functions of xanthones, natural structures can serve as inspiration for the synthesis of new xanthones with diverse molecular functions. Therefore, the molecules discussed below have the potential for use in natura or may serve as a basis for future synthetic derivations.

6. Coumarins

Coumarins are chemical compounds of natural or synthetic origin; they are stable and of low molecular weight, called benzo-α-pyrone, and have the isomeric chemical structure of chromones (benzo-γ-pyrone). As natural products, they can be produced by fungi and bacteria and in the secondary metabolism of plants [116,117]. The first report of isolation occurred in 1820 from the seeds of *Dipteryx odorata* species, popularly known as tonka bean, by A. Vogel, a regular member of the Royal Academy of Science in Munich [118]. Currently, the isolation of coumarins in hundreds of plant species is already known, with greater occurrence in the families *Apiaceae*, *Rutaceae*, *Asteraceae*, *Fabaceae*, *Oleaceae*, *Moraceae*, and *Thymelaceacea* [119]. Although coumarins are reported to occur in all parts of plants, they are most commonly found in fruits, roots, stems, and leaves [120].

Coumarins are compounds derived from the ortho-hydroxylation reaction of cinnamic acid, forming 2-coumaric acid, which undergoes cis–trans isomerization followed by lactonization, giving rise to the base structure of coumarin [37,120]. Coumarins can be divided into four major subgroups, which include single coumarins, pyrone-substituted coumarins, furanocoumarins, and pyranocoumarins [121]. From a pharmacological point of view, coumarin compounds and their derivatives are of great importance in the prevention and treatment of diseases. Some are used as anticoagulants, antitumor agents, antispasmodics, choleretic drugs, and antibiotics, such as the drug novobiocin, a potent inhibitor of Gram-positive bacteria [122,123].

The coumarins tanizin (**30**) and gravellifenore (**31**) (Figure 6) are examples of coumarin compounds with good inhibitory potential against bacterial strains. Isolated from the bark of *Helietta apiculata* species, both compounds showed MIC values $\leq$ 50 µg/mL against Gram-positive and Gram-negative bacteria (Table 1) [28]. Structurally, the two substances differ only in the C-5 and C-7 carbon positions, which may be reflected in the different MIC values seen in Table 1. Studies indicate that the coumarin substitution patterns are related to their pharmacological and biochemical activities and therapeutic applications [119].

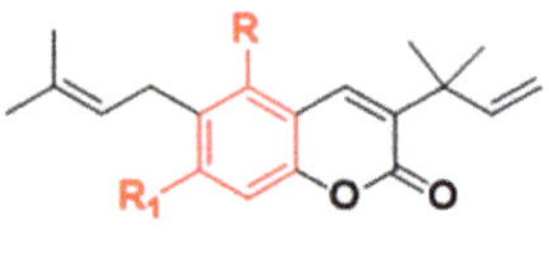

30 R = OH; R_1 = OCH$_3$
31 R = H; R_1 = OH

Figure 6. Coumarins with antibacterial activity isolated from plants in Brazil. Tanizin (**30**), and gravellifenore (**31**).

In coumarins, it was seen that the addition of polar or nonpolar groups and their respective locations can interfere with the antibacterial activity. The introduction of a hydroxyl group at the C-7 position of the aromatic ring can reduce the antibacterial activity. However, the addition of two methoxyl groups at the C-7 and C-8 positions can make the compounds more active against Gram-positive and Gram-negative microorganisms [124]. Other studies have revealed that, in mono-oxygenated coumarins, the addition of methoxyl or methyl groups at the C-6 and C-7 positions may decrease the antibacterial activity against Gram-positive strains. It is suggested that the lipophilic character and planar structure of coumarins are factors that may confer reasonably high antibacterial activities on these structures. [125]. In a preliminary study of the structure–activity relationship, it was shown that the inclusion of biphenyl groups at the C-3 position can confer strong activity against the DNA helicases of Gram-positive bacteria. However, when ester functionality was tested at this position, inactive compounds were found. Furthermore, it was seen that the change of substituents at the C-7 position influences the effectiveness against both DNA helicases [126]. Additionally, some coumarins are potential efflux pump inhibitors. In a study with seven coumarins, it was observed that phenyl groups at C-4,2-methylbutanoyl at C-6 and prenyl at C-8 seem to contribute to the inhibition of the efflux pump in the clinical strain *S. aureus* 1199B [127].

7. Phenolic Acids

Phenolic acids (phenolcarboxylic acids) are ubiquitous compounds in plants and are frequently reported in fruits, vegetables, spices, and herbs, in addition to being found in fungi and bacteria. In food, phenolic acids are associated with nutritional, antioxidant, and organoleptic properties, while in the survival mechanism of plants, these compounds contribute to protein synthesis, photosynthesis, nutrient absorption, and allelopathy [128,129]. Several phenolic acids have been described to have many biological activities, including antifungal, antioxidant, antibacterial, and anti-inflammatory activities [130–138]

Structurally, phenolic acids are characterized by having an aromatic ring directly linked to a hydroxyl group (phenolic hydroxyl) and a carboxyl group and can be found in nature conjugated to esters, ethers, simple sugars, vegetable polymers, organic acids, or polyphenols [139,140]. They are compounds produced in greater numbers in the shikimic acid pathway through the precursors L-phenylalanine or L-tyrosine. These phenolic compounds have two distinct structures: hydroxycinnamic and hydroxybenzoic. Hydroxycinnamic acids or cinnamic acids (C6-C3) can be produced by all plants through the deamination of L-phenylalanine; however, the formation of cinnamic acids from L-tyrosine is restricted to some plants. Hydroxybenzoics (C6-C1), on the other hand, can be formed at the beginning of the shikimate pathway through intermediates and through alternative routes by derivatives of hydroxycinnamic acids [37,141–143]. Similar to other compounds previously discussed in the present study, phenolic acids affect bacteria through damage to the cell membrane wall. This effect leads to a change in cell surface hydrophobicity and charge, with consequent leakage of cytoplasmic content [144]. In a study of the antibacterial mechanism using rosmarinic acid, inhibition of Gram-positive and Gram-negative bacteria was estimated by the destruction of bacterial cells and cellular proteins in addition to inhibition of Na+/K+-ATP-ase activity [145]. However, although the mechanism of action on the cell wall of bacteria has been elucidated, as well as coumarins, phenolic acids can also act on fungi through the interaction between caffeic acid derivatives and the 1,3-β-glucan synthase fraction [146].

Gallic (**32**) and protocatechuic (**33**) acids (Figure 7) were tested for their antibacterial activities against Gram-positive and Gram-negative strains. Gallic acid isolated from *Himatanthus sucuuba* species in northern Brazil showed inhibitory activity against *S. aureus* (MRSA), *S. epidermidis*, *P. mirabilis*, *S. haemolyticus*, and *E. coli*, with values ranging between 31 and 125 μg/mL (Table 1). The highest efficacy values were against methicillin-resistant *S. aureus* and *P. mirabilis* [29]. The protocatechuic acid isolated from the aerial parts of *Calophyllum brasiliense* differs from gallic acid only by the absence of hydroxyl in the C-5 position of the aromatic ring, showing MIC against all tested strains with values from 200 to 700 μg/mL (Table 1). In the same study, gallic acid was tested against the same bacterial strains but did not show activity against the tested microorganisms up to a concentration of 1000 μg/mL [27].

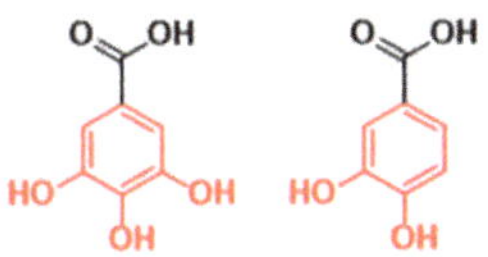

Figure 7. Phenolic acids with antibacterial activity isolated from plants in Brazil. Gallic acid (**32**) and protocatechuic acid (**33**).

As with other phenolic compounds, the antimicrobial activity of phenolic acids depends on the chemical structure of the compounds. It has been seen that the number and positions of the substituent groups on the aromatic ring, the length of the side chain, and the unsaturations influence the increase or decrease of the antibacterial activity of phenolic acids [147]. For example, in studies of the inhibitory action of caffeic acid and caffeic acid alkyl esters against *S. aureus* and *E. coli*, it was observed that compounds with long alkyl side chains were more efficient in inhibiting Gram-positive bacteria, and medium alkyl chains were more potent in inhibiting Gram-negative bacteria [148].

The lipophilicity of phenolic acids is one of the determining factors for the antimicrobial potential; the greater the lipophilicity of these compounds, the greater the inhibitory capacity. Studies have observed that as hydroxyl groups are replaced by methoxyl, there is unsaturation in the molecules, in addition to a decrease in pH, which causes acidification in the plasmatic membrane of pathogens; thus, there is an increase in the lipophilicity of phenolic acids, consequently improving the antibacterial activity [14,15].

8. Other Phenolic Compounds

Other types of phenolic compounds were also found in plants from Brazil and tested for their inhibitory action against microorganisms. The neolignan, dihydro-dehydrodiconiferyl alcohol (**34**), and the lignan, Lyoniresinol (**35**) (Figure 8), were active against *S. aureus* strains. Compound (**34**), isolated from *Styrax ferrugineus* species, showed potent inhibitory action against *S. aureus* ATCC 12228 (MIC 20 µg/mL) [30]. Lignan (**35**) from *Clusia burlemarxii* leaves showed a MIC value of 25 µg/mL against the *S. aureus* strain (ATCC 6538) [19]. Lignans and neolignans are natural products formed by two phenylpropane units (C6C3) through oxidative dimerization [149,150]. When the formation of the molecule occurs by the β, β′ bond, the term used is lignan; in the absence of this bond, the molecule formed by two C6-C3 units is neolignan [151]. Regarding the structure activities of the lignans, the substituents and the absolute configuration in the hydrofuran rings seem to affect the antibacterial activity in these compounds, and the presence of methoxyl groups in the C-9 and C-9′ positions can cause the inactivation of the compound against Gram-positive bacteria [152]. Although there are lignans that are reported to have antibacterial activity [153–156], studies relating to the structure-activity of these compounds are scarce.

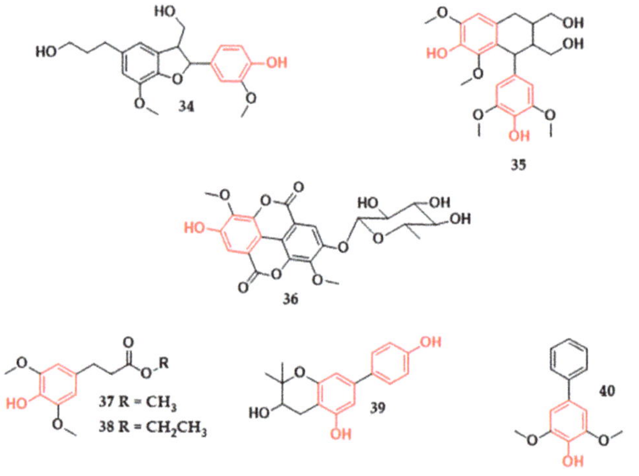

Figure 8. Other classes of phenolic compounds with antibacterial activity isolated from plants in Brazil. Dihydrodehydrodiconiferyl alcohol (**34**), lyoniresinol (**35**), Acid 3,3′-dimethoxyellagic acid-4-*O*-α-rhamnopyranoside (**36**), (E)-methyl-4-hydroxy-3,5-dimethoxycinnamate (**37**), (E)-ethyl-4-hydroxy-3,5-dimethoxycinnamate (**38**), 2,2-dimethyl-3,5-dihydroxy-7-(4-hydroxyphenyl)chromane (**39**), aucuparin (**40**).

The ellagic acid derivative (**36**) (Figure 8) isolated from the roots of *Euphorbia tirucalli* exhibited inhibitory activity against *S. aureus* (ATCC 6538) and *E. coli* (ATCC 8739), being more active against *S. aureus* (Table 1) [16]. The cinnamic acid derivatives (E)-methyl-4-hydroxy-3,5-dimethoxycinnamate (**37**) and (E)-ethyl-4-hydroxy-3,5-dimethoxycinnamate (**38**) (Figure 8) from the bark of *Hellietta apiculata* species were active against Gram-positive and Gram-negative bacteria (Table 1). In particular, compound (**37**) showed a greater inhibitory effect than compound (**38**) against *B. cereus* (ATCC 33019) and *E. aerogenes* (ATCC 13048), with a minimum inhibitory concentration value up to four-times lower [28]. Compounds (**37**) and (**38**) structurally differ only in the methyl and ethyl substituents of the side chain of the compound.

Biphenyl 2,2-dimethyl-3,5-dihydroxy-7-(4-hydroxyphenyl) chromane was reported from *Clusia burlemarxii* species (**39**) (Figure 8) and showed antibacterial activity against four pathogens (Table 1). The higher activity was against the microorganisms *M. luteus* ATCC 10240 (MIC = 25 µg/mL) and *S. aureus* ATCC 6538 (MIC = 50 µg/mL) [19]. Another biphenyl, known as aucuparin (**40**) (Figure 8), isolated from *Kielmeyera coriacea* has shown antibacterial activity against *S. aureus* (ATCC 25922), *E. coli* (ATCC 25922), *B. subtilis* (ATCC 6623), and *P. aeruginosa* (ATCC 15442) (Table 1). It is more active against Gram-positive bacteria, with minimum inhibitory concentrations of 3.12 µg/mL and 12.5 µg/mL against *B. subtilis* and *S. aureus*, respectively [25]. Studies indicate that the formation of biphenyl aucuparin in plants occurs in response to attacks by microorganisms [157].

9. Phenolic Compounds and a Possible Farm-to-Fork Influence

The use of natural compounds comprises an important strategy in the control of pathogens, from the farm to the health of the consumer. Studies have evaluated that, in cattle fed a forage-based diet (containing a high presence of phenolic compounds), there was less elimination of *E. coli* O157:H7 in the animals' feces [158]. The load of *E. coli* O157:H7 shed in bovine feces is of special importance since super-shedding events have already been reported and are responsible for the high presence of this pathogen in animal feces [159]. The super-shedding event was related to the presence of 80% hide contamination in all animals present in a feedlot [160]. Thus, the action of phenolic compounds has the potential to be explored in animal feed to reduce the concentration of bacteria in feces, which consequently could reduce contamination in animal hide and its prevalence in the industry.

In addition, the application of phenolic compounds in food production has been extensively discussed recently. In a study performed by Zamuz et al. [161], the authors mentioned the potential of using phenolic compounds in the fight against *Listeria monocytogenes*, an important human pathogen that presents characteristics of microbial growth under refrigeration and potential for biofilm formation. The production of microbial biofilms in the food industry represents one of the main contamination factors during the food process [162]. This microbial strategy allows the formation of layers of cells that help in the persistence of the microorganism during food production, since the application of sanitizers will act on the outermost layers, protecting the deeper layers of microorganisms from the action of the substance [163]. In several studies, flavonoids, such as naringin, were indicated as agents with the potential to be nonspecific inhibitors of autoinducer-mediated cell–cell signaling in *E. coli* bacteria [164]. This interference can have a direct impact on bacterial biofilm formation since it is estimated that communication between bacterial cells plays a crucial role in biofilm formation. Nevertheless, other studies have identified the ability of flavans to inhibit the formation of *S. aureus* biofilms, both with natural molecules and with synthetic forms [165]. This fact allows the substance to be obtained in larger quantities to enable large-scale application. Additionally, the presence of microbial biofilms can lead to resistance to these compounds, since subinhibitory amounts of sanitizer could lead to selective pressure on the isolates and the consequent development of resistance mechanisms [166]. As previously mentioned, the use of synthetic xanthones to combat multidrug resistance has already been addressed by Duraes et al. [114]. The authors highlight the role of seven xanthone compounds in decreasing the efflux of ethidium bromide, which can be translated as an efflux pump inhibitor [114]. Drug efflux is the main mechanism of resistance in Gram-negative bacteria [167]. This allows the bacteria to regulate the cell's internal contents and enable the removal of toxic substances, including antimicrobial compounds [168]. Thus, the use of phenolic compounds could be another ally in combating biofilm contamination and antimicrobial resistance.

Remaining in food production, the use of phenolic compounds can be a good strategy to extend the shelf life of a product, acting as an active component in the packaging [169]. In a study performed by Gaikwad [170], the authors verified that the use of pyrogallol coated in a polymeric film obtained an antimicrobial effect in both Gram-positive (*S. aureus*) and Gram-negative (*E. coli*) bacteria. Such results highlight the power of phenolic compounds since the molecule affected different bacterial groups. An important point to be emphasized in the use of phenolic compounds in packaging is that these substances, in addition to their antimicrobial action, also have an antioxidant effect, which contributes to increasing the shelf life and maintaining the chemical quality of the product. In a study performed by Kalogianni [171], the author discussed important aspects of the use of phenolic compounds in the meat industry, such as the direct application of the compounds in the meat and inclusion in the animal diet.

In addition to animal and food production, the use of phenolic compounds through diet has been widely studied in human health. Although most studies have focused on the antioxidant properties of these substances, some studies have aimed to investigate the role of phenolic compounds in the gut microbiota. It is important to note that in recent years, several studies have linked diseases with the intestinal microbiota through dysbiosis [172], and a diet composed of polyphenols could be a key factor in helping this problem, since phenolic compounds are also related to the expression of prebiotic properties, leading to antimicrobial activity against microbial pathogens [172]. Furthermore, in a study performed by Tuohu et al. [173], the authors pointed out that, although the fundamentals about the role of polyphenols are not clear in the gut microbiota, the anti-age, antimicrobial, and anti-inflammatory properties point to a potential benefit when consumed in adequate amounts. The benefits of a polyphenol-based diet have been evaluated in vivo; for example, in the study performed in Brazil by Gris et al. [174], the authors verified that the administration of red wines with a high phenolic content promoted high antioxidant activity in mice. Furthermore, in another study in the country, the authors highlighted that extracts from murici and gabiroba (Brazilian Cerrado fruits) obtained protective effects against genotoxic and mutagenic inducers in mice [175]. In addition, in a study performed by Siqueira et al. [176], the authors verified that extracts derived from *Spondias tuberosa* (a fruit from a plant native to northeast Brazil) presented the molecules of chlorogenic acid, caffeic acid, rutin, and isoquercitrin that showed therapeutic potential in inflammatory conditions in mice. Therefore, as mentioned throughout the text, phenolic compounds have wide potential for use, from the farm in animal production to the quality of human health, and represent a strong ally in the search for natural compounds that help in the fight against microbial contamination and, at the same time, help in other activities such as cellular antioxidation and anti-inflammatory properties.

10. Conclusions and Future Perspectives

As discussed throughout the manuscript, there are several phenolic compounds with different properties, such as antimicrobial, antiaging, and anti-inflammatory properties. In the present study, we discuss substances isolated from plants in Brazil, with a focus on their antimicrobial capacity. We emphasize that, although several studies showed promising values of microbial inactivation, the studies are still in the initial stage "in vitro", and their application on a large scale is still a limiting factor in the application of these compounds. Furthermore, several studies have analyzed the properties of polyphenols in *Staphylococcus aureus* and *Escherichia coli*, and, although these two microorganisms represent two distinct classes (Gram-positive and negative, respectively), several other pathogens still need to be evaluated to test the effectiveness of the compounds, as well as a possible cross effect in a bacterial community. However, the results presented shed light on promising compounds that can help in food production and human health in the coming years, as we are experiencing the era of multiresistant microorganisms, which, in addition to costing countless lives, also has a high global economic cost [177].

Author Contributions: Conceptualization, A.C.K. and V.S.C.; methodology, A.C.K.; formal analysis A.C.K. and V.S.C.; investigation, A.C.K. and V.S.C.; data curation, A.C.K.; writing—original draft preparation, A.C.K. and V.S.C.; writing—review and editing, A.C.K. and V.S.C.; supervision, V.S.C. All authors have read and agreed to the published version of the manuscript.

Funding: This research received no external funding.

Institutional Review Board Statement: Not applicable.

Informed Consent Statement: Not applicable.

Data Availability Statement: This study consisted of a literature review, and we discussed results previously published in scientific manuscripts. References are included in Table 1.

Conflicts of Interest: The authors declare no conflict of interest.

References

1. Lister, J. On the Antiseptic Principle in the Practice of Surgery. *BMJ* **1867**, *2*, 246–248. [CrossRef] [PubMed]
2. Pitt, D.; Aubin, J.-M. Joseph Lister: Father of modern surgery. *Can. J. Surg.* **2012**, *55*, E8–E9. [CrossRef] [PubMed]
3. Tan, S.Y.; Tatsumura, Y. Alexander Fleming (1881–1955): Discoverer of penicillin. *Singap. Med. J.* **2015**, *56*, 366–367. [CrossRef] [PubMed]
4. Karam, G.; Chastre, J.; Wilcox, M.H.; Vincent, J.-L. Antibiotic strategies in the era of multidrug resistance. *Crit. Care* **2016**, *20*, 136. [CrossRef]
5. Guerra, A.; Reis, L.K.; Borges, F.L.G.; Ojeda, P.T.A.; Pineda, D.A.M.; Miranda, C.O.; de Lima Maidana, D.P.F.; dos Santos, T.M.R.; Shibuya, P.S.; Marques, M.C.; et al. Ecological restoration in Brazilian biomes: Identifying advances and gaps. *For. Ecol. Manag.* **2020**, *458*, 117802. [CrossRef]
6. Lattanzio, V. Phenolic Compounds: Introduction. In *Natural Products: Phytochemistry, Botany and Metabolism of Alkaloids, Phenolics and Terpenes*; Ramawat, K.G., Mérillon, J.M., Eds.; Springer: Berlin/Heidelberg, Germany, 2013; pp. 1543–1580.
7. Al Mamari, H.H. Phenolic Compounds: Classification, Chemistry, and Updated Techniques of Analysis and Synthesis. In *Phenolic Compounds: Chemistry, Synthesis, Diversity, Non-Conventional Industrial, Pharmaceutical and Therapeutic Applications*; Badria, F.A., Ed.; IntechOpen: London, UK, 2021; pp. 73–94.
8. Minatel, I.O.; Borges, C.V.; Ferreira, I.M.; Gomez, H.A.G.; Chen, C.O.; Lima, G.P.P. Phenolic Compounds: Functional Properties, Impact of Processing and Bioavailability. In *Phenolic Compounds—Biological Activity*, 1st ed.; Soto-Hernandez, M., Palma-Tenango, M., Garcia-Mateos, M.D.R., Eds.; IntechOpen: Rijeka, Croatia, 2017; pp. 1–23.
9. González-Burgos, E.; Gómez-Serranillos, M.P. Effect of Phenolic Compounds on Human Health. *Nutrients* **2021**, *13*, 3922. [CrossRef]
10. Fang, Y.; Lu, Y.; Zang, X.; Wu, T.; Qi, X.; Pan, S.; Xu, X. 3D-QSAR and docking studies of flavonoids as potent *Escherichia coli* inhibitors. *Sci. Rep.* **2016**, *6*, 23634. [CrossRef]
11. Damen, F.; Mpetga, J.D.S.; Demgne, O.M.F.; Çelik, I.; Wamba, B.E.N.; Tapondjou, L.A.; Beng, V.P.; Levent, S.; Kuete, V.; Tene, M. Roeperone A, a new tetraoxygenated xanthone and other compounds from the leaves of *Hypericum roeperianum* Schimp. (Hypericaceae). *Nat. Prod. Res.* **2020**, *36*, 2071–2077. [CrossRef]
12. De Azevedo Maia, G.L.; Dos Santos Falcão-Silva, V.; Aquino, P.G.V.; De Araújo-Júnior, J.X.; Tavares, J.F.; Da Silva, M.S.; Rodrigues, L.C.; De Siqueira-Júnior, J.P.; Barbosa-Filho, J.M. Flavonoids from Praxelis clematidea R.M. King and Robinson Modulate Bacterial Drug Resistance. *Molecules* **2011**, *16*, 4828–4835. [CrossRef]
13. Ávila, H.P.; Smânia, E.D.F.A.; Monache, F.D.; Smânia, A. Structure–activity relationship of antibacterial chalcones. *Bioorg. Med. Chem.* **2008**, *16*, 9790–9794. [CrossRef]
14. Sánchez-Maldonado, A.; Schieber, A.; Gänzle, M. Structure-function relationships of the antibacterial activity of phenolic acids and their metabolism by lactic acid bacteria. *J. Appl. Microbiol.* **2011**, *111*, 1176–1184. [CrossRef] [PubMed]
15. Cueva, C.; Moreno-Arribas, M.V.; Martín-Álvarez, P.J.; Bills, G.; Vicente, M.F.; Basilio, A.; Rivas, C.L.; Requena, T.; Rodríguez, J.M.; Bartolomé, B. Antimicrobial activity of phenolic acids against commensal, probiotic and pathogenic bacteria. *Res. Microbiol.* **2010**, *161*, 372–382. [CrossRef] [PubMed]
16. Lima, M.d.F.R.d.; Cavalcante, L.A.; Costa, E.C.T.D.A.; de Veras, B.O.; da Silva, M.V.; Cavalcanti, L.N.; Araújo, R.M. Bioactivity flavonoids from roots of *Euphorbia tirucalli* L. *Phytochem. Lett.* **2021**, *41*, 186–192. [CrossRef]
17. de Oliveira, D.M.; de Oliveira, D.B.C.; Nunes, Y.R.F.; Alves, T.M.D.A.; Kohlhoff, M.; Andrade, A.A.; Cota, B.B. Natural Occurring Phenolic Derivatives from *Mauritia flexuosa* (Buriti) Stems and Their Potential Antibacterial Activity against Methicillin-Resistant *Staphylococcus aureus* (MRSA). *Chem. Biodivers.* **2022**, *19*, e202100788. [CrossRef]

18. Çiçek, S.S.; Pérez, M.G.; Wenzel-Storjohann, A.; Bezerra, R.M.; Segovia, J.F.O.; Girreser, U.; Kanzaki, I.; Tasdemir, D. Antimicrobial Prenylated Isoflavones from the Leaves of the Amazonian Medicinal Plant *Vatairea guianensis* Aubl. *J. Nat. Prod.* **2022**, *85*, 927–935. [CrossRef] [PubMed]

19. Ribeiro, P.R.; Ferraz, C.G.; Guedes, M.L.; Martins, D.; Cruz, F.G. A new biphenyl and antimicrobial activity of extracts and compounds from Clusia burlemarxii. *Fitoterapia* **2011**, *82*, 1237–1240. [CrossRef]

20. Ferreira, E.O.; Salvador, M.J.; Pral, E.M.F.; Alfieri, S.C.; Ito, I.Y.; Dias, D.A. A New Heptasubstituted (E)-Aurone Glucoside and Other Aromatic Compounds of Gomphrena agrestis with Biological Activity. *Z. Naturforsch. C* **2004**, *59*, 499–505. [CrossRef]

21. Falcão-Silva, V.S.; Silva, D.A.; Souza, M.D.F.V.; Siqueira-Junior, J.P. Modulation of drug resistance in *staphylococcus aureus* by a kaempferol glycoside from *herissantia tiubae* (malvaceae). *Phytother. Res.* **2009**, *23*, 1367–1370. [CrossRef]

22. Costa, G.; Endo, E.; Cortez, D.; Nakamura, T.; Nakamura, C.; Filho, B.D. Antimicrobial effects of Piper hispidum extract, fractions and chalcones against Candida albicans and Staphylococcus aureus. *J. Med. Mycol.* **2016**, *26*, 217–226. [CrossRef]

23. Coqueiro, A.; Choi, Y.H.; Verpoorte, R.; Gupta, K.B.S.S.; De Mieri, M.; Hamburger, M.; Young, M.C.M.; Stapleton, P.; Gibbons, S.; Bolzani, V.D.S. Antistaphylococcal Prenylated Acylphoroglucinol and Xanthones from *Kielmeyera variabilis*. *J. Nat. Prod.* **2016**, *79*, 470–476. [CrossRef]

24. Pinheiro, L.; Nakamura, C.V.; Filho, B.P.D.; Ferreira, A.G.; Young, M.C.M.; Cortez, A.G. Antibacterial xanthones from Kielmeyera variabilis mart. (Clusiaceae). *Mem. Inst. Oswaldo Cruz.* **2003**, *98*, 549–552. [CrossRef] [PubMed]

25. Cortez, D.A.G.; Filho, B.A.A.; Nakamura, C.V.; Filho, B.P.D.; Marston, A.; Hostettmann, K. Antibacterial Activity of a Biphenyl and Xanthones from Kielmeyera coriacea. *Pharm. Biol.* **2002**, *40*, 485–489. [CrossRef]

26. Araújo, M.G.D.F.; Hilário, F.; Nogueira, L.G.; Vilegas, W.; Dos Santos, L.C.; Bauab, T.M. Chemical Constituents of the Methanolic Extract of Leaves of Leiothrix spiralis Ruhland and Their Antimicrobial Activity. *Molecules* **2011**, *16*, 10479–10490. [CrossRef] [PubMed]

27. Pretto, J.B.; Cechinel-Filho, V.; Noldin, V.F.; Sartori, M.R.K.; Isaias, D.E.B.; Cruz, A.B. Antimicrobial Activity of Fractions and Compounds from Calophyllum brasiliense (Clusiaceae/Guttiferae). *Z. Nat. C* **2004**, *59*, 657–662. [CrossRef]

28. Fernandes, T.S.; Copetti, D.; Carmo, G.D.; Neto, A.T.; Pedroso, M.; Silva, U.F.; Mostardeiro, M.A.; Burrow, R.E.; Dalcol, I.I.; Morel, A.F. Phytochemical analysis of bark from Helietta apiculata Benth and antimicrobial activities. *Phytochemistry* **2017**, *141*, 131–139. [CrossRef]

29. Silva, J.R.A.; Rezende, C.M.; Pinto, A.C.; Amaral, A.C.F. Cytotoxicity and Antibacterial Studies of iridoids and Phenolic Compounds Isolated from the Latex of Himatanthus sucuuba. *Afr. J. Biotechnol.* **2010**, *9*, 7357–7360.

30. Pauletti, P.M.; Araujo, A.R.; Young, M.C.M.; Giesbrecht, A.M.; Bolzani, V.D.S. nor-Lignans from the leaves of Styrax ferrugineus (Styracaceae) with antibacterial and antifungal activity. *Phytochemistry* **2000**, *55*, 597–601. [CrossRef] [PubMed]

31. Gani, M.A.; Shama, M. Phenolic Compounds. In *Bioactive Compounds: Biosynthesis, Characterization and Applications*, 1st ed.; Zepka, L.Q., Nascimento, T.C., Jacob-Lopes, E., Eds.; IntechOpen: London, UK, 2021; pp. 223–240.

32. Del Rio, D.; Rodriguez-Mateos, A.; Spencer, J.P.E.; Tognolini, M.; Borges, G.; Crozier, A. Dietary (Poly)phenolics in Human Health: Structures, Bioavailability, and Evidence of Protective Effects Against Chronic Diseases. *Antioxid. Redox Signal.* **2013**, *18*, 1818–1892. [CrossRef]

33. Sobiesiak, M. Chemical Structure of Phenols and Its Consequence for Sorption Processes. In *Phenolic Compounds: Natural Sources, importance and Applications*, 1st ed.; Soto-Hernandez, M., Palma-Tenango, M., Garcia-Mateos, M.D.R., Eds.; IntechOpen: Rijeka, Croatia, 2017; pp. 3–28.

34. Bravo, L. Polyphenols: Chemistry, Dietary Sources, Metabolism, and Nutritional Significance. *Nutr. Rev.* **1998**, *56*, 317–333. [CrossRef]

35. Alara, O.R.; Abdurahman, N.H.; Ukaegbu, C.I. Extraction of phenolic compounds: A review. *Curr. Res. Food Sci.* **2021**, *4*, 200–214. [CrossRef]

36. Dixon, R.A.; Paiva, N.L. Stress-Induced Phenylpropanoid Metabolism. *Plant Cell* **1995**, *7*, 1085–1097. [CrossRef] [PubMed]

37. Dewick, P.M. *Medicinal Natural Products*, 2nd ed.; John wiley & Sons, Ltd.: Chichester, UK, 2002; pp. 1–487.

38. Russell, W.; Duthie, G. Plant secondary metabolites and gut health: The case for phenolic acids. *Proc. Nutr. Soc.* **2011**, *70*, 389–396. [CrossRef]

39. Gao, J.; Yang, Z.; Zhao, C.; Tang, X.; Jiang, Q.; Yin, Y. A comprehensive review on natural phenolic compounds as alternatives to in-feed antibiotics. *Sci. China Life Sci.* **2022**, *65*, 1–17. [CrossRef] [PubMed]

40. Hussain, M.B.; Hassan, S.; Waheed, M.; Javed, A.; Farooq, M.A.; Tahir, A. Bioavailability and Metabolic Pathway of Phenolic Compounds. In *Plant Physiological Aspects of Phenolic Compounds*; Marcos, S.-H., Rosario, G.-M., Mariana, P.-T., Eds.; IntechOpen: Rijeka, Croatia, 2019.

41. Özçelik, B.; Orhan, I.; Toker, G. Antiviral and Antimicrobial Assessment of Some Selected Flavonoids. *Z. Naturforsch. C* **2006**, *61*, 632–638. [CrossRef]

42. Jung, H.-A.; Su, B.-N.; Keller, W.J.; Mehta, R.G.; Kinghorn, A.D. Antioxidant Xanthones from the Pericarp of Garcinia mangostana (Mangosteen). *J. Agric. Food Chem.* **2006**, *54*, 2077–2082. [CrossRef] [PubMed]

43. Zhu, H.; Xiao, Y.; Guo, H.; Guo, Y.; Huang, Y.; Shan, Y.; Bai, Y.; Lin, X.; Lu, H. The isoflavone puerarin exerts anti-tumor activity in pancreatic ductal adenocarcinoma by suppressing mTOR-mediated glucose metabolism. *Aging* **2021**, *13*, 25089–25105. [CrossRef]

44. Inoue, T.; Sugimoto, Y.; Masuda, H.; Kamei, C. Antiallergic Effect of Flavonoid Glycosides Obtained from *Mentha piperita* L. *Biol. Pharm. Bull.* **2002**, *25*, 256–259. [CrossRef]

45. A Fernández, M.; Sáenz, M.T.; García, M.D. Anti-inflammatory Activity in Rats and Mice of Phenolic Acids Isolated from *Scrophularia frutescens*. *J. Pharm. Pharmacol.* **1998**, *50*, 1183–1186. [CrossRef]

46. Kim, Y.-M.; Lee, C.-H.; Kim, H.-G.; Lee, H.-S. Anthraquinones Isolated from *Cassia tora* (Leguminosae) Seed Show an Antifungal Property against Phytopathogenic Fungi. *J. Agric. Food Chem.* **2004**, *52*, 6096–6100. [CrossRef]

47. Ani, V.; Varadaraj, M.C.; Naidu, K.A. Antioxidant and antibacterial activities of polyphenolic compounds from bitter cumin (*Cuminum nigrum* L.). *Eur. Food Res. Technol.* **2006**, *224*, 109–115. [CrossRef]

48. Swan, E.P. Health Hazards Associated with Extractives. In *Natural Products of Woody Plants I*; Rowe, J.W., Ed.; Springer Science & Business Media: New York, NY, USA, 1989; pp. 931–946.

49. Pietta, P.-G. Flavonoids as Antioxidants. *J. Nat. Prod.* **2000**, *63*, 1035–1042. [CrossRef]

50. Winkel, B.S.J. The Biosynthesis of flavonoids. In *The Science of Flavonoids*; Grotewold, E., Ed.; Springer: New York, NY, USA, 2006; pp. 71–96.

51. Iwashina, T. The Structure and Distribution of the Flavonoids in Plants. *J. Plant Res.* **2000**, *113*, 287–299. [CrossRef]

52. Wu, T.; He, M.; Zang, X.; Zhou, Y.; Qiu, T.; Pan, S.; Xu, X. A structure–activity relationship study of flavonoids as inhibitors of *E. coli* by membrane interaction effect. *Biochim. Biophys. Acta* **2013**, *1828*, 2751–2756. [CrossRef] [PubMed]

53. Andersen Oyvind, M.; Markham Kenneth, R. *Flavonoids: Chemistry, Biochemistry and Applications*; CRC Press: Boca Raton, FL, USA, 2005.

54. Swain, T. (Ed.) *Biochemistry of Plant Phenolics*; Springer Science & Business Media: Berlin/Heidelberg, Germany, 2013.

55. Ullah, A.; Munir, S.; Badshah, S.L.; Khan, N.; Ghani, L.; Poulson, B.G.; Emwas, A.-H.; Jaremko, M. Important Flavonoids and Their Role as a Therapeutic Agent. *Molecules* **2020**, *25*, 5243. [CrossRef] [PubMed]

56. Wu, T.; Zang, X.; He, M.; Pan, S.; Xu, X. Structure–Activity Relationship of Flavonoids on Their Anti-*Escherichia coli* Activity and Inhibition of DNA Gyrase. *J. Agric. Food Chem.* **2013**, *61*, 8185–8190. [CrossRef] [PubMed]

57. Górniak, I.; Bartoszewski, R.; Króliczewski, J. Comprehensive review of antimicrobial activities of plant flavonoids. *Phytochem. Rev.* **2018**, *18*, 241–272. [CrossRef]

58. Reygaert, W.C. The antimicrobial possibilities of green tea. *Front. Microbiol.* **2014**, *5*, 434. [CrossRef]

59. Fathima, A.; Rao, J.R. Selective toxicity of Catechin—A natural flavonoid towards bacteria. *Appl. Microbiol. Biotechnol.* **2016**, *100*, 6395–6402. [CrossRef]

60. Silva, L.N.; Da Hora, G.C.A.; Soares, T.A.; Bojer, M.S.; Ingmer, H.; Macedo, A.J.; Trentin, D.S. Myricetin protects Galleria mellonella against Staphylococcus aureus infection and inhibits multiple virulence factors. *Sci. Rep.* **2017**, *7*, 2823. [CrossRef]

61. Veiko, A.G.; Olchowik-Grabarek, E.; Sekowski, S.; Roszkowska, A.; Lapshina, E.A.; Dobrzynska, I.; Zamaraeva, M.; Zavodnik, I.B. Antimicrobial Activity of Quercetin, Naringenin and Catechin: Flavonoids Inhibit *Staphylococcus aureus*-Induced Hemolysis and Modify Membranes of Bacteria and Erythrocytes. *Molecules* **2023**, *28*, 1252. [CrossRef]

62. Tiza, N.U.; Thato, M.; Raymond, D.; Jeremy, K.; Burtram, C.F. Additive antibacterial activity of naringenin and antibiotic combinations against multidrug resistant Staphylococcus aureus. *Afr. J. Microbiol. Res.* **2015**, *9*, 1513–1518. [CrossRef]

63. Wen, Q.-H.; Wang, R.; Zhao, S.-Q.; Chen, B.-R.; Zeng, X.-A. Inhibition of Biofilm Formation of Foodborne *Staphylococcus aureus* by the Citrus Flavonoid Naringenin. *Foods* **2021**, *10*, 2614. [CrossRef] [PubMed]

64. Wang, L.-H.; Wang, M.-S.; Zeng, X.-A.; Xu, X.-M.; Brennan, C.S. Membrane and genomic DNA dual-targeting of citrus flavonoid naringenin against Staphylococcus aureus. *Integr. Biol.* **2017**, *9*, 820–829. [CrossRef] [PubMed]

65. Křížová, L.; Dadáková, K.; Kasparovska, J.; Kašparovský, T. Isoflavones. *Molecules* **2019**, *24*, 1076. [CrossRef]

66. Inbaraj, B.S.; Chen, B.H. Isoflavones in Foods and Ingestion in the Diet. In *Isoflavones: Chemistry, Analysis, Function and Effects*; Preedy, V.R., Ed.; The Society of Chemistry: Cambridge, UK, 2013; pp. 28–43.

67. Mukne, A.P.; Viswanathan, V.; Phadatare, A.G. Structure pre-requisites for isoflavones as effective antibacterial agents. *Pharmacogn. Rev.* **2011**, *5*, 13–18. [CrossRef]

68. Sato, M.; Tanaka, H.; Tani, N.; Nagayama, M.; Yamaguchi, R. Different antibacterial actions of isoflavones isolated from Erythrina poeppigiana against methicillin-resistant Staphylococcus aureus. *Lett. Appl. Microbiol.* **2006**, *43*, 243–248. [CrossRef]

69. Sadgrove, N.J.; Oliveira, T.B.; Khumalo, G.P.; van Vuuren, S.F.; van Wyk, B.E. Antimicrobial isoflavones and de-rivatives from Erythrina (Fabaceae): Structure activity perspective (Sar & Qsar) on experimental and mined values against Staphylococcus aureus. *Antibiotics* **2020**, *9*, 223.

70. Cruz, B.G.; dos Santos, H.S.; Bandeira, P.N.; Rodrigues, T.H.S.; Matos, M.G.C.; Nascimento, M.F.; Carvalho, G.; Filho, R.B.; Teixeira, A.M.; Tintino, S.R.; et al. Evaluation of antibacterial and enhancement of antibiotic action by the flavonoid kaempferol 7-O-β-D-(6″-O-cumaroyl)-glucopyranoside isolated from Croton piauhiensis müll. *Microb. Pathog.* **2020**, *143*, 104144. [CrossRef]

71. Wang, Q.; Wang, H.; Xie, M. Antibacterial mechanism of soybean isoflavone on Staphylococcus aureus. *Arch. Microbiol.* **2010**, *192*, 893–898. [CrossRef]

72. Hazni, H.; Ahmad, N.; Hitotsuyanagi, Y.; Takeya, K.; Choo, C.Y. Phytochemical constituents from Cassia alata with inhibition against methicillin-resistant Staphylococcus aureus (MRSA). *Planta Med.* **2008**, *74*, 1802–1805. [CrossRef]

73. Mandalari, G.; Bennett, R.; Bisignano, G.; Trombetta, D.; Saija, A.; Faulds, C.; Gasson, M.; Narbad, A. Antimicrobial activity of flavonoids extracted from bergamot (Citrus bergamia Risso) peel, a byproduct of the essential oil industry. *J. Appl. Microbiol.* **2007**, *103*, 2056–2064. [CrossRef] [PubMed]

74. Lopes, L.A.A.; Rodrigues, J.B.D.S.; Magnani, M.; de Souza, E.L.; de Siqueira-Júnior, J.P. Inhibitory effects of flavonoids on biofilm formation by Staphylococcus aureus that overexpresses efflux protein genes. *Microb. Pathog.* **2017**, *107*, 193–197. [CrossRef] [PubMed]

75. Zhuang, C.; Zhang, W.; Sheng, C.; Zhang, W.; Xing, C.; Miao, Z. Chalcone: A Privileged Structure in Medicinal Chemistry. *Chem. Rev.* **2017**, *117*, 7762–7810. [CrossRef]

76. Dimmock, J.; Elias, D.; Beazely, M.; Kandepu, N. Bioactivities of Chalcones. *Curr. Med. Chem.* **1999**, *6*, 1125–1149. [CrossRef]

77. Tran, T.D.; Do, T.H.; Tran, N.C.; Ngo, T.D.; Tran, C.D.; Thai, K.M. Synthesis and anti Methicillin resistant Staphylococcus aureus activity of substituted chalcones alone and in combination with non-beta-lactam antibiotics. *Bioorg. Med. Chem. Lett.* **2012**, *22*, 4555–4560. [CrossRef] [PubMed]

78. Alcaráz, L.; Blanco, S.; Puig, O.; Tomás, F.; Ferretti, F. Antibacterial Activity of Flavonoids Against Methicillin-resistant Staphylococcus aureus strains. *J. Theor. Biol.* **2000**, *205*, 231–240. [CrossRef]

79. Batovska, D.; Parushev, S.; Stamboliyska, B.; Tsvetkova, I.; Ninova, M.; Najdenski, H. Examination of growth inhibitory properties of synthetic chalcones for which antibacterial activity was predicted. *Eur. J. Med. Chem.* **2009**, *44*, 2211–2218. [CrossRef]

80. Fair, R.J.; Tor, Y. Antibiotics and Bacterial Resistance in the 21st Century. *Perspect. Med. Chem.* **2014**, *6*, S14459. [CrossRef]

81. World Health Organization. *Antimicrobial Resistance: Global Report on Surveillance*; WHO Press: Geneva, Switzerland, 2014; Available online: http://apps.who.int/iris/bitstream/10665/112642/1/9789241564748_eng.pdf (accessed on 4 February 2023).

82. Abreu, A.C.; McBain, A.J.; Simões, M. Plants as sources of new antimicrobials and resistance-modifying agents. *Nat. Prod. Rep.* **2012**, *29*, 1007–1021. [CrossRef]

83. Diniz-Silva, H.T.; Magnani, M.; de Siqueira, S.; de Souza, E.L.; de Siqueira-Júnior, J.P. Fruit flavonoids as modulators of norfloxacin resistance in Staphylococcus aureus that overexpresses norA. *LWT* **2017**, *85*, 324–326. [CrossRef]

84. Wang, S.-Y.; Sun, Z.-L.; Liu, T.; Gibbons, S.; Zhang, W.-J.; Qing, M. Flavonoids from *Sophora moorcroftiana* and their Synergistic Antibacterial Effects on MRSA. *Phytother. Res.* **2013**, *28*, 1071–1076. [CrossRef] [PubMed]

85. Lan, J.-E.; Li, X.-J.; Zhu, X.-F.; Sun, Z.-L.; He, J.-M.; Zloh, M.; Gibbons, S.; Mu, Q. Flavonoids from *Artemisia rupestris* and their synergistic antibacterial effects on drug-resistant *Staphylococcus aureus*. *Nat. Prod. Res.* **2019**, *35*, 1881–1886. [CrossRef] [PubMed]

86. Kumarihamy, M.; Tripathi, S.K.; Khan, S.; Muhammad, I. Schottiin, a new prenylated isoflavones from *Psorothamnus schottii* and antibacterial synergism studies between methicillin and fremontone against methicillin-resistant *Staphylococcus aureus* ATCC 1708. *Nat. Prod. Res.* **2021**, *36*, 2984–2992. [CrossRef] [PubMed]

87. Morel, C.; Stermitz, F.R.; Tegos, G.; Lewis, K. Isoflavones as Potentiators of Antibacterial Activity. *J. Agric. Food Chem.* **2003**, *51*, 5677–5679. [CrossRef]

88. Yuan, G.; Guan, Y.; Yi, H.; Lai, S.; Sun, Y.; Cao, S. Antibacterial activity and mechanism of plant flavonoids to gram-positive bacteria predicted from their lipophilicities. *Sci. Rep.* **2021**, *11*, 10471. [CrossRef]

89. Roberts, J.C. Naturally Occurring Xanthones. *Chem. Rev.* **1961**, *61*, 591–605. [CrossRef]

90. Vieira, L.M.M.; Kijjoa, A. Naturally-occurring xanthones: Recent developments. *Curr. Med. Chem.* **2005**, *12*, 2413–2446. [CrossRef] [PubMed]

91. Huang, Q.; Wang, Y.; Wu, H.; Yuan, M.; Zheng, C.; Xu, H. Xanthone Glucosides: Isolation, Bioactivity and Synthesis. *Molecules* **2021**, *26*, 5575. [CrossRef] [PubMed]

92. Negi, J.S.; Bisht, V.K.; Singh, P.; Rawat, M.S.M.; Joshi, G.P. Naturally Occurring Xanthones: Chemistry and Biology. *J. Appl. Chem.* **2013**, *2013*, 621459. [CrossRef]

93. Diderot, N.T.; Silvere, N.; Etienne, T. Xanthones as therapeutic agents: Chemistry and pharmacology. *Adv. Phytomed.* **2006**, *2*, 273–298. [CrossRef]

94. Gottlieb, O.R. Evolution of Xanthones in Gentianaceae and Guttiferae. In *Micromolecular Evolution, Systematics and Ecology*; Springer: Berlin/Heidelberg, Germany, 1982; pp. 89–95. [CrossRef]

95. Panda, S.; Chand, M.; Sakhuja, R.; Jain, S. Xanthones as potential antioxidants. *Curr. Med. Chem.* **2013**, *20*, 4481–4507. [CrossRef] [PubMed]

96. Sukandar, E.R.; Kaennakam, S.; Raab, P.; Nöst, X.; Rassamee, K.; Bauer, R.; Siripong, P.; Ersam, T.; Tip-Pyang, S.; Chavasiri, W. Cytotoxic and Anti-Inflammatory Activities of Dihydroisocoumarin and Xanthone Derivatives from *Garcinia picrorhiza*. *Molecules* **2021**, *26*, 6626. [CrossRef]

97. Nhan, N.-T.; Nguyen, P.-H.; Tran, M.-H.; Nguyen, P.-D.; Tran, D.-T.; To, D.-C. Anti-inflammatory xanthone derivatives from *Garcinia delpyana*. *J. Asian Nat. Prod. Res.* **2020**, *23*, 414–422. [CrossRef]

98. Seesom, W.; Jaratrungtawee, A.; Suksamrarn, S.; Mekseepralard, C.; Ratananukul, P.; Sukhumsirichart, W. Antileptospiral activity of xanthones from Garcinia mangostanaand synergy of gamma-mangostin with penicillin G. *BMC Complement. Altern. Med.* **2013**, *13*, 182. [CrossRef] [PubMed]

99. El-Nashar, H.A.S.; El-Labbad, E.M.; Al-Azzawi, M.A.; Ashmawy, N.S. A New Xanthone Glycoside from *Mangifera indica* L.: Physicochemical Properties and In Vitro Anti-Skin Aging Activities. *Molecules* **2022**, *27*, 2609. [CrossRef] [PubMed]

100. Wu, J.; Dai, J.; Zhang, Y.; Wang, J.; Huang, L.; Ding, H.; Li, T.; Zhang, Y.; Mao, J.; Yu, S. Synthesis of Novel Xanthone Analogues and Their Growth Inhibitory Activity Against Human Lung Cancer A549 Cells. *Drug Des. Dev. Ther.* **2019**, *13*, 4239–4246. [CrossRef]

101. Miladiyah, I.; Jumina, J.; Haryana, S.M.; Mustofa, M. Biological activity, quantitative structure–activity relationship analysis, and molecular docking of xanthone derivatives as anticancer drugs. *Drug Des. Dev. Ther.* **2018**, *12*, 149–158. [CrossRef]

102. Dineshkumar, B.; Mitra, A.; Manjunatha, M. Studies on the anti-diabetic and hypolipidemic potentials of mangiferin (Xanthone Glucoside) in streptozotocin-induced Type 1 and Type 2 diabetic model rats. *Int. J. Adv. Pharm. Sci.* **2010**, *1*, 75–85. [CrossRef]
103. Nguyen, C.N.; Trinh, B.T.D.; Tran, T.B.; Nguyen, L.-T.T.; Jäger, A.K.; Nguyen, L.-H.D. Anti-diabetic xanthones from the bark of Garcinia xanthochymus. *Bioorg. Med. Chem. Lett.* **2017**, *27*, 3301–3304. [CrossRef]
104. Pinto, D.C.; Fuzzati, N.; Pazmino, X.C.; Hostettmann, K. Xanthone and antifungal constituents from Monnina obtusifolia. *Phytochemistry* **1994**, *37*, 875–878. [CrossRef]
105. Cane, H.P.C.A.; Saidi, N.; Mustanir, M.; Darusman, D.; Idroes, R.; Musman, M. Evaluation of Antibacterial and Antioxidant Activities of Xanthone Isolated from Orophea corymbosa Leaf. *Rasayan J. Chem.* **2020**, *13*, 2215–2222. [CrossRef]
106. Iinuma, M.; Tosa, H.; Tanaka, T.; Asai, F.; Kobayashl, Y.; Shimano, R.; Miyauchi, K.-I. Antibacterial Activity of Xanthones from Guttiferaeous Plants against Methicillin-resistant *Staphylococcus aureus*. *J. Pharm. Pharmacol.* **1996**, *48*, 861–865. [CrossRef]
107. Auranwiwat, C.; Trisuwan, K.; Saiai, A.; Pyne, S.G.; Ritthiwigrom, T. Antibacterial tetraoxygenated xanthones from the immature fruits of Garcinia cowa. *Fitoterapia* **2014**, *98*, 179–183. [CrossRef] [PubMed]
108. Wang, W.; Liao, Y.; Huang, X.; Tang, C.; Cai, P. A novel xanthone dimer derivative with antibacterial activity isolated from the bark of *Garcinia mangostana*. *Nat. Prod. Res.* **2017**, *32*, 1769–1774. [CrossRef]
109. Dharmaratne, H.; Sakagami, Y.; Piyasena, K.; Thevanesam, V. Antibacterial activity of xanthones from *Garcinia mangostana* (L.) and their structure–activity relationship studies. *Nat. Prod. Res.* **2013**, *27*, 938–941. [CrossRef] [PubMed]
110. Zou, H.; Koh, J.J.; Li, J.; Qiu, S.; Aung, T.T.; Lin, H.; Lakshminarayanan, R.; Dai, X.; Tang, C.; Lim, F.H.; et al. Design and synthesis of amphiphilic xanthone-based, membrane-targeting antimicrobials with improved membrane selectivity. *J. Med. Chem.* **2013**, *56*, 2359–2373. [CrossRef]
111. Boonnak, N.; Karalai, C.; Chantrapromma, S.; Ponglimanont, C.; Fun, H.-K.; Kanjana-Opas, A.; Chantrapromma, K.; Kato, S. Anti-Pseudomonas aeruginosa xanthones from the resin and green fruits of Cratoxylum cochinchinense. *Tetrahedron* **2009**, *65*, 3003–3013. [CrossRef]
112. Miladiyah, I.; Rachmawaty, F.J. Potency of Xanthone Derivatives as antibacterial agent against Methicillin-Resistant Staphylococcus Aureus (MRSA). *J. Kedokt. Kesehat. Indones.* **2017**, *8*, 124–135. [CrossRef]
113. Yan, L.; Guan, T.; Wang, S.; Zhou, C.; Wang, M.; Wang, X.; Zhang, K.; Han, X.; Lin, J.; Tang, Q.; et al. Novel xan-thone antibacterials: Semi-synthesis, biological evaluation, and the action mechanisms. *Bioorg. Med. Chem.* **2023**, *83*, 117232.
114. Durães, F.; Resende, D.; Palmeira, A.; Szemerédi, N.; Pinto, M.; Spengler, G.; Sousa, E. Xanthones Active against Multidrug Resistance and Virulence Mechanisms of Bacteria. *Antibiotics* **2021**, *10*, 600. [CrossRef]
115. Pinto, M.M.M.; Palmeira, A.; Fernandes, C.; Resende, D.I.S.P.; Sousa, E.; Cidade, H.; Tiritan, M.E.; Correia-da-Silva, M.; Cravo, S. From Natural Products to New Synthetic Small Molecules: A Journey through the World of Xanthones. *Molecules* **2021**, *26*, 431. [CrossRef] [PubMed]
116. Mishra, S.; Pandey, A.; Manvati, S. Coumarin: An emerging antiviral agent. *Heliyon* **2020**, *6*, e03217. [CrossRef] [PubMed]
117. Akkol, E.K.; Genç, Y.; Karpuz, B.; Sobarzo-Sánchez, E.; Capasso, R. Coumarins and Coumarin-Related Compounds in Pharmacotherapy of Cancer. *Cancers* **2020**, *12*, 1959. [CrossRef]
118. Abernethy, J.L. The historical and current interest in coumarin. *J. Chem. Educ.* **1969**, *46*, 561. [CrossRef]
119. Matos, M.J.; Santana, L.; Uriarte, E.; Abreu, O.A.; Molina, E.; Yordi, E.G. Coumarins—Na Important Class of Phyto-chemicals. In *Phytochemicals: Isolation, Characterisation and Role in Human Health*, 1st ed.; Rao, A.V., Rao, L.G., Eds.; IntechOpen: Rijeka, Croatia, 2015; pp. 113–140.
120. Jain, P.K.; Joshi, H. Coumarin: Chemical and Pharmacological Profile. *J. Appl. Pharm. Sci.* **2012**, *2*, 236–240.
121. Lacy, A.; O'Kennedy, R. Studies on Coumarins and Coumarin-Related Compounds to Determine their Therapeutic Role in the Treatment of Cancer. *Curr. Pharm. Des.* **2004**, *10*, 3797–3811. [CrossRef]
122. Stefanachi, A.; Leonetti, F.; Pisani, L.; Catto, M.; Carotti, A. Coumarin: A Natural, Privileged and Versatile Scaffold for Bioactive Compounds. *Molecules* **2018**, *23*, 250. [CrossRef] [PubMed]
123. Riveiro, M.; De Kimpe, N.; Moglioni, A.; Vazquez, R.; Monczor, F.; Shayo, C.; Davio, C. Coumarins: Old Compounds with Novel Promising Therapeutic Perspectives. *Curr. Med. Chem.* **2010**, *17*, 1325–1338. [CrossRef]
124. Kayser, O.; Kolodziej, H. Antibacterial Activity of Simple Coumarins: Structural Requirements for Biological Activity. *Z. Naturforsch. C* **1999**, *54*, 169–174. [CrossRef] [PubMed]
125. De Souza, S.M.; Monache, F.D.; Smânia, A., Jr. Antibacterial Activity of Coumarins. *Z. Nat. C* **2005**, *60*, 693–700. [CrossRef]
126. Li, B.; Pai, R.; Di, M.; Aiello, D.; Barnes, M.H.; Butler, M.M.; Tashjian, T.F.; Peet, N.P.; Bowlin, T.L.; Moir, D.T. Couma-rin-based inhibitors of Bacillus anthracis and Staphylococcus aureus replicative DNA helicase: Chemical optimization, bi-ological evaluation, and antibacterial activities. *J. Med. Chem.* **2012**, *55*, 10896–10908. [CrossRef] [PubMed]
127. Roy, S.K.; Kumari, N.; Pahwa, S.; Agrahari, U.C.; Bhutani, K.K.; Jachak, S.M.; Nandanwar, H. NorA efflux pump in-hibitory activity of coumarins from Mesua ferrea. *Fitoterapia* **2013**, *90*, 140–150. [CrossRef] [PubMed]
128. Robbins, R.J. Phenolic Acids in Foods: An Overview of Analytical Methodology. *J. Agric. Food Chem.* **2003**, *51*, 2866–2887. [CrossRef] [PubMed]
129. Maga, J.A.; Katz, I. Simple phenol and phenolic compounds in food flavor. *Crit. Rev. Food Sci. Nutr.* **1978**, *10*, 323–372. [CrossRef]
130. Graf, E. Antioxidant potential of ferulic acid. *Free Radic. Biol. Med.* **1992**, *13*, 435–448. [CrossRef]
131. Gülçin, I. Antioxidant activity of caffeic acid (3,4-dihydroxycinnamic acid). *Toxicology* **2006**, *217*, 213–220. [CrossRef]

132. Chen, J.H.; Ho, C.-T. Antioxidant Activities of Caffeic Acid and Its Related Hydroxycinnamic Acid Compounds. *J. Agric. Food Chem.* **1997**, *45*, 2374–2378. [CrossRef]

133. Zhang, S.; Gai, Z.; Gui, T.; Chen, J.; Chen, Q.; Li, Y. Antioxidant Effects of Protocatechuic Acid and Protocatechuic Aldehyde: Old Wine in a New Bottle. *Evid. Based Complement. Altern. Med.* **2021**, *2021*, 6139308. [CrossRef]

134. Maurya, D.K.; Devasagayam, T.P.A. Antioxidant and prooxidant nature of hydroxycinnamic acid derivatives ferulic and caffeic acids. *Food Chem. Toxicol.* **2010**, *48*, 3369–3373. [CrossRef]

135. Su, M.; Liu, F.; Luo, Z.; Wu, H.; Zhang, X.; Wang, D.; Zhu, Y.; Sun, Z.; Xu, W.; Miao, Y. The Antibacterial Activity and Mechanism of Chlorogenic Acid Against Foodborne Pathogen *Pseudomonas aeruginosa*. *Foodborne Pathog. Dis.* **2019**, *16*, 823–830. [CrossRef]

136. Lou, Z.; Wang, H.; Zhu, S.; Ma, C.; Wang, Z. Antibacterial Activity and Mechanism of Action of Chlorogenic Acid. *J. Food Sci.* **2011**, *76*, M398–M403. [CrossRef] [PubMed]

137. Hussin, N.M.; Muse, R.; Ahmad, S.; Ramli, J.; Mahmood, M.; Sulaiman, M.R.; Shukor, M.Y.A.; Rahman, M.F.A.; Aziz, K.N.K. Antifungal Activity of Extracts and Phenolic Compounds from *Barringtonia racemose* L. (lecythidaceae). *Afr. J. Biotechnol.* **2009**, *8*, 2835–2842.

138. Liu, H.; Ma, S.; Xia, H.; Lou, H.; Zhu, F.; Sun, L. Anti-inflammatory activities and potential mechanisms of phenolic acids isolated from Salvia miltiorrhiza f. alba roots in THP-1 macrophages. *J. Ethnopharmacol.* **2018**, *222*, 201–207. [CrossRef] [PubMed]

139. Abotaleb, M.; Liskova, A.; Kubatka, P.; Büsselberg, D. Therapeutic Potential of Plant Phenolic Acids in the Treatment of Cancer. *Biomolecules* **2020**, *10*, 221. [CrossRef]

140. Mandal, S.M.; Chakraborty, D.; Dey, S. Phenolic acids act as signaling molecules in plant-microbe symbioses. *Plant Signal. Behav.* **2010**, *5*, 359–368. [CrossRef] [PubMed]

141. Lorigooini, Z.; Jamshidi-Kia, F.; Hosseini, Z. Analysis of aromatic acids (phenolic acids and hydroxycinnamic acids). In *Recent Advances in Natural Products Analysis*; Silva, A.S., Nabavi, S.F., Saeedi, M., Nabavi, S.M., Eds.; Elsevier: Amsterdam, The Netherlands, 2020; pp. 199–219. [CrossRef]

142. Gross, G.G. Phenolic Acids. In *Secondary Plant Products: A Comprehensive Treatise*; Conn, E.E., Ed.; Academic Press, Inc.: New York, NY, USA, 1981; Volume 7, pp. 301–315.

143. Marchiosi, R.; Dos Santos, W.D.; Constantin, R.P.; De Lima, R.B.; Soares, A.R.; Finger-Teixeira, A.; Mota, T.R.; de Oliveira, D.M.; Foletto-Felipe, M.D.P.; Abrahão, J.; et al. Biosynthesis and metabolic actions of simple phenolic acids in plants. *Phytochem. Rev.* **2020**, *19*, 865–906. [CrossRef]

144. Borges, A.; Ferreira, C.; Saavedra, M.J.; Simões, M. Antibacterial Activity and Mode of Action of Ferulic and Gallic Acids Against Pathogenic Bacteria. *Microb. Drug Resist.* **2013**, *19*, 256–265. [CrossRef] [PubMed]

145. Zhang, J.; Cui, X.; Zhang, M.; Bai, B.; Yang, Y.; Fan, S. The antibacterial mechanism of perilla rosmarinic acid. *Biotechnol. Appl. Biochem.* **2021**, *69*, 1757–1764. [CrossRef]

146. Ma, C.M.; Abe, T.; Komiyama, T.; Wang, W.; Hattori, M.; Daneshtalab, M. Synthesis, anti-fungal and 1, 3-β-d-glucan synthase inhibitory activities of caffeic and quinic acid derivatives. *Bioorg. Med. Chem.* **2010**, *18*, 7009–7014. [CrossRef]

147. Kumar, N.; Goel, N. Phenolic acids: Natural versatile molecules with promising therapeutic applications. *Biotechnol. Rep.* **2019**, *24*, e00370. [CrossRef]

148. Bouarab-Chibane, L.; Forquet, V.; Lantéri, P.; Clément, Y.; Léonard-Akkari, L.; Oulahal, N.; Degraeve, P.; Bordes, C. Anti-bacterial properties of polyphenols: Characterization and QSAR (Quantitative structure–activity relationship) models. *Front. Microbiol.* **2019**, *10*, 829. [CrossRef] [PubMed]

149. Barker, D. Lignans. *Molecules* **2019**, *24*, 1424. [CrossRef]

150. Gottlied, O.R. Chemistry of Neolignans with Potential Biological Activity. In *New Natural Products and Plant Drugs with Pharmacological, Biological or Therapeutical Activity*; Wagner, H., Wolff, P., Eds.; Springer: Berlin/Heidelberg, Germany, 1977; pp. 227–262.

151. Moss, G.P. Nomenclature of Lignans and Neolignans (IUPAC Recommendations 2000). *Pure Appl. Chem.* **2000**, *72*, 1493–1523. [CrossRef]

152. Maruyama, M.; Yamauchi, S.; Akiyama, K.; Sugahara, T.; Kishida, T.; Koba, Y. Antibacterial activity of a vir-gatusin-related compound. *Biosci. Biotechnol. Biochem.* **2007**, *71*, 677–680. [CrossRef]

153. Favela-Hernández, J.M.J.; García, A.; Garza-González, E.; Rivas-Galindo, V.M.; Camacho-Corona, M.R. Antibacterial and Antimycobacterial Lignans and Flavonoids from *Larrea tridentata*. *Phytother. Res.* **2012**, *26*, 1957–1960. [CrossRef]

154. Mi, Q.-L.; Liang, M.-J.; Gao, Q.; Song, C.-M.; Huang, H.-T.; Xu, Y.; Wang, J.; Deng, L.; Yang, G.-Y.; Guo, Y.-D.; et al. Arylbenzofuran Lignans from the Seeds of Arctium lappa and Their Bioactivity. *Chem. Nat. Compd.* **2020**, *56*, 53–57. [CrossRef]

155. Yang, G.-Z.; Hu, Y.; Yang, B.; Chen, Y. Lignans from the Bark of *Zanthoxylum planispinum*. *Helv. Chim. Acta* **2009**, *92*, 1657–1664. [CrossRef]

156. Kumarasamy, Y.; Nahar, L.; Cox, P.J.; Dinan, L.N.; Ferguson, C.A.; Finnie, D.A.; Jaspars, M.; Sarker, S.D. Biological Activity of Lignans from the Seeds of Centaurea scabiosa. *Pharm. Biol.* **2003**, *41*, 203–206. [CrossRef]

157. Watanabe, K.; Ishiguri, Y.; Nonaka, F.; Morita, A. Isolation and identification of aucuparin as a phytoalexin from *Eriobotrya japonica* L. *Agric. Biol. Chem.* **1982**, *46*, 567–568. [CrossRef]

158. Wells, J.E.; Berry, E.D.; Varel, V.H. Effects of Common Forage Phenolic Acids on *Escherichia coli* O157:H7 Viability in Bovine Feces. *Appl. Environ. Microbiol.* **2005**, *71*, 7974–7979. [CrossRef]

159. Castro, V.S.; Figueiredo, E.; McAllister, T.; Stanford, K. Farm to fork impacts of super-shedders and high-event periods on food safety. *Trends Food Sci. Technol.* **2022**, *127*, 129–142. [CrossRef]

160. Arthur, T.M.; Keen, J.E.; Bosilevac, J.M.; Brichta-Harhay, D.M.; Kalchayanand, N.; Shackelford, S.D.; Wheeler, T.L.; Nou, X.; Koohmaraie, M. Longitudinal study of *Escherichia coli* O157:H7 in a beef cattle feedlot and role of high-level shed-ders in hide contamination. *Appl. Environ. Microbiol.* **2009**, *75*, 6515–6523. [CrossRef] [PubMed]
161. Zamuz, S.; Munekata, P.E.; Dzuvor, C.K.; Zhang, W.; Sant'Ana, A.S.; Lorenzo, J.M. The role of phenolic compounds against Listeria monocytogenes in food. A review. *Trends Food Sci. Technol.* **2021**, *110*, 385–392. [CrossRef]
162. Shi, X.; Zhu, X. Biofilm formation and food safety in food industries. *Trends Food Sci. Technol.* **2009**, *20*, 407–413. [CrossRef]
163. Beauchamp, C.S.; Dourou, D.; Geornaras, I.; Yoon, Y.; Scanga, J.A.; Belk, K.E.; Smith, G.C.; Nychas, G.J.E.; Sofos, J.N. Sanitizer efficacy against *Escherichia coli* O157: H7 biofilms on inadequately cleaned meat-contact surface materials. *Food Prot. Trends* **2012**, *32*, 173–182.
164. Vikram, A.; Jayaprakasha, G.K.; Jesudhasan, P.R.; Pillai, S.D.; Patil, B.S. Suppression of bacterial cell–cell signalling, bio-film formation and type III secretion system by citrus flavonoids. *J. Appl. Microbiol.* **2010**, *109*, 515–527. [CrossRef]
165. Manner, S.; Skogman, M.; Goeres, D.; Vuorela, P.; Fallarero, A. Systematic Exploration of Natural and Synthetic Flavonoids for the Inhibition of Staphylococcus aureus Biofilms. *Int. J. Mol. Sci.* **2013**, *14*, 19434–19451. [CrossRef]
166. Martinez-Suárez, J.V.; Ortiz, S.; López-Alonso, V. Potential impact of the resistance to quaternary ammonium disinfect-ants on the persistence of Listeria monocytogenes in food processing environments. *Front. Microbiol.* **2016**, *7*, 638. [CrossRef]
167. Soto, S.M. Role of efflux pumps in the antibiotic resistance of bacteria embedded in a biofilm. *Virulence* **2013**, *4*, 223–229. [CrossRef]
168. Pearson, J.P.; Van Delden, C.; Iglewski, B.H. Active efflux and diffusion are involved in transport of Pseudomonas aeru-ginosa cell-to-cell signals. *J. Bacteriol.* **1999**, *181*, 1203–1210. [CrossRef]
169. Costa, D.C.; Costa, H.S.; Albuquerque, T.G.; Ramos, F.; Castilho, M.C.; Sanches-Silva, A. Advances in phenolic com-pounds analysis of aromatic plants and their potential applications. *Trends Food Sci. Technol.* **2015**, *45*, 336–354. [CrossRef]
170. Gaikwad, K.K.; Singh, S.; Lee, Y.S. Antimicrobial and improved barrier properties of natural phenolic compound-coated polymeric films for active packaging applications. *J. Coat. Technol. Res.* **2018**, *16*, 147–157. [CrossRef]
171. Kalogianni, A.I.; Lazou, T.; Bossis, I.; Gelasakis, A.I. Natural Phenolic Compounds for the Control of Oxidation, Bacterial Spoilage, and Foodborne Pathogens in Meat. *Foods* **2020**, *9*, 794. [CrossRef]
172. Kasprzak-Drozd, K.; Oniszczuk, T.; Stasiak, M.; Oniszczuk, A. Beneficial effects of phenolic compounds on gut microbi-ota and metabolic syndrome. *Int. J. Mol. Sci.* **2021**, *22*, 3715. [CrossRef] [PubMed]
173. Tuohy, K.M.; Conterno, L.; Gasperotti, M.; Viola, R. Up-regulating the Human Intestinal Microbiome Using Whole Plant Foods, Polyphenols, and/or Fiber. *J. Agric. Food Chem.* **2012**, *60*, 8776–8782. [CrossRef] [PubMed]
174. Gris, E.F.; Mattivi, F.; Ferreira, E.A.; Vrhovsek, U.; Pedrosa, R.C.; Bordignon-Luiz, M.T. Phenolic profile and effect of regu-lar consumption of Brazilian red wines on in vivo antioxidant activity. *J. Food Compost. Anal.* **2013**, *31*, 31–40. [CrossRef]
175. Malta, L.G.; Ghiraldini, F.G.; Reis, R.; Oliveira, M.D.V.; Silva, L.B.; Pastore, G.M. In vivo analysis of antigenotoxic and antimuta-genic properties of two Brazilian Cerrado fruits and the identification of phenolic phytochemicals. *Food Res. Int.* **2012**, *49*, 604–611. [CrossRef]
176. da Silva Siqueira, E.M.; Félix-Silva, J.; de Araújo, L.M.L.; Fernandes, J.M.; Cabral, B.; Gomes, J.A.D.S.; de Araújo Roque, A.; Tomaz, J.C.; Lopes, N.P.; de Freitas Fernandes-Pedrosa, M.; et al. Spondias tuberosa (Anacardiaceae) leaves: Profiling phenolic compounds by HPLC-DAD and LC–MS/MS and in vivo anti-inflammatory activity. *Biomed. Chromatogr.* **2016**, *30*, 1656–1665. [CrossRef]
177. Hofer, U. The cost of antimicrobial resistance. *Nat. Rev. Genet.* **2018**, *17*, 3. [CrossRef]

Article

Investigation of the Effect of pH on the Adsorption–Desorption of Doxycycline in Feed for Small Ruminants

Rositsa Mileva, Tsvetelina Petkova, Zvezdelina Yaneva and Aneliya Milanova *

Department of Pharmacology, Animal Physiology, Biochemistry and Chemistry, Faculty of Veterinary Medicine, Trakia University, 6000 Stara Zagora, Bulgaria
* Correspondence: aneliya.milanova@trakia-uni.bg

Abstract: Orally administered tetracycline antibiotics interact with feed, which may impact their bioavailability and efficacy. Therefore, the pH-dependent adsorption of doxycycline and its interaction with feed for ruminants was studied in vitro. Adsorption experiments on animal feed (135 and 270 mg) with initial doxycycline concentrations of 35, 75, and 150 µg/mL were performed. Desorption experiments were conducted by agitation of a predetermined mass of doxycycline-loaded animal feed in PBS, at pH = 3.0, 6.0, and 7.4, to simulate changes in the gastrointestinal tract. Antibiotic concentrations were determined by LC-MS/MS analysis. The adsorption/desorption of doxycycline was described by mathematical models. Chemisorption with strong intermolecular interactions between the active functional groups of doxycycline and the organic biomass was found. The experimental release curve comprised three sections: initial prolonged 27–30% release (pH = 6.0), followed by moderate 56–59% release (pH = 3.0), and final 63–74% release (pH = 7.4). The sigmoidal model showed a considerable role of diffusion with an initial prevalence of desorption and a decreased desorption rate thereafter. The Weibull equation revealed an initial release stage followed by a lag time section and sustained release. The study of doxycycline adsorption by the animal feed proved a maximum 80% encapsulation efficiency and revealed initial diffusion followed by chemisorption. The highest release efficiency of 74% suggests high bioavailability of doxycycline after oral administration in ruminants.

Keywords: adsorption; desorption; doxycycline; pH dependence; small ruminant feed

Citation: Mileva, R.; Petkova, T.; Yaneva, Z.; Milanova, A. Investigation of the Effect of pH on the Adsorption–Desorption of Doxycycline in Feed for Small Ruminants. *Antibiotics* **2023**, *12*, 268. https://doi.org/10.3390/antibiotics12020268

Academic Editor: Dóra Kovács

Received: 12 January 2023
Revised: 24 January 2023
Accepted: 26 January 2023
Published: 28 January 2023

1. Introduction

In modern animal husbandry, large numbers of animals are concentrated on a small area, which is a prerequisite for the occurrence of bacterial infections. Although many actions have been taken to improve animal welfare and replace antibiotics, infectious diseases occur which require adequate therapy with antibacterial drugs [1]. The problem of the selection and spread of bacterial resistance against antibiotics in industrial animal husbandry requires knowledge of the factors affecting the antibiotics' bioavailability, which can decrease their efficacy. Oral administration of antibiotics with drinking water is the most often recommended route because the animals drink water even when they refuse to consume feed [2]. While subcutaneous or intramuscular administration of antibiotics is in most cases associated with predictable concentrations in the body, the oral route leads to complex interactions of the antibiotics with the microbiome and the contents of the gastrointestinal tract [3–5]. Therefore, precise use of antibiotics requires knowledge of a number of factors that may affect their absorption and the achievement of effective concentrations at the site of action, which is essential for the treatment of systemic infections [3].

Tetracyclines continue to be a group of antibiotics with significant use in veterinary practice, and new EU regulations define them as drugs with limited negative impact on resistance development and spread in the terms of co-selection [6]. The cited document emphasizes the requirements for their responsible use. They are most often applied orally, preferably with drinking water, in the mass treatment of farm animals [7]. Doxycycline is one of the preferable tetracyclines in the treatment of bacterial infections in animals. Its salts dissolve in water, and their solutions are stable for longer periods at acidic pH, while in alkaline pH they precipitate [8]. Chelation of tetracyclines with metal ions such as Ca^{2+}, Al^{3+}, and others reduces their bioavailability and efficacy. Compared to the older members of the group, doxycycline has a lower affinity for metal ions [9]. However, its ability to form chelated complexes with them has been observed [10]. This interaction is a prerequisite for the excretion of significant concentrations of tetracyclines in the environment through fecal masses and for their long retention in water and soil. The interaction of tetracyclines with feed masses and their adsorption on them in the gastrointestinal tract is also very important. It hinders the absorption and effectiveness of these antibiotics [3,4]. The non-specific binding rates of doxycycline to poultry feed were found to be 87.9 to 88.8%, respectively, at pH 2.5 and 6.5 [11]. In the cited study, it was found that the mixing of mycotoxin binders with the feed at effective doses did not affect the adsorption of doxycycline. An absence of interaction of doxycycline with mycotoxin binders and lack of effect on the bioavailability of the antibiotic were observed in pigs in another experimental setting [12]. These data suggest that the presence of nutrients in the gastrointestinal tract can significantly reduce the bioavailability of doxycycline in a wide pH range from 2.5 to 6.5. The presence of large amounts of nutrients in the rumen of ruminants could have a significant effect on the bioavailability of doxycycline. Additionally, the slow passage of food through the gastrointestinal tract of ruminants can affect their absorption. Taking into account the literature data for other animal species, the importance of doxycycline's interaction with feed can be defined as significant, but the available literature about the interaction of doxycycline with feed for ruminants, depending on pH, is scarce.

Therefore, the aim of the present study was to investigate the adsorption of doxycycline, as a representative of the tetracyclines, on feed for ruminants and to characterize the interaction by simulating release conditions in different parts of the gastrointestinal tract via changing the pH. Mathematical models were used to describe the adsorption and desorption behaviors of doxycycline.

2. Results and Discussion

In the current investigations, the effect of initial doxycycline concentration and animal feed mass on the adsorption efficiency of the solid phase was investigated (Figure 1a,b). The kinetics experimental data revealed a direct relationship between the antibiotic initial concentration and adsorption capacity, which is due to the greater number of organic molecules saturating the active sites of the biomaterial (Figure 1a). An increased mass of feed, however, was associated with lower capacity as a result of the smaller number of molecules occupying a greater number of adsorption sites (Figure 1b).

The goals of mathematical modelling of sorption/desorption processes are to provide an opportunity for prediction of the sorption behavior of a given system at different conditions or varying system parameters, to define the rate-limiting stage, and to reveal the mechanism of the adsorption/desorption process.

2.1. Adsorption of Doxycycline

The kinetics experimental data of doxycycline adsorption on the animal feed were described by the pseudo-second-order kinetics model and the diffusion–chemisorption model. The values of the characteristic model parameters and correlation coefficients determined by linear regression analyses are presented in Table 1.

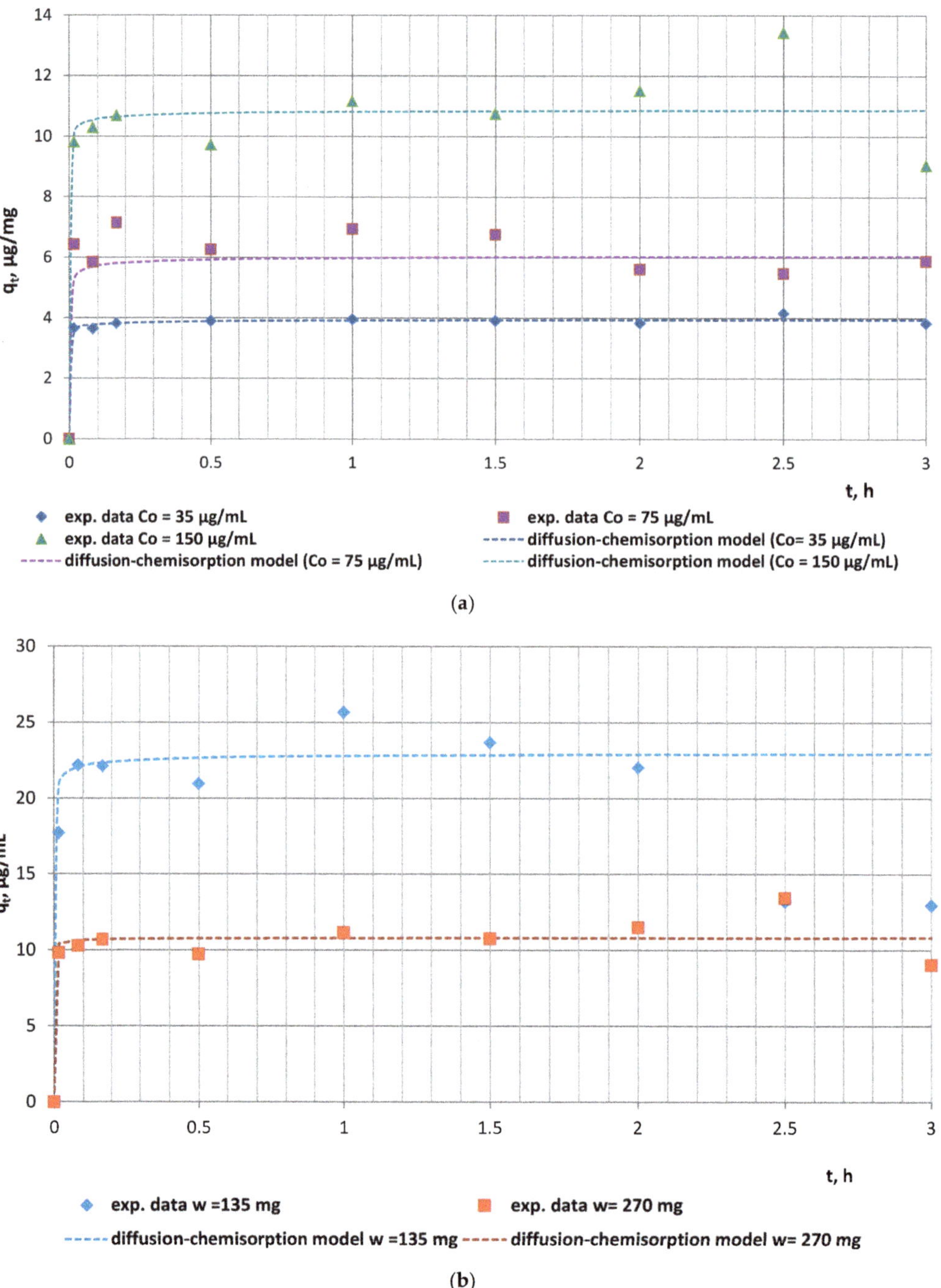

Figure 1. Experimental encapsulation kinetics results vs. diffusion–chemisorption model data: (**a**) effect of initial concentration (w = 270 mg); (**b**) effect of animal feed mass (C_0 = 150 µg/mL).

Table 1. Values of the model parameters and error functions of the applied adsorption kinetics models to the system doxycycline/animal feed.

Mathematical Model	Model Parameters		
	Effect of Initial Concentration (w = 270 mg)		
	$C_o = 35$ µg/mL	$C_o = 75$ µg/mL	$C_o = 150$ µg/mL
Pseudo-second-order kinetics model $\frac{1}{q_t} = \frac{1}{k_2 \times q_e^2} + \frac{1}{q_e} \times t$	$q_{e2} = 3.934$ µg/mg $k_2 = 129.87$ $h = 510.91$ µg/(mg·min) $R^2 = 0.9972$	$q_{e2} = 5.685$ µg/mg $k_2 = 3.291$ $h = 18.71$ µg/(mg·min) $R^2 = 0.9915$	$q_{e2} = 10.515$ µg/mg $k_2 = 3.015$ $h = 31.70$ µg/(mg·min) $R^2 = 0.9464$
Diffusion–chemisorption model $\frac{t^{0.5}}{q_t} = \frac{1}{q_e} \times t^{0.5} + \frac{1}{K_{DC}}$	$K_{DC} = 357.14$ $q_e^{DC} = 3.962$ µg/mg $R^2 = 0.9968$	$K_{DC} = 277.78$ $q_e^{DC} = 6.11$ µg/mg $R^2 = 0.9781$	$K_{DC} = 2000.00$ $q_e^{DC} = 10.846$ µg/mg $R^2 = 0.9403$
	Effect of animal feed mass ($C_0 = 150$ µg/mL)		
	w = 135 mg		w = 270 mg
Pseudo-second-order kinetics model	$q_{e2} = 15.748$ µg/mg $k_2 = 0.4337$ $h = 6.83$ µg/(mg·min) $R^2 = 0.8853$		$q_{e2} = 10.515$ µg/mg $k_2 = 3.015$ $h = 31.70$ µg/(mg·min) $R^2 = 0.9464$
Diffusion–chemisorption model	$K_{DC} = 1666.66$ $q_e^{DC} = 23.095$ µg/mg $R^2 = 0.9916$		$K_{DC} = 2000.00$ $q_e^{DC} = 10.846$ µg/mg $R^2 = 0.9403$

k_2: pseudo-second-order rate constant; h: initial rate of adsorption, µg/(mg.min); KDC: rate constant in the diffusion–chemisorption model.

The high values of the correlation coefficients ($R^2 > 0.9403$) and the significantly close values of the experimental and model adsorption capacities q_{e2} and q_e^{DC} for all studied doxycycline/animal feed experimental series determined the applicability of both models to describe the kinetics of the antibiotic adsorption by the animal feed. The series with doxycycline concentration $C_0 = 35$ µg/mL and fodder mass w = 270 mg was characterized by the highest initial rate of adsorption (h = 510.91 µg/(mg.min)), which could be explained by the absence of competition between sorbate molecules due to their low number combined with the greater number of vacant active sites on/within fodder particles. Obviously, the increased initial doxycycline concentration and the reduced fodder mass were prerequisites for the lower initial adsorption rate of the experimental series with $C_0 = 75$ µg/mL and 150 µg/mL and with w = 270 mg fodder mass.

Considering the theoretical assumptions of the pseudo-second-order kinetics model that chemisorption is the operative reaction mechanism, the latter observations undoubtedly outline the significant role of chemical processes related to the formation of strong intermolecular interactions between the active functional groups of doxycycline and those of the organic biomass, as the rate-limiting stage of the process.

Fodder, as a biosorbent, is characterized by inherent complex physical, chemical, and biological characteristics, making it necessary to test multiple kinetic models to achieve the best possible simulation. Such a multiple model is the diffusion–chemisorption model in which the rate of solid-phase concentration change (q_t, µg/mg) is a function of the rate of mass transfer of the organic molecules from the fluid phase to the biosorption site, characterized by the rate constant (K_{DC}, µg/(mg·t$^{0.5}$)), the equilibrium sorption capacity (q_e, µg/mg), and the square root of time. The comparative analyses of the calculated rate constants (Table 1) reveal an increase in the diffusion–chemisorption rate with increasing doxycycline initial concentration up to 150 µg/mL. This trend is expected due to the increased concentration gradient developed between the inner and outer regions of the sorbent particles, which is the driving force for diffusion. Thus, the significant applicability of the diffusion–chemisorption model indicates that the remarkable role of film/intraparticle diffusion of the antibiotic molecules through the boundary layer surrounding the fodder particulates or within the internal pores, especially at the initial stages of the adsorption process, cannot be neglected. Based on these data, it can be concluded that the adsorption kinetics of doxycycline on fodder are expected to depend mainly on diffusion-limited processes, as affected by the heterogeneous distributions of active sites, functional groups, and pore sizes, and continual partitioning of antibiotic molecules between the dissolved state and fixed state of adsorption [13].

2.2. Desorption of Doxycycline

The in vitro desorption experiments aimed to simulate the behavior of doxycycline-loaded feed in the gastrointestinal tract of ruminants, which, in turn, could allow the assessment of the negative health effects arising from the consumption of antibiotic-containing fodder. The desorption processes were modelled in an attempt to explicate the rate-limiting steps and to select an appropriate release model which can quantify the effect of changing solution and sorbent parameters and can aid in the eventual development of predictive models that would enable process design and in vitro behavior analyses with minimal experimentation. As the accuracy of fit is paramount in the development of predictive models, the error functions SSE, MSE, and RMSE and correlation coefficients were also determined.

The experimental doxycycline in vitro release from doxycycline-loaded fodder was studied in a simulated physiological medium to assess the effect of pH on the extent of antibiotic desorption in the gastrointestinal tract of ruminants. Two series of experiments were conducted. The first series was run with the mass of doxycycline-loaded fodder w = 133.3 mg containing 11.03 µg/mg of the antibiotic in the solid phase (Figure 2a). The second series was performed with the mass of doxycycline-loaded fodder w = 137.8 mg containing 41.6 µg/mg of the antibiotic in the solid phase (Figure 2b). The solid-phase concentrations of doxycycline were calculated on the basis of the quantity of antibiotic encapsulated by the feed. All the data were derived from the experiments performed in the current study. The extent of the in vitro antibiotic desorption at pH = 6.0 (simulating the pH in the rumen), at pH = 3.0 (simulating the pH in the abomasium), and at pH = 7.4 (simulating the pH in the small intestine) was investigated.

The experimental kinetics release curves, characterized by a pulsatile release mode, comprised three well-defined major sections (Figure 2a,b). The initial part resembled the conditions in the rumen and showed prolonged 27–30% release at pH = 6.0 for 1.5 h of incubation. Moderate release, up to 56–59%, was observed at pH = 3.0, which simulated the pH in the abomasium. The last part of the curve presented the final 63–74% release at pH = 7.4, close to the pH in the small intestines. The suggested concept for the release mechanism in the last stage is the infiltration of tiny antibiotic-loaded feed particles into the mucus, resulting in gradual disintegration, facilitated doxycycline release, and subsequent permeation through the paracellular pathway to the bloodstream.

Four single-, two-, and three-parameter release kinetics mathematical models—Higuchi, Korsmeyer–Peppas, Weibull, and the sigmoidal model—were applied to describe the experimental desorption results. The values of the calculated model parameters and error functions obtained by linear and nonlinear regression analyses are presented in Table 2.

The Higuchi model assumes that the drug release occurs predominantly through Fickian diffusion and has typically been observed in drug-carrier systems with constituents of a hydrophobic nature. The low value of R2 and the high values of the other error functions in the present study indicated the limited applicability of this model for the current experimental data.

The Korsmeyer–Peppas model describes simultaneously the following release mechanisms: water diffusion into the biomatrix, matrix swelling, and dissolution/relaxation of the matrix [14]. The release exponent n of the Korsmeyer–Peppas model describes the drug release mechanism: Fick diffusion for n = 0.5 and non-Fickian diffusion for 0.5 < n < 1. The values of the parameter in this study (n = 0.815–0.982) indicate anomalous or non-Fickian antibiotic release including diffusion and relaxation effects; thus, an appropriate model such as the sigmoidal model, which could describe such complex behavior, was required. The release (transport) constant (a) provides information on the drug formulation, such as the structural characteristics of the carrier. The identical characteristics of the fodder explain the commensurable values of the parameter a for both studied series.

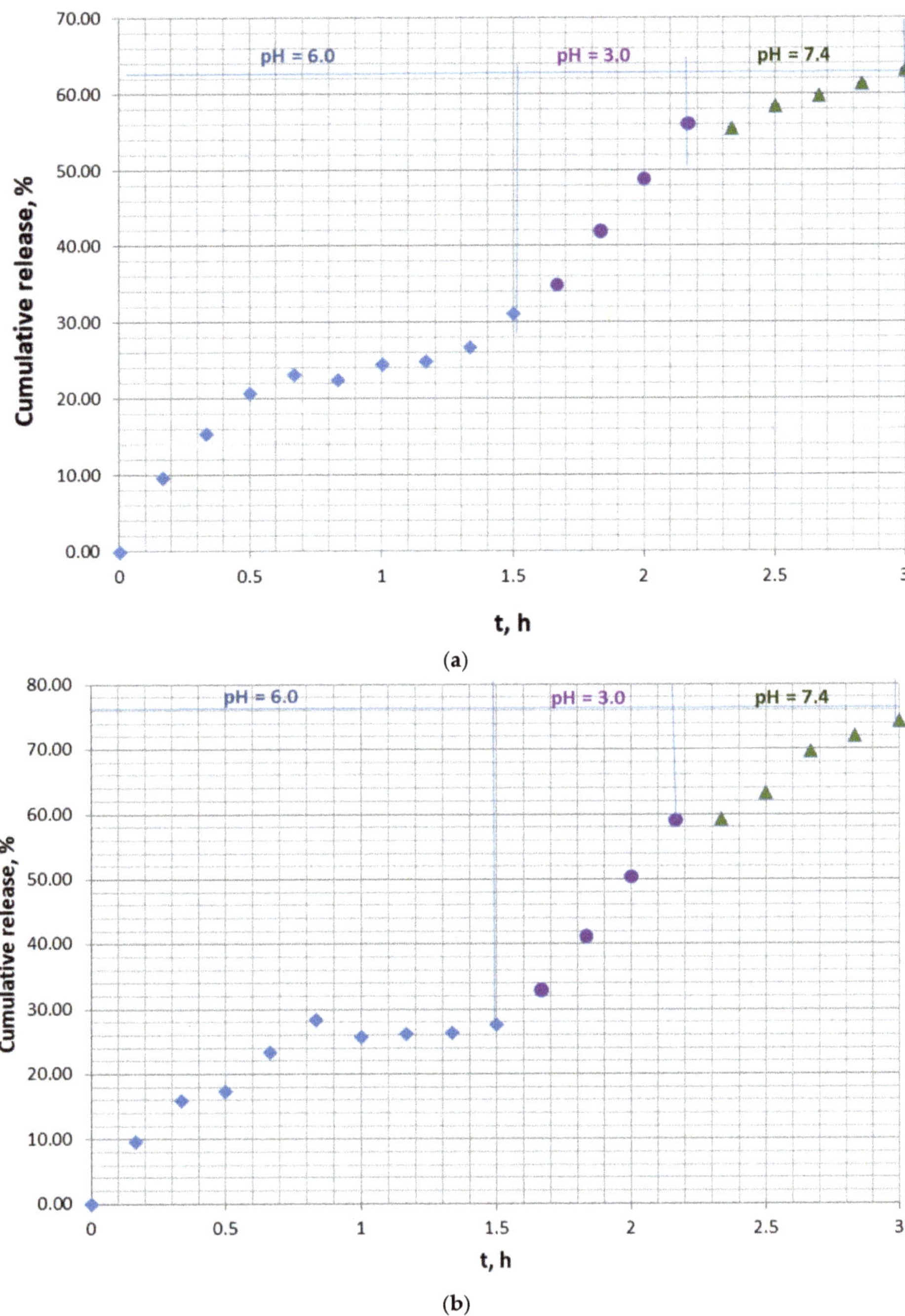

Figure 2. In vitro kinetics release curve of doxycycline from animal feed in simulated gastrointestinal and stomach medium: (**a**) solid doxycycline concentration 11.03 μg/mg, doxycycline-loaded fodder w = 133.3 mg; (**b**) solid doxycycline concentration 41.6 μg/mg, doxycycline-loaded fodder w = 137.8 mg.

Table 2. Values of the model parameters and error functions of the applied in vitro release kinetics models to the doxycycline/animal feed system in simulated gastrointestinal and stomach medium.

Mathematical Model	Model Parameters		Error Functions		Regression Analyses
	Series 1	Series 2	Series 1	Series 2	
Higuchi $C_t = k_H \times t^{0.5}$	$k_H = 4.33$	$k_H = 18.56$	$R^2 = 0.9139$	$R^2 = 0.8672$	Linear regression
Korsmeyer–Peppas $\frac{C_t}{C_o} = a \times t^n$	$a = 0.083$ $n = 0.815$	$a = 0.079$ $n = 0.982$	$R^2 = 0.9139$ $SSE = 0.003$ $MSE = 1.6 \times 10^{-4}$ $RMSE = 0.013$	$R^2 = 0.9139$ $SSE = 0.004$ $MSE = 2.3 \times 10^{-4}$ $RMSE = 0.015$	Nonlinear regression
Weibull $C_t = C_o \times [1 - e^{-\frac{(t-T)^b}{a}}]$	at b = 1 $C_o = 97.19$ $a_w = 0.022$ $T = 0.291$	at b = 1 $C_o = 21.6$ $a_w = 0.26$ $T = 0.469$	$R^2 = 0.962$ $SSE = 3.101$ $MSE = 0.194$ $RMSE = 0.440$	$R^2 = 0.960$ $SSE = 63.085$ $MSE = 3.943$ $RMSE = 1.986$	Nonlinear regression
Sigmoidal $\frac{C_t}{C_e} = k_{s1} \times (t - l)^{n_s} + k_{s2} \times (t - l)^{n_s}$	Region 1 $k_{s1} = 27.063$ $n_s = 0.004$ $k_{s2} = 27.446$ Region 2 $k_{s1} = 11.011$ $n_s = 0.057$ $k_{s2} = 11.267$	Region 1 $k_{s1} = 130.31$ $n_s = 0.001$ $k_{s2} = 130.657$ Region 2 $k_{s1} = 0.190$ $n_s = 2.111$ $k_{s2} = 0.009$	Region 1 $R^2 = 0.958$ $SSE = 0.002$ $MSE = 0.001$ $RMSE = 0.025$ Region 2 $R^2 = 0.964$ $SSE = 0.020$ $MSE = 0.002$ $RMSE = 0.047$	Region 1 $R^2 = 0.904$ $SSE = 0.005$ $MSE = 0.001$ $RMSE = 0.033$ Region 2 $R^2 = 0.984$ $SSE = 0.009$ $MSE = 0.001$ $RMSE = 0.033$	Nonlinear regression

k_H: Higuchi release rate constant; a: release constant in the Korsmeyer–Peppas model; aw: time process parameter in the Weibull model; T: lag time; b: shape parameter in the Weibull model, characterizing the curve as exponential at b = 1; k_{s1}, k_{s2}: sigmoidal model release constants; n_{s1}, n_{s2}: sigmoidal model exponents.

In the Weibull equation, the parameter T is a location parameter denoting the lag time before the onset of the drug release procedure, while b describes the shape of the dissolution curve progression. For b = 1, the shape of the curve corresponds to an exponential profile, which coincides with the mode of the experimental release curve. Obviously, the modes of the experimental release curves generally include an initial release stage followed by a lag-time section and then sustained release. Lag time in pharmacokinetics is associated with the finite time necessary for a drug to enter the central circulation after extravascular administration. This parameter describes the absorption phase and depends on the drug dissolution/release processes from the delivery system and on drug migration to the surface of the sorbent [15,16].

Due to the two clearly defined separate regions (region 1: until 1.5 h and region 2: from 1.5 h to 3 h) on both experimental kinetics release curves, a modified modelling methodology was undertaken. The sigmoidal model was applied separately to each of the regions due to the deviations in the release behavior of the antibiotic in the rumen as compared to that in the abomasum and the small intestine, which provoked different modes of the respective curves. The comparative analyses of the experimental and model data and the values of error functions proved the better applicability of the sigmoidal model, as it had the highest average R2 and the lowest average SSE, MSE, and RMSE values for both experimental series. Moreover, the model release curves practically coincided with the experimental ones for both studied series (Figure 3a,b).

The values of the diffusional constant k_{s1}, the relaxation constant k_{s2}, and the diffusional exponent n_s enable the assessment of the contributions of relaxation and diffusion mechanisms within the different stages of the release process. The data in Table 2 outline that the values of these constants for each of the regions are commensurable, which, in turn, indicates the relative contribution of both mechanisms within each region. However, the significantly higher values of the kinetic constants k_{s1} for region 1 as compared to k_{s2} for region 2 are indicative of the considerable role of diffusion in simulated rumen medium. Obviously, the role of relaxation as a limiting mechanism of doxycycline release from fodder is also significant. This could be explained by the chemical composition of fodder, namely, the high content of a variety of biopolymers (starch, hemicellulose, cellulose, lignins, etc.) that are susceptible to polymer chain relaxation caused by mechanical stress (agitation) and/or pH changes. The inflexion points registered at approximately 28% and 31% equilibrium release outline an initial stage where the rate of desorption exceeds that of sorption and a second region characterized by a decreased desorption rate. The probable formation of compact supramolecular networks comprising doxycycline molecules and water dipoles via intermolecular hydrogen bonding could also evoke sigmoidal behavior during the in vitro release process [17].

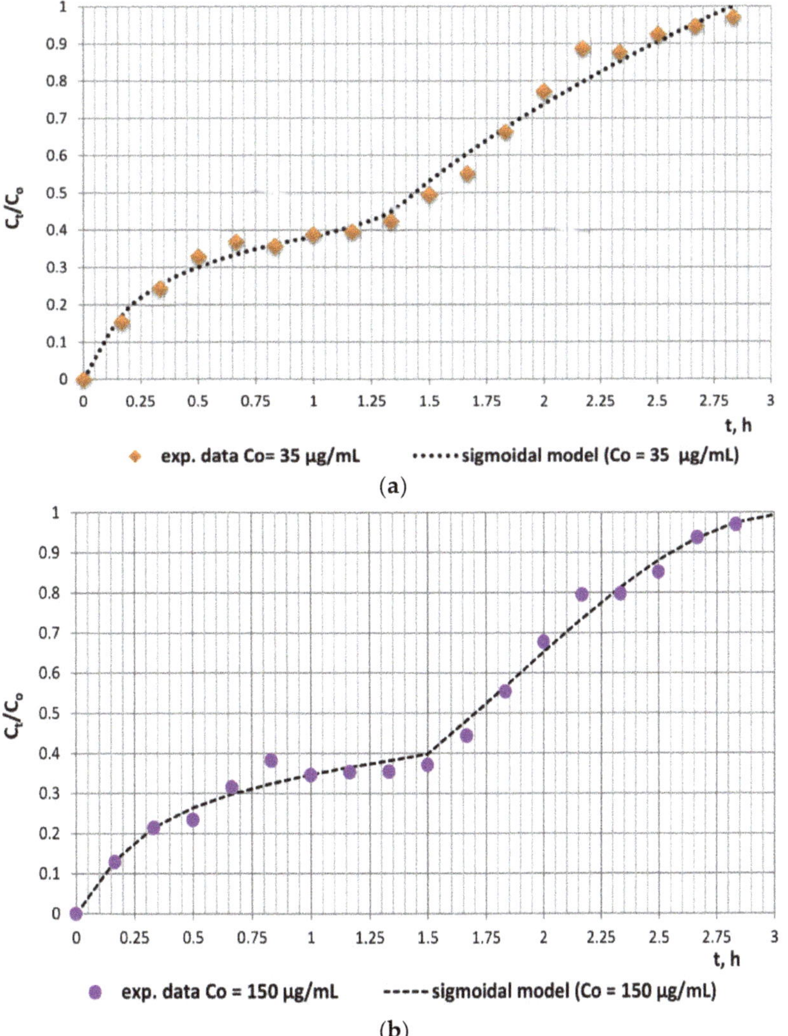

Figure 3. Applicability of the sigmoidal model to the experimental in vitro release data: (**a**) solid doxycycline concentration 11.03 µg/mg, doxycycline-loaded fodder w = 133.3 mg; (**b**) solid doxycycline concentration 41.6 µg/mg, doxycycline-loaded fodder w = 137.8 mg.

In conclusion, the adsorption study of doxycycline by animal feed proved a maximum 80% encapsulation efficiency of the fodder towards the antibiotic. Similar results were reported for interactions between doxycycline and feed for broiler chickens and pigs [11,12]. High non-specific binding of doxycycline (around 88%) to the components of the feed matrix was observed, which could have impact on its bioavailability [11]. The capacity for its adsorption on the feed was studied in in vitro experiments, conducted in buffered solutions with or without feed. These tests were performed at a specific acidic or alkaline pH and more often at neutral pH [18,19]. Our results, for the first time, describe interactions between doxycycline and feed for small ruminants, and they are in accordance with the observed dependence of the processes of adsorption and desorption of polar substances on the pH of the medium, which varies in different parts of the gastrointestinal tract [20,21]. Orally administered drugs pass through different parts of the digestive tract, where the values of pH vary from acidic in the stomach (3.5–5–6.5, depending on the species) to alkaline in the gut (7.5–7.8). Data from the modeling of doxycycline adsorption/desorption demonstrated the need to study these processes at different pH in order to adequately assess the presence of free drug molecules available for absorption. The limitations of the current investigation concern the use of feed specific for lactating sheep, while some changes in the proportions and the type of feeding of the animals can lead to variations in doxycycline–feed interaction. Although the metabolism of doxycycline is negligible, it would be of interest to discover how the addition of specific enzymes to the simulated medium affects the adsorption/desorption of the antibiotic in the feed for small ruminants.

3. Materials and Methods

3.1. Drugs and Reagents

Doxycycline hyclate with purity $\geq$98% (HPLC grade, Sigma-Aldrich, St. Louis, MO, USA) and crystalline oxytetracycline hydrochloride $\geq$95% were used as analytical standards. HydroDoxx 500 mg/g Oral Powder (Huvepharma, Sofia, Bulgaria) was used to prepare solutions for adsorption experiments. The following reagents were used for extraction and for the further analysis of the drugs: trifluoroacetic acid (99.5%) (Fisher Chemical, Waltham, MA, USA), LC/MS grade acetonitrile OPTIMA® (Fisher Chemical), LC-MS grade methanol (CHROMASOLV LC-MS, Honeywell, Seelze, Germany), ~98% formic acid for mass spectrometry (Honeywell Fluka™, Seelze, Germany), and water for chromatography (LC-MS Grade, LiChrosolv®, Merck KGaA, Darmstadt, Germany). The reagents PBS (pH = 7.4) and HCl (ACS reagent, 37%) used for the in vitro release experiments were supplied by Sigma.

3.2. Experimental Procedure and Adsorption/Desorption Study

The feed used in the current experiments consisted of 80% hay (72% meadow hay and 8% alfalfa hay) and 20% concentrated feed for lactating domestic sheep (*Ovis aries*) with 18% crude protein (HL-TopMix, Sliven, Bulgaria). The concentrated feed contained maize, wheat, soybean meal, sunflower meal, maize fodder, protected fat, vitamin premix, and micro- and macro-elements, without copper. The analytical content was 18% crude protein, 7.5% crude fiber, and 3.2% crude fat. The amounts of coarse fodder and concentrated feed were selected according to the practices for feeding lactating sheep. The feed was milled and well-mixed before the experiments. The amount of doxycycline hyclate used to prepare the experimental solutions was calculated on the basis of a sheep with a body weight of 55 kg and an oral dose of 20 mg/kg doxycycline, according to the data from our previous study [22]. The concentrations of the antibiotic solutions were calculated by taking into account the total dose of the antibiotic and the volume of the contents of the rumen in small ruminants [23,24]. Based on these calculations, doxycycline solutions were prepared in Milli-Q (Evoqua Water Technologies, Pittsburgh, PA, USA) water at 35, 75, and 150 µg/mL, corresponding to low, medium, and high concentrations, respectively. The values of pH at different segments of the gastrointestinal tract were taken into account, and experiments were performed at pH 3.0, 6.0, and 7.4 [23].

Five series of adsorption experiments with initial doxycycline concentrations C_0 = 35, 75, and 150 µg/mL, mass of animal feed (coarse/concentrated fodder = 3/2 *w/w*) w = 135 and 270 mg, and solution volume V = 50 mL at pH = 6 and T = 37 °C were conducted in batch mode for 3 h. The adsorption capacity of the feed (qt, μg/mg) was determined by the formula

$$q_t = \frac{(Co - Ct) \cdot w}{V} \tag{1}$$

where Ct is the doxycycline concentration in the liquid phase at time t, μg/mL.

Desorption experiments were conducted by agitation of a predetermined mass of doxycycline-loaded animal feed in simulated gastrointestinal medium without enzymes, comprising PBS pH = 7.4 adjusted to pH = 3.0 and 6.0 with 1 M HCl, for 3 h at T = 37 °C in a Digital Waterbath WNB 22 (Memmert GmbH, Büchenbach, Germany). Samples were taken at predetermined time intervals, and fresh medium of an equal volume was added to restore the total amount of the medium. The release efficiency (E, %) was calculated by

$$E = \frac{\text{amount of released substance}}{\text{total amount of encapsulated substance}} \times 100\% \tag{2}$$

Blanks containing no antibiotic and replicates of each adsorption/desorption point were used for each series of experiments. The concentration of doxycycline in the liquid phase was determined by LC-MS/MS (Agilent 6400c, Agilent Technologies, Santa Clara, CA, USA). Standard solutions were prepared in the same matrix as blank samples. Blank matrix was used for dilution of the samples.

3.3. LC-MS/MS Analysis of Doxycycline in Samples Containing Animal Feed

The extraction of doxycycline from samples containing 135 or 270 mg animal feed (coarse/concentrated fodder = 3/2 *w/w*) in 50 mL simulated gastrointestinal fluid was carried out by the method previously described [25]. Dilutions of the samples were prepared in the same matrix which did not contain the antibiotic. Several serial dilutions 1:10 *v:v* were prepared from every sample so that the suitable dilution was used for the determination of the concentrations. The extraction was performed as follows: 500 μL samples were added to 10 μL oxytetracycline as an internal standard (at final concentration 0.1 μg/mL) and 65 μL trifluoroacetic acid (TFA). The samples were vortexed for 1 min and centrifuged at 10,800× *g* for 10 min at 22 °C. After this step, the supernatant was filtered through 0.22 μm syringe filters and transferred into injection vials. Then, 5 μL of the filtrate was injected into the LC-MS/MS system. The calibration curve was prepared in the same matrix. The final concentrations of doxycycline were 0, 0.01, 0.05, 0.1, 0.25, 0.5, 0.75, and 1 μg/mL.

The LC-MS/MS method was used to determine the concentration of doxycycline. An Agilent 6460c triple-quadrupole mass spectrometer with AJS technology was used for the analysis. The liquid chromatography (LC) system consisted of a 1260 Infinity II quaternary pump and a 1260 Infinity II Vial Sampler. A Poroshell 120 EC C18 column (4.6 mm i.d. 100 mm, 2.7 m, Agilent Technologies, Santa Clara, CA, USA) was used at a temperature of 40 °C for chromatographic separation of the tetracyclines, using doxycycline as the target compound and oxytetracycline as an internal standard. The mobile phases for LC-MS analysis were a) 0.1% formic acid in ultrapure water and b) 100% acetonitrile. Gradient elution was used for separation of the compounds at a flow rate of 0.3 mL.min-1. The gradient was as follows: 0–0.5 min (90% A, 10% B), 0.50–8 min (from 90% A, 10% B to 2% A, 98% B), and 8–12 min (2% A, 98% B). The injection volume was 5μL. The total run time was 12 min with a post-run of 5 min. The gas temperature was 350 °C, the drying gas (nitrogen) flow rate was 12 L/min, the nebulizer gas (nitrogen) pressure was 45 psi, the sheath gas (nitrogen) temperature was 400 °C, and the sheath flow rate was 12 L/min. The capillary voltage, nozzle voltage, and dwell time were 4000 V, 500 V, and 200 ms, respectively. The qualifying ion for doxycycline was 445.1 m/z, and that for oxytetracycline was 461.1 m/z. The quantifying ions for these substances were 410/428.1 m/z and 443.1/444 m/z, respectively. The standard curve was linear (R^2 = 0.9992) between 0.01 and 1 μg/mL. The values of mean accuracy and inter-day and intra-day precision were 95.30 ± 9.17%, 1.16–11.23, and 0.58–2.32%, respectively. The values of LOD and LOQ were 1.12 ng/mL and 3.41 ng/mL.

3.4. Mathematical Modelling

The experimental adsorption data were described by the pseudo-second-order kinetics and diffusion–chemisorption models [26]. The desorption behavior of the studied system was modelled by four release kinetics models: Higuchi, Korsmeyer–Peppas, Weibull, and the sigmoidal model [27]. Linear and nonlinear regression analyses were applied.

3.5. Statistical Analysis

The statistical analyses of the experimental results, the values of the correlation coefficients R2, and the error functions (sum of the squared errors (SSE), mean squared error (MSE), and root mean square error (RMSE)) were determined by linear/nonlinear regression analyses using XLSTAT statistical software for Excel (Microsoft Corporation, Washington, USA).

4. Conclusions

The data in the current study, for the first time, describe the interactions between doxycycline and feed for small ruminants, with the aim to obtain more information about possible effects on the bioavailability of the antibiotic after its oral administration. The results from our proposed mechanism of adsorption based on the applied mathematical models comprise initial diffusion followed by chemisorption. The in vitro release behavior of doxycycline followed a sigmoidal mode comprising three sections: initial prolonged release in the rumen, moderate release in the abomasium, and final release in the small intestine. The highest release efficiency of 74% in simulated physiological medium indicates high bioavailability of the drug after oral administration in small ruminants. Initial adsorption of the antibiotic on the fodder was followed by almost complete desorption at pH 7.4, which could contribute to significant absorption of doxycycline and the achievement of therapeutic levels in central circulation.

Author Contributions: Conceptualization, A.M. and Z.Y.; methodology, A.M. and Z.Y.; software, A.M. and Z.Y.; validation, R.M. and T.P.; formal analysis, T.P.; investigation, T.P. and Z.Y.; resources, R.M.; data curation, T.P.; writing—original draft preparation, A.M. and Z.Y.; writing—review and editing, A.M. and Z.Y.; visualization, Z.Y.; supervision, A.M. and Z.Y.; project administration, A.M.; funding acquisition, A.M. All authors have read and agreed to the published version of the manuscript.

Funding: This research was funded by Trakia University, grant number 01/2020.

Institutional Review Board Statement: Not applicable.

Informed Consent Statement: Not applicable.

Data Availability Statement: Not applicable.

Conflicts of Interest: The authors declare no conflict of interest.

References

1. Sharma, C.; Rokana, N.; Chandra, M.; Singh, B.P.; Gulhane, R.D.; Gill, J.P.S.; Ray, P.; Puniya, A.K.; Panwar, H. Antimicrobial resistance: Its surveillance, impact, and alternative management strategies in dairy animals. *Front Vet Sci.* **2018**, *4*, 237. [CrossRef] [PubMed]
2. Martinez, M.; Blondeau, J.; Cerniglia, C.E.; Fink-Gremmels, J.; Guenther, S.; Hunter, R.P.; Li, X.Z.; Papich, M.; Silley, P.; Soback, S.; et al. Workshop report: The 2012 antimicrobial agents in veterinary medicine: Exploring the consequences of antimicrobial drug use: A 3-D approach. *J. Vet. Pharm. Therap.* **2014**, *37*, e1–e16. [CrossRef] [PubMed]
3. Marasanapalle, V.P.; Li, X.; Jasti, B.R. Effects of food on drug absorption. In *Oral Bioavailability. Basic Principles, Advanced Concepts, and Applications*; Hu, M., Li, X., Eds.; John Wiley & Sons, Inc.: Hoboken, NJ, USA, 2011; pp. 221–231.
4. Del Castillo, J.R.E. Tetracyclines. In *Antimicrobial Therapy in Veterinary Medicine*, 5th ed.; Giguère, S., Prescott, J.F., Dowling, P.M., Eds.; Wiley Blackwell: Hoboken, NJ, USA, 2013; pp. 257–268.
5. Mulder, M.; Radjabzadeh, D.; Kiefte-de Jong, J.C.; Uitterlinden, A.G.; Kraaij, R.; Stricker, B.H.; Verbon, A. Long-term effects of antimicrobial drugs on the composition of the human gut microbiota. *Gut Microbes* **2020**, *12*, 1. [CrossRef]
6. European Medicine Agency. Categorisation of Antibiotics in the European Union. EMA/CVMP/CHMP/682198/2017. 2019. Available online: https://www.ema.europa.eu/en/documents/report/categorisation-antibiotics-european-union-answer-request-european-commission-updating-scientific_en.pdf (accessed on 11 January 2023).

7. Collignon, P.J.; McEwen, S.A. One health-its importance in helping to better control antimicrobial resistance. *Trop. Med. Infect. Dis.* **2019**, *4*, 22. [CrossRef]
8. Marx, J.O.; Vudathala, D.; Murphy, L.; Rankin, S.; Hankenson, F.C. Antibiotic administration in the drinking water of mice. *J. Am. Assoc. Lab. Anim. Sci.* **2014**, *53*, 301–306. [PubMed]
9. Yang, F.; Sun, N.; Zhao, G.Z.S.; Wang, Y.; Wang, M.F. Pharmacokinetics of doxycycline after a single intravenous, oral or intramuscular dose in Muscovy ducks (*Cairina moschata*). *Br. Poult. Sci.* **2015**, *56*, 137–142. [CrossRef]
10. Smith, V.A.; Cook, S.D. Doxycycline-a role in ocular surface repair. *Br. J. Ophthalmol.* **2004**, *88*, 619–625. [CrossRef]
11. De Baere, S.; De Mil, T.; Antonissen, G.; Devreese, M.; Croubels, S. In vitro model to assess the adsorption of oral veterinary drugs to mycotoxin binders in a feed- and aflatoxin B1-containing buffered matrix. *Food Addit. Contam. Part A* **2018**, *35*, 1728–1738. [CrossRef]
12. De Mil, T.; Devreese, M.; De Saeger, S.; Eeckhout, M.; De Backer, P.; Croubels, S. Influence of mycotoxin binders on the oral bioavailability of doxycycline in pigs. *J. Agric. Food Chem.* **2016**, *64*, 2120–2126. [CrossRef]
13. Hubbe, M.A.; Azizian, S.; Douven, S. Implications of apparent pseudo-second-order adsorption kinetics onto cellulosic materials: A review. *BioResources* **2019**, *14*, 7582–7626. [CrossRef]
14. Permanadewi, I.; Kumoro, A.C.; Wardhani, D.H.; Aryanti, N. Modelling of controlled drug release in gastrointestinal tract simulation. *J. Phys.Conf. Ser.* **2019**, *1295*, 012063. [CrossRef]
15. Nerella, N.G.; Block, L.H.; Noonan, P.K. The impact of lag time on the estimation of pharmacokinetic parameters. I. One-compartment open model. *Pharm. Res.* **1993**, *10*, 1031–1036. [CrossRef] [PubMed]
16. Ruiz-Garcia, A.; Tan, W.; Li, J.; Haughey, M.; Masters, J.; Hibma, J.; Lin, S. Pharmacokinetic models to characterize the absorption phase and the influence of a proton pump Inhibitor on the overall exposure of dacomitinib. *Pharmaceutics* **2020**, *12*, 330. [CrossRef] [PubMed]
17. Legendre, A.O.; Silva, L.R.R.; Silva, D.M.; Rosa, I.M.L.; Azarias, L.C.; de Abreu, P.J.; de Araújo, M.B.; Neves, P.P.; Torres, C.; Martins, F.T.; et al. Solid state chemistry of the antibiotic doxycycline: Structure of the neutral monohydrate and insights into its poor water solubility. *Cryst. Eng. Comm.* **2012**, *14*, 2532–2540. [CrossRef]
18. Kurtbay, H.M.; Becki, Z.; Merdivan, M.; Yurdakoc, K. Reduction of ochratoxin A levels in red wine by bentonite, modified bentonites, and chitosa. *J. Agric. Food Chem.* **2008**, *56*, 2541–2545. [CrossRef]
19. Cavret, S.; Laurent, N.; Videmann, B.; Mazallon, M.; Lecoeur, S. Assessment of deoxynivalenol (DON) adsorbents and characterization of their efficacy using complementary in vitro tests. *Food Addit. Contam.* **2010**, *27*, 43–53. [CrossRef]
20. Santos, R.R.; Vermeulen, S.; Haritova, A.; Fink-Gremmels, J. Isotherm modeling of organic activated bentonite and humic acid polymer used as mycotoxin adsorbents. *Food Addit. Contam.* **2011**, *28*, 1578–1589. [CrossRef]
21. Telma dos Santos, C.; Hervé, R.; da Silva, P.R. Adsorption capacity of phenolic compounds onto; cellulose and xylan. *Food Sci. Technol.* **2015**, *35*, 314–320. [CrossRef]
22. Mileva, R.; Subev, S.; Gehring, R.; Milanova, A. Oral doxycycline pharmacokinetics: Lambs in comparison with sheep. *J. Vet. Pharm. Therap.* **2020**, *43*, 268–275. [CrossRef]
23. Miltko, R.; Bełżecki, G.; Kowalik, B.; Skomiał, J. Presence of carbohydrate-digesting enzymes throughout the digestive tract of sheep. *Turk. J. Vet. Anim. Sci.* **2016**, *40*, 271–277. [CrossRef]
24. Imamidoost, R.; Cant, J.P. Non-steady-state modeling of effects of timing and level of concentrate supplementation on ruminal pH and forage intake in high-producing, grazing ewes. *J. Anim. Sci.* **2005**, *83*, 1102–1115. [CrossRef] [PubMed]
25. Mileva, R. Determination of free doxycycline concentrations in the plasma and milk of sheep and in the plasma of rabbits by using the HPLC method. *Maced. Vet. Rev.* **2019**, *42*, 123–130. [CrossRef]
26. Yaneva, Z.; Georgieva, N.V.; Bekirska, L.L.; Lavrova, S. Drug mass transfer mechanism, thermodynamics, and in vitro release kinetics of antioxidant-encapsulated zeolite microparticles as a drug carrier system. *Chem. Biochem. Eng. Q.* **2018**, *32*, 281–298. [CrossRef]
27. Yaneva, Z.; Georgieva, N. Physicochemical and morphological characterization of pharmaceutical nanocarriers and mathematical modeling of drug encapsulation/release mass transfer processes. In *Nanoscale Fabrication, Optimization, Scale-Up and Biological Aspects of Pharmaceutical Nanotechnology*, 1st ed.; Grumezescu, A., Andrew, W., Eds.; Elsevier: Amsterdam, The Netherlands, 2018; Chapter 5; pp. 174–218.

Article

Population Pharmacokinetic Analysis Proves Superiority of Continuous Infusion in PK/PD Target Attainment with Oxacillin in Staphylococcal Infections

Irena Murínová [1,2], Martin Švidrnoch [3], Tomáš Gucký [3], Jan Hlaváč [4], Pavel Michálek [5], Ondřej Slanař [4] and Martin Šíma [4,*]

[1] Department of Clinical Pharmacy, Military University Hospital Prague, 16902 Prague, Czech Republic
[2] Department of Applied Pharmacy, Faculty of Pharmacy, Masaryk University, 60177 Brno, Czech Republic
[3] Laboratory of Pharmacology and Toxicology, AGEL Laboratories, 74101 Novy Jicin, Czech Republic
[4] Department of Pharmacology, First Faculty of Medicine, Charles University and General University Hospital in Prague, 12800 Prague, Czech Republic
[5] Department of Anaesthesia and Intensive Care, First Faculty of Medicine, Charles University and General University Hospital in Prague, 12808 Prague, Czech Republic
* Correspondence: martin.sima@lf1.cuni.cz

Abstract: Considering its very short elimination half-life, the approved oxacillin dosage might not be sufficient to maintain the pharmacokinetic/pharmacodynamics (PK/PD) target of time-dependent antibiotics. This study aimed to describe the population pharmacokinetics of oxacillin and to explore the probability of PK/PD target attainment by using various dosing regimens with oxacillin in staphylococcal infections. Both total and unbound oxacillin plasma concentrations retrieved as a part of routine therapeutic drug-monitoring practice were analyzed using nonlinear mixed-effects modeling. Monte Carlo simulations were used to generate the theoretical distribution of unbound oxacillin plasma concentration–time profiles at various dosage regimens. Data from 24 patients treated with oxacillin for staphylococcal infection have been included into the analysis. The volume of distribution of oxacillin in the population was 11.2 L, while the elimination rate constant baseline of $0.73\ h^{-1}$ increased by $0.3\ h^{-1}$ with each $1\ mL/s/1.73\ m^2$ of the estimated glomerular filtration rate (eGFR). The median value of oxacillin binding to plasma proteins was 86%. The superiority of continuous infusion in achieving target PK/PD values was demonstrated and dosing according to eGFR was proposed. Daily oxacillin doses of 9.5 g, 11 g, or 12.5 g administered by continuous infusion have been shown to be optimal for achieving target PK/PD values in patients with moderate, mild, or normal renal function, respectively.

Keywords: antibiotics; nonlinear mixed-effects modeling; glomerular filtration rate; dosing regimen; oxacillin; Monte Carlo simulations

Citation: Murínová, I.; Švidrnoch, M.; Gucký, T.; Hlaváč, J.; Michálek, P.; Slanař, O.; Šíma, M. Population Pharmacokinetic Analysis Proves Superiority of Continuous Infusion in PK/PD Target Attainment with Oxacillin in Staphylococcal Infections. *Antibiotics* **2022**, *11*, 1736. https://doi.org/10.3390/antibiotics11121736

Academic Editor: Dóra Kovács

Received: 9 November 2022
Accepted: 29 November 2022
Published: 1 December 2022

Publisher's Note: MDPI stays neutral with regard to jurisdictional claims in published maps and institutional affiliations.

1. Introduction

Oxacillin is a narrow-spectrum semisynthetic penicillin that belongs to the beta-lactam antibiotics used to treat infections caused by Gram-positive cocci. The isoxazole group, which is attached to the beta 6 position on the penicillin core, ensures penicillinase resistance [1]. This is the reason why it is used in clinical practice, especially for moderate-to-severe penicillin-resistant staphylococcal infections. It is mainly used for the treatment of infections of the skin and soft tissues, osteomyelitis, and endocarditis. It can also be used for the treatment of CNS, urinary, or respiratory infections [1,2]. Oxacillin replaced its predecessor methicillin and offers a better safety profile, especially limiting the risk of interstitial nephritis. Methicillin and oxacillin share similar pharmacokinetic/pharmacodynamic (PK/PD) properties. Only a year after the introduction of methicillin into practice, methicillin-resistant Staphylococcus aureus was discovered and its occurrence in the hospital environment is quite common today [3]. Its weighted mean population prevalence in the European economic area was 16.7% in 2020 but varies significantly with a clear north to south/east gradient [4].

After the administration of oxacillin (nowadays only parenterally), it quickly distributes into most tissues and fluids of the body, including cerebrospinal fluid and bones, and crosses the placenta. It has a low volume of distribution (Vd), which is approximately equivalent to extracellular fluid, and has a high level of protein binding (~90%) [1]. A part of the oxacillin dose is metabolized into active 5-hydroxymethyl and inactive penicilloic acid derivatives in the liver [5]. Oxacillin, same as its metabolites, is excreted primarily in the urine via tubular secretion and glomerular filtration; only a minor part is eliminated by bile or maybe in breast milk. The elimination half-life is approximately 0.5 h and could be prolonged in patients with renal impairment or by concomitant use of probenecid due to competitive inhibition of renal tubular secretion [1,2].

The approved dosage of oxacillin for common infections is 0.25–1 g every 4 or 6 h, administered intravenously or intramuscularly. For severe infections the dose can be escalated up to 12 g/day [2,6].

The effect of beta-lactam antibiotics is time-dependent. The time when unbound oxacillin plasma concentrations are maintained above the minimal inhibitory concentration (fT > MIC) is considered as the most appropriate PK/PD target. The lower-limit threshold is set to fT > MIC = 40–70% (% of time between dosages) and the optimum is fT > MIC = 100%, but some studies have argued for an even more forceful target of fT > 4 × MIC = 100% [7,8].

Given the very short elimination half-life of oxacillin and its time-dependent antibiotic effect, it seems that approved dosages administered via standard infusion might not be sufficient to achieve this PK/PD target. The issue of achieving the optimal PK/PD target has already been addressed by other studies with the halogenated oxacillin derivatives cloxacillin and flucloxacillin, and they agreed that extended (3 h) or continuous infusion is more efficient than the standard 0.5 h infusion [9,10].

This study aimed to describe the pharmacokinetics of oxacillin using a population approach and to explore the probability of PK/PD target attainment using various dosing regimens with oxacillin in staphylococcal infections.

2. Results

A total of 24 patients (14 males, 10 females) were enrolled in the analysis. Their demographic and laboratory characteristics are summarized in Table 1. Patients received oxacillin to treat staphylococcal infections of the central nervous system (n = 7), sepsis (n = 3), orthopedic (n = 10), or other infections (n = 4) (e.g., bacteriuria, bacteremia, or endocarditis). Oxacillin doses ranged between 1 g and 3 g every 4 h and were administered via a 0.5 h intravenous infusion. Only one patient was given oxacillin via an extended 3 h infusion. Infection was caused by methicillin-susceptible *Staphylococcus aureus* in 21 cases, by *Staphylococcus hominis* in 2 cases, and by *Staphylococcus warneri* in one case. The median (IQR) value of MIC was 0.25 (0.25–0.5) mg/L (i.e., the most frequent MIC value was 0.25 mg/L).

Table 1. Demographic and laboratory characteristics of patients.

Characteristics	Median	IQR	Range
Age (years)	55	45–72	26–84
Body weight (kg)	84	74–96	57–145
Height (cm)	174	168–182	153–195
Body mass index (kg/m^2)	28	23–33	20–39
Serum creatinine (μmol/L)	82	65–92	50–151
eGFR (mL/s/1.73 m^2)	1.53	0.98–1.71	0.59–2.10
Serum urea (mmol/L)	4.5	3.3–5.6	1.8–11.8
Serum albumin (g/L)	32	26–37	22–44

eGFR is estimated glomerular filtration rate according to the CKD-EPI formula. IQR is interquartile range.

A total of 32 oxacillin plasma concentrations were included in the PK analysis, with an average of 1.33 concentrations per patient. Twenty-four concentration points were taken as trough level, while eight samples were taken as peak level (sample collection after infusion completion). The median (IQR) value of oxacillin binding to plasma proteins was 86% (83–88%). Oxacillin binding to plasma proteins was not significantly associated with any of the covariates tested (total oxacillin plasma level, serum albumin, sex, age, body weight, height, BMI, and eGFR).

2.1. Population PK Analysis

Total plasma oxacillin concentration–time data were best fitted using a one-compartmental model with linear elimination kinetics. A proportional error model was the most accurate for residual and interpatient variability. The population model was parametrized using volume of distribution (Vd) and elimination rate constant (K_e). The population PK estimates for the oxacillin final model are summarized in Table 2. Among the investigated variables, the most significant covariate was eGFR for oxacillin K_e. The population Vd of oxacillin was 11.2 L, while K_e started at a baseline of 0.73 h^{-1} and increased by 0.3 h^{-1} with each 1 mL/s/1.73 m^2 of eGFR.

Table 2. Estimates of the final oxacillin population pharmacokinetic model.

Parameter	Estimate	R.S.E. (%)
Fixed effects		
Vd_pop (L)	11.2	30.4
K_e_pop (h^{-1})	0.73	26.1
β_K_e_eGFR (h^{-1} per each 1 mL/s/1.73 m^2 of eGFR)	0.3	50.8
Standard deviation of the random effects		
Ω_Vd	0.7	22.5
Ω_K_e	0.11	37.5
Error model parameters		
b	0.4	19.8

Vd is oxacillin volume of distribution, K_e is oxacillin elimination rate constant, and eGFR is estimated glomerular filtration rate.

The final equations describing the relationships between oxacillin PK parameters and their covariate are as follows:

$$\text{Log(Vd)} = \log(\text{Vd\_pop}) + \eta\_\text{Vd}$$

$$\text{Log}(K_e) = \log(K_e\_\text{pop}) + \beta\_K_e\_\text{eGFR} \times \text{eGFR} + \eta\_K_e$$

where pop represents the typical value of the parameter, β represents the covariate effect on the parameter, and η represents a random effect variable.

The GOF plots for the final covariate model for oxacillin showed no substantial deviations (Figure 1). The R.S.E. values documented that the PK parameters in the population model for oxacillin were correctly estimated (Table 2). The VPC plot of the final oxacillin model showed that the predictions corresponded to the observed data, confirming the validity of the model to predict the PK data (Figure 2).

2.2. Monte Carlo Simulations

Figure 3 shows the simulated unbound plasma concentration profiles of oxacillin over time (500 replicates of all subjects in the dataset) following intravenous administration of oxacillin at a dose of 1 g every 6 h by 0.5 h infusion, 1 g every 4 h by 0.5 h infusion, 3 g every 6 h by 0.5 h infusion, 1 g every 4 h by 3 h infusion, 6 g daily by continuous infusion, and 12 g daily by continuous infusion.

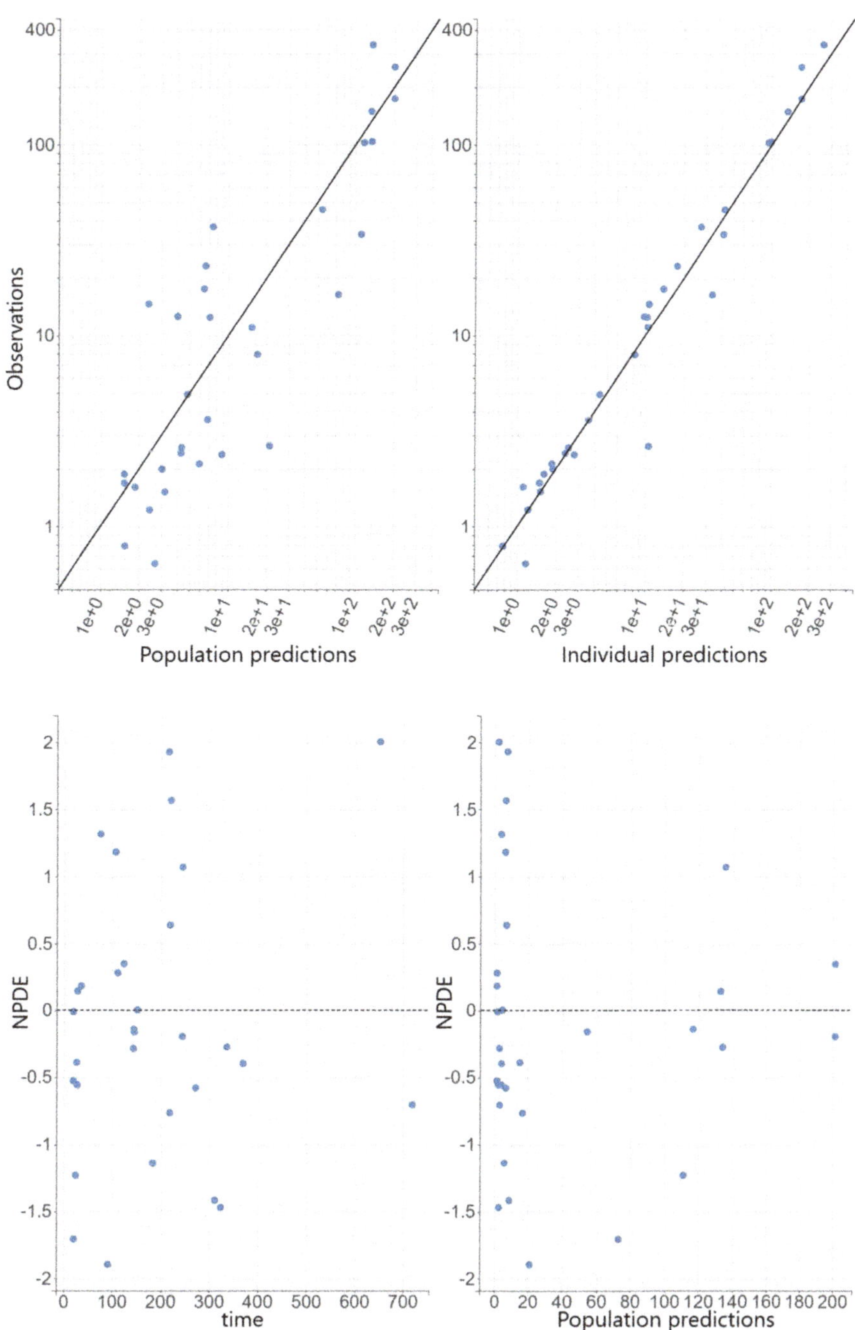

Figure 1. Population and individual predictions of oxacillin versus observed concentrations (log–log scale), and normalized prediction distribution errors (NPDE) versus time and population predictions.

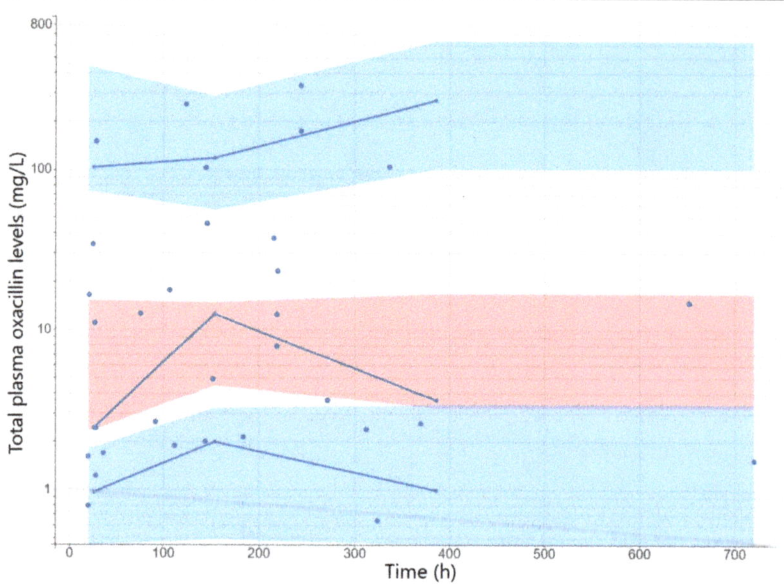

Figure 2. Semilogarithmically plotted visual predictive check and the observed data of the total plasma oxacillin concentrations in time for the final model. Solid lines represent the 10th, 50th, and 90th percentiles of the observed data. Shaded regions represent 90% confidence interval around the 10th (below blue region), 50th (pink region), and 90th (above blue region) percentiles of the simulated data.

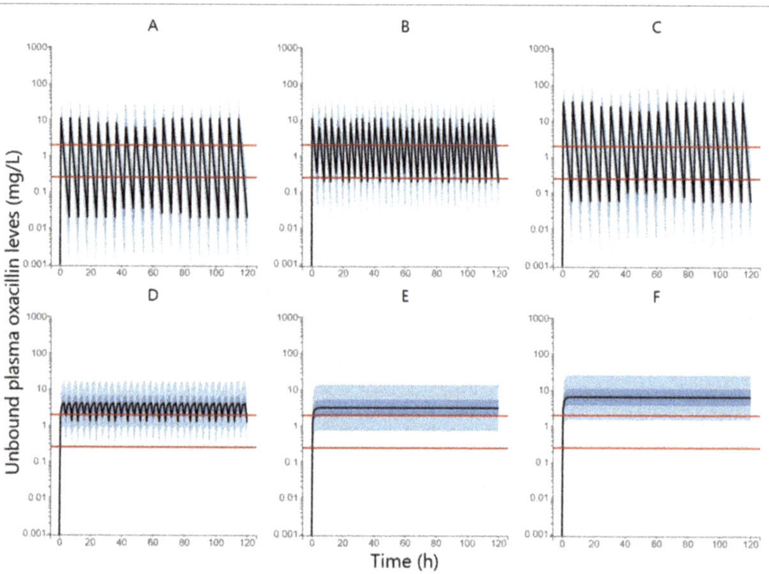

Figure 3. Simulated unbound plasma concentration profiles of oxacillin over time following intravenous administration of oxacillin at a dose of 1 g every 6 h by 0.5 h infusion (**A**), 1 g every 4 h by 0.5 h infusion (**B**), 3 g every 6 h by 0.5 h infusion (**C**), 1 g every 4 h by 3 h infusion (**D**), 6 g daily by continuous infusion (**E**), and 12 g daily by continuous infusion (**F**). The curve represents the median, and the four bands represent percentiles (5–27.5%, 27.5–50%, 50–72.5%, and 72.5–95%) of the 90% distribution of simulated concentrations. Red lines represent two different MIC values (0.25 and 2 mg/L).

Table 3 summarized the PTA values of all above-mentioned oxacillin dosing regimens at various MIC values (0.25 and 2 mg/L) and various PK/PD targets (fT > MIC = 100% and fT > MIC = 50%). The dosing regimen was considered to be successful when ≥ 90% of patients reached the PK/PD target. The standard 0.5 h infusion was sufficient neither by shortening the dosing interval to 4 h nor by increasing the dose size to 3 g. Extended 3 h infusion can only be used at low MIC values (e.g., 0.25 mg/L), whereas continuous infusion has been shown to be effective in achieving PK/PD outcomes even when targeting higher MIC values. However, if we want to cover the PK/PD target up to the EUCAST epidemiological cut-off value for oxacillin of 2 mg/L, the dose needs to be increased to 12 g per day, administered via continuous infusion.

Table 3. Probability of target attainment at steady-state after intravenous administration of oxacillin in various dosing regimens (1 g every 6 h via 0.5 h infusion, 1 g every 4 h via 0.5 h infusion, 3 g every 6 h via 0.5 h infusion, 1 g every 4 h via 3 h infusion, 6 g per day via continuous infusion, and 12 g per day via continuous infusion) at various MIC values (0.25 and 2 mg/L) and various PK/PD targets (fT > MIC = 100% and fT > MIC = 50%).

	Probability of Target Attainment (%)			
PK/PD Target	**fT > MIC = 100%**		**fT > MIC = 50%**	
MIC	0.25 mg/L	2 mg/L	0.25 mg/L	2 mg/L
1 g every 6 h (0.5 h infusion)	4.35	0.05	60.83	5.55
1 g every 4 h (0.5 h infusion)	41.52	2.60	60.25	27.69
3 g every 6 h (0.5 h infusion)	17.56	0.64	87.77	27.48
1 g every 4 h (3 h infusion)	95.30	32.23	99.93	72.40
6 g/day (continuous infusion)	99.82	69.95	99.82	69.95
12 g/day (continuous infusion)	99.97	90.72	99.97	90.72
p-Value	<0.0001	<0.0001	<0.0001	<0.0001

In a simulation of oxacillin administered by continuous infusion at a dose graded according to eGFR (daily dose of 9.5 g, 11 g, and 12.5 g in patients with moderate renal impairment, mild renal impairment, and normal renal function, respectively), the PTA was 90.18%; therefore, this dosing can be considered successful in achieving the optimal PK/PD target of fT > MIC = 100% with the EUCAST epidemiological cut-off value for oxacillin of 2 mg/L using the lowest possible doses.

3. Discussion

Oxacillin has been used for many years as one of the first-choice drugs in the treatment of staphylococcal infections [11]. Nevertheless, considering its very short elimination half-life and time-dependent antibiotic effect, the approved dosage administered via standard 0.5 h infusion might not be sufficient to maintain the PK/PD target. This widespread practice may then be the cause of frequent staphylococcal resistance [12]. To prevent the development of this unfavorable trend, antibiotic dosing needs to be optimized to achieve the PK/PD target in the majority of the population. Therefore, therapeutic drug monitoring of oxacillin has been implemented into clinical routines in our hospital and optimized oxacillin dosing has been proposed based on a one-year follow-up population analysis of the acquired data.

Based on our analysis, the population volume of the distribution of oxacillin was 11.2 L, which, at median a body weight of 84 kg, corresponds to a body-weight-normalized Vd of 0.13 L/kg. However, the patients in our study were overweight (median BMI of 28 kg/m^2) and if we converted their weight to a normal BMI value of 22 kg/m^2, then the body-weight-normalized Vd comes out to 0.17 L/kg. This value is consistent with the volume of extracellular water into which the beta-lactam antibiotics were distributed [13]. The only significant covariate of oxacillin pharmacokinetics in our population model was eGFR for oxacillin K_e, which started at a baseline of 0.73 h^{-1} and increased by 0.3 h^{-1} with each 1 mL/s/1.73 m^2 of eGFR. In patients with an eGFR of 1.5 mL/s/1.73 m^2 (normal renal function), the oxacillin K_e was therefore 1.18 h^{-1}, which corresponds to the elimination half-life of 0.6 h. This value is fully consistent with that reported by the summary of product characteristics [2] and it is also the reason for the difficulty in achieving the PK/PD target at standard dosing.

While experimental data reported the antibiotic effects of beta-lactams at fT > MIC = 40–70%, clinical studies showed that a higher threshold of fT > MIC = 100 should be considered [7,8]. In our study, we explored the probability of target attainment for two pharmacological outcomes: fT > MIC = 100% was defined as an optimal PK/PD target, while fT > MIC = 50% was a minimal PK/PD target. However, considering the severity of some staphylococcal infections (e.g., endocarditis, sepsis, osteomyelitis, or pneumonia), the risk of developing resistance, and that therapeutic beta-lactam monitoring is usually not routinely implemented, we consider a higher PK/PD target to be more appropriate for the optimal dosage proposal. Both PK/PD target attainments were also explored for two different MIC values, i.e., 0.25 mg/L as the most frequent MIC value in our study and 2 mg/L as the EUCAST epidemiological cut-off value for oxacillin [14]. Again, we recommend assuming a higher MIC value in clinical routines if we do not have exactly measured strain-specific MIC values.

It should also be kept in mind that only the unbound fraction of the drug can exert an antibiotic effect. Since oxacillin is highly bound to plasma proteins, its unbound fraction should be measured, or at least estimated, from the total level using a correction factor. In our study, oxacillin binding to plasma proteins ranged from 74 to 97%, with a median value of 86%. This value is only slightly lower than that reported for healthy volunteers (89–94%) [2]. Since oxacillin binds primarily to albumin, an association between the unbound fraction and serum albumin level would be expected. However, we did not observe any significant relation. This can be explained by the fact that there were no patients with severe hypoalbuminemia in our study in whom the effect of increasing the unbound fraction of oxacillin would be most pronounced.

Monte Carlo simulations based on the oxacillin population pharmacokinetic model confirmed our assumption that the approved dosage administered via standard 0.5 h infusion is not sufficient to reach the PK/PD target. To achieve a higher PTA, dose intensification is needed, which can be done by several approaches—increasing the dose sizes, shortening the dosing interval, or extending the duration of the infusion. Increasing the dose sizes leads to only a small increase in PTA, which is logical for an antibiotic with a time-dependent effect. A shortening of the dosing interval would certainly be appropriate, but a dosing interval of less than 4 h would already be highly impractical for clinical routines. Therefore, extending the duration of the infusion is the most appropriate approach. As we can see in Table 3, an extended 3 h infusion is sufficient to cover the PK/PD target only at lower MIC values (0.25 mg/L); therefore, continuous infusion seems to be optimal. Another advantage of continuous infusion is that it shows the same PTA rate for both lower (fT > MIC = 50%) and higher (fT > MIC = 100%) PK/PD targets, i.e., if we reach the lower PK/PD target, we spontaneously reach the higher one. Thus, our findings confirmed our hypothesis that oxacillin should ideally be administered via continuous infusion, which is also consistent with the dosing recommendations for flucloxacillin and cloxacillin from the other studies that have addressed this issue [9,10]. However, if we want to cover the PK/PD target up to the EUCAST epidemiological cut-off value for oxacillin of 2 mg/L in the whole population, the daily dose administered via continuous infusion needs to be increased up to 12 g per day. To cover this PK/PD target with the lowest possible daily doses, we proposed a dosage graded by eGFR as a covariate of oxacillin elimination. This intention is best met by daily doses of 9.5 g, 11 g, and 12.5 g in patients with moderate renal impairment, mild renal impairment, and normal renal function, respectively.

We acknowledge several limitations of our study. Firstly, we evaluated only the achievement of the PK/PD target, not the real clinical outcomes. Secondly, since our analysis was based on TDM data from clinical routines, the majority of samples were taken as a trough concentration, and thus the distribution phase in our population model may not be exhaustively described. On the other hand, diagnostics of our model did not show any inaccuracies throughout the monitoring time course. Moreover, for oxacillin as a time-dependent antibiotic, the elimination phase is crucial for achieving the PK/PD target. Therefore, our dosing proposal should not be biased by the sampling limitation. Finally, since patients with severe renal impairment or with augmented renal function were not present in our study cohort, we are unable to suggest dosing for these extreme cases.

4. Materials and Methods

4.1. Study Design

A retrospective observational pharmacokinetic study was conducted in adults (age $\geq$ 18 years) treated with oxacillin intravenous infusion for staphylococcal infection admitted to mixed wards of the Military University Hospital in Prague from June 2021 to June 2022. Patients were included in the study if they had at least one measurement of oxacillin plasma level during treatment. Exclusion criteria were extracorporeal life support and renal replacement therapy. The study was approved by the local ethics committee of the Military University Hospital in Prague under the registration number 108/17–98/2022, and followed the principles established by the Declaration of Helsinki. Due to the retrospective nature of this study, which involved only analysis of routine clinical data, study-specific informed consent for was waived. The collection and processing of anonymized data is in the public interest.

4.2. Data Retrieval

The clinical records of all included patients were reviewed to collect information on age, sex, height, and body weight, as well as serum creatinine, urea, and albumin levels. Body mass index (BMI) was computed as the ratio of body weight (kg) to the square of height (m). For each patient, the glomerular filtration rate (eGFR) was calculated according to the chronic kidney disease epidemiology collaboration formula. Oxacillin dosing regimens, including administration times and infusion rates, were recorded. Both total and unbound oxacillin serum concentrations were determined as a routine part of the therapeutic drug-monitoring procedure. Sampling times were also collected. Minimal inhibitory concentration (MIC) value of oxacillin for isolated staphylococcal strain was also recorded in each patient.

4.3. Bioanalytical Assay

4.3.1. Chemicals and Reagents

Acetonitrile (ACN), methanol, and water (all of them HPLC-MS grade) were supplied by Chem-Lab NV (Zedelgem, Belgium). Ammonium acetate ($\geq$99.9%), formic acid ($\geq$98%), and oxacillin sodium salt monohydrate were purchased from Sigma-Aldrich (St. Louis, MO, USA). Amoxicillin-d4 ($\geq$98%, internal standard) was purchased from Toronto Research Chemicals (Toronto, Canada). Human sera from more than 10 volunteers of different sexes and ages were pooled and used as matrix. The drug-free status was verified by measuring a blank sample of the pooled serum. Calibration samples of oxacillin were prepared by direct spiking of the verified serum to final concentrations of 1, 5, 10, 50, and 100 mg/L.

4.3.2. Instrumentation

The LC-MS/MS analyses were performed using a Shimadzu LC-MS system Nexera X2 LC-30AD coupled with a LCMS-8060 triple quadrupole mass spectrometer. All separations were carried out on an Kinetex® C18 column (2.6 µm, 3 × 50 mm; Phenomenex, Torrance, CA, USA) thermostated to 40 °C. The LC system consisted of two binary pumps, a solvent rack equipped with a degasser, thermostated autosampler, and a column oven. The separation conditions used were as follows: sample injection volume 1 µL; mobile phases (MF): (A) consisted of water/ACN (9:1, v/v) containing 0.1% (v/v) of formic acid and 10 mM ammonium acetate and (B) consisted of water containing 0.1% (v/v) of formic acid and 10 mM ammonium acetate. The separation was conducted in an isocratic mode with a flow rate of 0.2 mL/min (60% MF A and 40% MF B). The LC-MS/MS system was operated with an electrospray ionization probe in a positive mode. The MS/MS detection was performed using a multiple-reaction monitoring mode (MRM) with the following MRM transitions: m/z 402.2→144.0 (quantifier) and 402.2→186.0 (qualifier) for oxacillin and m/z 368.2→227.2 for amoxicillin-d4.

4.3.3. Sample Preparation and Quantification

A very high fraction of oxacillin is bound to plasma proteins; thus, an ultrafiltration step was applied to isolate the free fraction of oxacillin. An amount of 500 µL of serum aliquot was transferred to the centrifugal filter (Centrifree® PL Regenerated Cellulose, 30 kDa; Sigma-Aldrich, St. Louis, MO, USA) and centrifuged at 1000 rcf for 15 min. To 30 µL of sample (serum and ultrafiltrate) were added 10 µL of internal standard solution (amoxicillin-d4) and 60 µL of ACN. The solution was vortexed and centrifuged at $20,000 \times g$ for 5 min.

Supernatants were transferred to plastic inserts, placed in vials, and analyzed. A five-point calibration method was used for quantification of the clinical samples.

4.4. Statistics

Descriptive parameter medians and interquartile ranges (IQRs) were calculated using MS Excel 2013 (Microsoft Corporation, Redmond, WA, USA). Mann–Whitney U-test and linear regression model were used to evaluate the relationships between oxacillin binding to plasma proteins and categorical and continuous variables, respectively. GraphPad Prism software version 8.2.1 (GraphPad Inc., La Jolla, CA, USA) was used for all comparisons, and p-levels less than 0.05 were considered statistically significant.

4.5. Population PK Analysis

Oxacillin plasma concentrations against time were analyzed using nonlinear mixed-effects modeling. The model parameters were assumed to be log-normally distributed and were estimated by maximum likelihood using the stochastic approximation expectation maximization algorithm within the Monolix Suite software version 2021R1 (Lixoft SAS, Antony, France). The model was built in three steps.

(1) *Base model*

One- and two-compartmental models with first-order or Michaelis–Menten elimination were tested for the structural model. Log-normally distributed interindividual variability terms with estimated variance were tested on each PK parameter. Proportional, additive, and combination error models were tested for the residual error model. The most appropriate model was selected based on the minimum objective function value (OFV), adequacy of the goodness-of-fit (GOF) plots, and low relative standard errors (R.S.E.) of the estimated PK parameters.

(2) *Covariate model*

Age, bodyweight, height, body mass index, and serum creatinine, urea, and albumin levels, as well as eGFR were tested as continuous covariates (characteristics predictive of interindividual variability), while sex and the reason for oxacillin treatment were tested as categorical covariates. Preliminary graphical assessment and univariate associations using Pearson's correlation test for the effects of covariates on PK estimates was performed. Covariates with $p < 0.05$ were considered for the covariate model. A stepwise covariate modeling procedure was then performed. For model selection, a decrease in OFV of >3.84 points between nested models ($p < 0.05$) was considered statistically significant, assuming a χ^2-distribution. Other criteria for model selection were reasonably low R.S.E. values of the structural model parameter estimates, physiological plausibility of the parameter values obtained, and absence of bias in the GOF plots.

(3)   *Model evaluation*

The adequacy of the model was assessed using GOF graphs. Observed values were plotted against individual and population prediction values. The normalized prediction distribution errors (NPDE) were plotted against time to assess randomness around the line of unity. The visual predictive check (VPC) was conducted to assess the predictability of the final model. For this purpose, 1000 replicate values of the original dataset were simulated using the final model parameter estimates, and the simulated distribution was compared with the distribution from the observations. From all replicates, 90% prediction intervals for the 10th, 50th, and 90th percentile of the simulations were computed and graphically presented.

4.6. Monte Carlo Simulations

Monte Carlo simulations (500 replicates of all the individuals in dataset) based on a final population PK model of oxacillin were performed to generate theoretical distribution of unbound oxacillin plasma concentration profiles over time using Simulx version 2021 (Lixoft SAS, Antony, France). The following oxacillin dosing regimens were simulated: 1 g every 6 h by 0.5 h infusion, 1 g every 4 h by 0.5 h infusion, 3 g every 6 h by 0.5 h infusion, 1 g every 4 h by 3 h infusion, 6 g daily by continuous infusion, and 12 g daily by continuous infusion.

As an optimal PK/PD target was considered if unbound oxacillin plasma concentrations are maintained above the minimal inhibitory concentration for the entire dosing interval (fT > MIC = 100%), but we also tested the achievement of fT > MIC = 50% as a minimal PK/PD target in beta-lactam antibiotics. Probability of target attainment (PTA) at steady-state was calculated for all dosage regimens and different MIC values–0.25 mg/L (as the most frequent MIC value in our study) and 2 mg/L (as an EUCAST epidemiological cut-off value for oxacillin) [14]. Chi-square test was used for evaluation of differences in PTA between various dosing regimens.

In order to propose an optimal dosing regimen that would cover the PK/PD target with the EUCAST epidemiological cut-off value for oxacillin of 2 mg/L using the lowest possible dose, administration of oxacillin continuous infusion in dose scaled by eGFR as the main covariate of oxacillin PK was subsequently simulated. Patients with moderate renal impairment (eGFR = 0.5–1.0 mL/s/1.73 m^2) received oxacillin daily dose of 9.5 g, patients with mild renal impairment (eGFR = 1.0–1.5 mL/s/1.73 m^2) received daily dose of 11 g, and patients with normal renal function (eGFR > 1.5 mL/s/1.73 m^2) received daily dose of 12.5 g.

5. Conclusions

We described oxacillin population pharmacokinetics in patients with staphylococcal infection. The only significant covariate was eGFR for oxacillin clearance. The median value of oxacillin binding to plasma proteins was 86%. We proved the superiority of continuous infusion in the attainment of the PK/PD target and proposed an eGFR-scaled dosing. If we want to target the optimal PK/PD target of fT > MIC = 100%, with a EUCAST epidemiological cut-off value for oxacillin of 2 mg/L, the daily oxacillin doses of 9.5 g, 11 g, and 12.5 g should be administered via continuous infusion in patients with moderate renal impairment, mild renal impairment, and normal renal function, respectively. As this study only assessed the PK/PD target achievement, confirmatory trials exploring real clinical outcomes would be needed.

Author Contributions: Conceptualization, I.M., M.Š. (Martin Šíma), and O.S.; methodology, M.Š. (Martin Šíma), M.Š. (Martin Švidrnoch), and T.G.; software, M.Š. (Martin Šíma); validation, M.Š. (Martin Šíma), O.S., M.Š. (Martin Švidrnoch), and T.G.; formal analysis, M.Š. (Martin Šíma), M.Š. (Martin Švidrnoch), and T.G.; investigation, I.M.; resources, P.M.; data curation, I.M. and J.H.; writing—original draft preparation, I.M., M.Š. (Martin Šíma), J.H., M.Š. (Martin Švidrnoch), and T.G.; writing—review and editing, P.M. and O.S.; visualization, M.Š. (Martin Šíma); supervision, O.S.; project administration, I.M.; funding acquisition, P.M. All authors have read and agreed to the published version of the manuscript.

Funding: This study was supported by the Ministry of Health of the Czech Republic (project MH CZ-DRO-VFN64165) and by Charles University project Cooperatio (research area PHAR).

Institutional Review Board Statement: The study was conducted in accordance with the Declaration of Helsinki and approved by the local ethics committee of the Military University Hospital in Prague under the registration number 108/17-98/2022 on 17 October 2022.

Informed Consent Statement: Due to the retrospective nature of this study, which involved only analysis of routine clinical data, study-specific informed consent was waived. The collection of anonymized data and its processing are in the public interest.

Data Availability Statement: The data that support the findings of this study are available from the corresponding author upon reasonable request.

Acknowledgments: We are very grateful to the physicians, nurses, and other clinical pharmacists of Military University Hospital Prague for their cooperation in the careful collection of samples for this study.

Conflicts of Interest: The authors declare no conflict of interest.

References

1. Enna, S.J.; Bylund, D.B. Elsevier Science (Firm). In *XPharm: The Comprehensive Pharmacology Reference*; Elsevier: Amsterdam, The Netherlands, 2008; ISBN 0080552323/9780080552323.
2. State Institute for Drug Control of Czech Republic. Prostaphylin—Summary of Product Characteristics. Available online: https://www.sukl.eu/ (accessed on 19 September 2022).
3. Larsen, J.; Raisen, C.L.; Ba, X.; Sadgrove, N.J.; Padilla-Gonzalez, G.F.; Simmonds, M.S.J.; Loncaric, I.; Kerschner, H.; Apfalter, P.; Hartl, R.; et al. Emergence of methicillin resistance predates the clinical use of antibiotics. *Nature* **2022**, *602*, 135–141. [CrossRef] [PubMed]
4. World Health Organization. Antimicrobial Resistance Surveillance in Europe. Available online: https://www.ecdc.europa.eu/sites/default/files/documents/Joint-WHO-ECDC-AMR-report-2022.pdf (accessed on 22 September 2022).
5. Thijssen, H.H. Analysis of isoxazolyl penicillins and their metabolites in body fluids by high-performance liquid chromatography. *J. Chromatogr.* **1980**, *183*, 339–345. [CrossRef] [PubMed]
6. Schlossberg, D.L.; Rafik, S. *Antibiotics Manual: A Guide to Commonly Used Antimicrobials*; John Wiley & Sons: Hoboken, NJ, USA, 2017; ISBN 1119220750/9781119220756.
7. Wong, G.; Briscoe, S.; McWhinney, B.; Ally, M.; Ungerer, J.; Lipman, J.; Roberts, J.A. Therapeutic drug monitoring of beta-lactam antibiotics in the critically ill: Direct measurement of unbound drug concentrations to achieve appropriate drug exposures. *J. Antimicrob. Chemother.* **2018**, *73*, 3087–3094. [CrossRef] [PubMed]
8. Scharf, C.; Liebchen, U.; Paal, M.; Taubert, M.; Vogeser, M.; Irlbeck, M.; Zoller, M.; Schroeder, I. The higher the better? Defining the optimal beta-lactam target for critically ill patients to reach infection resolution and improve outcome. *J. Intensive Care* **2020**, *8*, 86. [CrossRef] [PubMed]
9. Wilkes, S.; van Berlo, I.; Ten Oever, J.; Jansman, F.; Ter Heine, R. Population pharmacokinetic modelling of total and unbound flucloxacillin in non-critically ill patients to devise a rational continuous dosing regimen. *Int. J. Antimicrob. Agents* **2019**, *53*, 310–317. [CrossRef] [PubMed]
10. Courjon, J.; Garzaro, M.; Roger, P.M.; Ruimy, R.; Lavrut, T.; Chelli, M.; Raynier, J.L.; Chirio, D.; Demonchy, E.; Cabane, L.; et al. A Population Pharmacokinetic Analysis of Continuous Infusion of Cloxacillin during Staphylococcus aureus Bone and Joint Infections. *Antimicrob. Agents Chemother.* **2020**, *64*, e01562-20. [CrossRef] [PubMed]
11. Klein, J.O.; Sabath, L.D.; Steinhauer, B.W.; Finland, M. Oxacillin Treatment of Severe Staphylococcal Infections. *N. Engl. J. Med.* **1963**, *269*, 1215–1225. [CrossRef] [PubMed]
12. Foster, T.J. Antibiotic resistance in Staphylococcus aureus. Current status and future prospects. *FEMS Microbiol. Rev.* **2017**, *41*, 430–449. [CrossRef] [PubMed]
13. Sinnollareddy, M.G.; Roberts, M.S.; Lipman, J.; Roberts, J.A. beta-lactam pharmacokinetics and pharmacodynamics in critically ill patients and strategies for dose optimization: A structured review. *Clin. Exp. Pharmacol. Physiol.* **2012**, *39*, 489–496. [CrossRef] [PubMed]
14. European Committee on Antimicrobial Susceptibility Testing. Breakpoint Tables for Interpretation of MICs and Zone Diameters. Available online: https://www.eucast.org/fileadmin/src/media/PDFs/EUCAST_files/Breakpoint_tables/v_12.0_Breakpoint_Tables.pdf (accessed on 19 September 2022).

antibiotics

Article

Individual Pharmacotherapy Management (IPM)—IV: Optimized Usage of Approved Antimicrobials Addressing Under-Recognized Adverse Drug Reactions and Drug-Drug Interactions in Polypharmacy

Ursula Wolf [1,*], Henning Baust [2], Rüdiger Neef [3] and Thomas Steinke [2,4]

1   Pharmacotherapy Management, University Hospital Halle (Saale), Martin Luther University Halle-Wittenberg, 06120 Halle (Saale), Germany
2   University Clinic for Anesthesiology and Operative Intensive Care Medicine, University Hospital Halle (Saale), Martin Luther University Halle-Wittenberg, 06120 Halle (Saale), Germany
3   Department of Orthopedics, Trauma and Reconstructive Surgery, Division of Geriatric Traumatology, University Hospital Halle (Saale), Martin Luther University Halle-Wittenberg, 06120 Halle (Saale), Germany
4   Clinic for Anesthesiology, Intensive Care Medicine and Pain Therapy, Carl-von Basedow-Klinikum Saalekreis, 06127 Merseburg, Germany
*   Correspondence: ursula.wolf@uk-halle.de

check for updates

Citation: Wolf, U.; Baust, H.; Neef, R.; Steinke, T. Individual Pharmacotherapy Management (IPM)—IV: Optimized Usage of Approved Antimicrobials Addressing Under-Recognized Adverse Drug Reactions and Drug-Drug Interactions in Polypharmacy. *Antibiotics* **2022**, *11*, 1381. https://doi.org/10.3390/antibiotics11101381

Academic Editor: Dóra Kovács

Received: 24 August 2022
Accepted: 5 October 2022
Published: 9 October 2022

Publisher's Note: MDPI stays neutral with regard to jurisdictional claims in published maps and institutional affiliations.

Abstract: Antimicrobial therapy is often a life-saving medical intervention for inpatients and outpatients. Almost all medical disciplines are involved in this therapeutic procedure. Knowledge of adverse drug reactions (ADRs) and drug-drug interactions (DDIs) is important to avoid drug-related harm. Within the broad spectrum of antibiotic and antifungal therapy, most typical ADRs are known to physicians. The aim of this study was to evaluate relevant pharmacological aspects with which we are not so familiar and to provide further practical guidance. Individual pharmacotherapy management (IPM) as a synopsis of internal medicine and clinical pharmacology based on the entirety of the digital patient information with reference to drug information, guidelines, and literature research has been continuously performed for over 8 years in interdisciplinary intensive care and trauma and transplant patients. Findings from over 52,000 detailed medication analyses highlight critical ADRs and DDIs, especially in these vulnerable patients with polypharmacy. We present the most relevant ADRs and DDIs in antibiotic and antifungal pharmacology, which are less frequently considered in relation to neurologic, hemostaseologic, hematologic, endocrinologic, and cardiac complexities. Constant awareness and preventive strategies help avoid life-threatening manifestations of these inherent risks and ensure patient and drug safety in antimicrobial therapy.

Keywords: antimicrobial; antibiotics; antifungals; adverse drug reaction (ADR); drug-drug interaction (DDI); polypharmacy; multimorbidity; intensive care patients; traumatology; elderly patients; organ failure; multi-organ failure; drug safety; patient safety

1. Introduction

Antimicrobial therapy often means a life-saving medical intervention for in-hospital and outpatients, and almost all medical disciplines are involved in this therapeutic regimen. Knowledge of adverse drug reactions (ADRs) and drug-drug interactions (DDIs) is important to avoid drug-related harm, as severe organ damage or life-threatening conditions can already occur with dosing due to drug-drug-inhibited metabolism. With respect to the expanding elderly patient population with polypharmacy, the almost rapidly increasing drug availabilities, and administration by different specialists in multimorbidity, ARDs and DDIs are turning into a major health problem worldwide. An ADR is "a response to a drug which is noxious and unintended and which occurs at doses normally used in man for prophylaxis, diagnosis, or therapy of disease or for the modification of physiologic function" (WHO 1972), which has already been defined by the WHO for 50 years [1]. A further, more clarified terminology refers to an ADR as "an appreciably harmful or unpleasant reaction resulting from an intervention related to the use of a medicinal product; adverse effects usually predict hazard from future administration and warrant prevention, or specific treatment, or alteration of the dosage regimen, or withdrawal of the product", and classical type A (augmented) and type B (bizarre) ADRs display variable patterns depending on age, sex, race, diseases, drug category, route of administration, and DDIs, as well as the individual genotypic profile, which is a further determinant for the manifestation of ADRs [2]. The classification into six ADR types (A, B, C, D, E, F) aims to differentiate dose-related (augmented), non-dose-related (bizarre), dose-related and time-related (chronic), time-related (delayed), withdrawal (end of use), and failure of therapy (failure) ADRs. According to the directive 2001/83/EC of the European Parliament and the 2001 Council on the Community code relating to medicinal products for human use, reported in the Guideline on good pharmacovigilance practices (GVP)-Annex I (Rev 4), an ADR is currently defined as: "A response to a medicinal product which is noxious and unintended. Response in this context means that a causal relationship between a medicinal product and an adverse event is at least a reasonable possibility. An adverse reaction, in contrast to an adverse event, is characterized by the fact that a causal relationship between a medicinal product and an occurrence is suspected. For regulatory reporting purposes, if an event is spontaneously reported, even if the relationship is unknown or unstated by the healthcare professional or consumer as primary source, it meets the definition of an adverse reaction. Therefore all spontaneous reports notified by healthcare professionals or consumers are considered suspected adverse reactions, since they convey the suspicions of the primary sources, unless the primary source specifically state that they believe the event to be unrelated or that a causal relationship can be excluded. Adverse reactions may arise from use of the product within or outside the terms of the marketing authorization or from occupational exposure. Use outside the marketing authorization includes off-label use, overdose, misuse, abuse and medication errors." [3].

As there are different definitions and categorizations for ADRs, which subsume DDIs [4–6], this may further complicate the predicament in the daily routine. Rather, the ADRs being increasingly reported post-marketing with varying drug co-medications may actually represent the result of so-far unknown pharmacokinetic or pharmacodynamic DDIs. In terms of our approach to management, the distinction between an inherently drug-associated ADR and a DDI resulting from combination therapy should always be clear for guiding physicians to avoid critical combinations, in particular. Any drug can cause ADRs to potentially manifest with a variety of clinical effects. Within the broad spectrum of antibiotic and antifungal therapy, most typical ADRs are or should be known to physicians. ADRs account for a high degree of hospitalization, up to 12%, and for a high rate of morbidity and mortality, ranking among the top 10 leading causes of death and illness in developed countries with enormous socioeconomic costs [7–11], irrespective of the presumably high number of unreported cases. This applies, especially, to antibiotics and intensive care patients [12,13]. In order not to confuse ADRs and DDIs with new disease symptoms and, thus, initiate an escalating spiral of therapeutic counter-regulation, the knowledge of specific ADRs is crucial, especially in intensive care medicine and in the polypharmacy of the elderly, who are particularly vulnerable in this context because of their frequent pre-existing severe or multimorbid conditions. Way back in 1997, Rochon and Gurwitz described the risks associated with the prescribing cascade and, in this context, called for the need to optimize drug treatment in the elderly patients most affected; in addition, a further distinction has been made bet- ween prescribing cascades that are sometimes appropriate, but must be regularly re-evaluated and documented, and problematic prescribing cascades [14,15]. Misinterpretation of an ADR as a new symptom or exacerbation of the underlying disease is the beginning of the prescribing cascade, prevalent in patients who already have extensive polypharmacy and impaired organ function and are, therefore, particularly susceptible to ADRs and DDIs. Today, 25 years later, the problem has become aggravated as the drugs available to treat an almost typical elderly multimorbid patient, e.g., with manifest hypertension, diabetes, heart failure, and COPD or urologic conditions, have become more numerous and diverse, physicians increasingly specialize, and the different disciplines do not even seem to be aware of the ADRs or DDIs of drugs from the other medical specialties. On top of this, changing organ functions of an intensively ill patient and, in addition, the complementary mechanical organ replacement procedures either temporarily or chronically influence the dosing regime and require very close monitoring of laboratory parameters in patients, often in life-threatening conditions.

Since it can be difficult to differentiate ADRs and DDIs from disease progression, for example, in the presence of pre-existing single- or multi-organ hypoxic failure, it is particularly important to focus on ADRs and DDIs from the onset of drug use, although in rare cases these may still occur even after drug discontinuation. The aim of this study was to highlight relevant pharmacological ADR and DDI aspects of antimicrobials administered in the hospital setting, which are not so widely known, and to provide further practical guidance from lessons learned based on long-term interdisciplinary experience with individual pharmacotherapy management (IPM).

2. Results

Findings from over 52,000 detailed medication analyses continuously performed for over 8 years in interdisciplinary intensive care, transplant, and trauma patients highlight critical ADR and DDI risks, especially in these vulnerable patients with polypharmacy and/or with concomitant organ deterioration. This is where our individual pharmacotherapy management (IPM) (Figure 1) has proven very reliable and valuable in identifying and targeting drug-associated risks at the earliest stage. The ability to have the most comprehensive digital view of the patient as the basis for IPM is the essential prerequisite for adjusting each drug precisely for the individual patient's medical condition as captured via metabolism and/or excretion depending on the specific mode of degradation of each drug, and always with the additional targeted focus on associated DDIs. Reading the Summary of Product Characteristics (SmPC) for each drug of an extensive drug list and, additionally, checking all pharmacokinetic and pharmacodynamic interactions between the drugs means an almost unmanageable procedure in the context of the time pressure of today's physicians' working conditions. We present the essence of often under-recognized but critical aspects of today's typical drug combinations related to antimicrobials to link the almost unmanageable set of risk aspects of ADRs and DDIs in extensive polypharmacy to everyday clinical realities and to enable the sharing of eight years of daily IPM experience for the most critical issues. We explicitly outline the relevant risks of ADRs and DDIs in the clinical pharmacology of antimicrobial agents that are less frequently addressed adequately in terms of neurologic, hemostaseologic, hematologic, endocrinologic, and cardiac complexities.

The IPM strategy, which aims to assess each drug as comprehensively as possible and which accounts for the entire patient's very acute clinical situation simultaneously, is shown in Figure 1. The first data on the preventative impact of this significantly effective IPM with a 90.2% reduction in delirium has been published previously [16]. Focusing on the entire patient condition with all of the information from the fully digitally available records proved to be a crucial and fundamental requirement for performing this extensive, individualized medication analysis, considering the respective, often abruptly changing, acute status in the ICU patients. The IPM process uncovers relevant complexities in almost all different groups of medications, and especially in polypharmacy, that have apparently been insufficiently addressed or not considered at all. Not infrequently, these manifested ADRs or DDIs were the reason for further deterioration of the clinical condition of the patient and his transfer to the ICU before application of IPM. Even therapeutic drug monitoring in antibiotics cannot guarantee drug safety, as it does not eliminate ARDs that are associated with regular dosing already or DDIs related to interacting co-medication. Despite knowledge of the often rapid alterations in renal function or metabolic rate due to

DDIs or hepatic dysfunction, especially in multimorbid elderly patients and in transplant and intensive care patients with polypharmacy, immediate concomitant dose adjustment of the antibiotic regimen to fine-tune the daily dose is required more often than it is practiced. For this purpose, and to differentiate from ADRs and DDIs, close laboratory monitoring of these organ functions is mandatory.

Individual Pharmacotherapy Management according to Wolf* - IPM

Details Procedure Medication Review (Ursula Wolf, MD, Internal Medicine)
Workload (6.5 minutes/patient=9.2 patients/hour)

→ 1. Actual patient condition ="digital overall view of the patient" capturing

1.1 All current medications including the admission medication and all transient changes
1.2 Entire diagnoses, surgeries and the current clinically relevant problems (note changes)
1.3 Laboratory parameters (organ functions liver, kidneys, myelopoiesis with differentiated anemias, coagulation, thyroid gland), electrolytes, serum albumin, prealbumin, immunoglobulins, inflammation parameters (note changes & different transient or long-term machine replacement therapies); electrocardiogram; body weight, BMI
1.4 Vital parameters and cardiopulmonary situation (course), acid-base balance
1.5 Acquisition of continuous patient history documentation from colleagues.

→ 2. Review all applied medications digitally via their Summary of Product Characteristics and their application to the updated patient situation (see above):

2.1 Correct indications? Evidence for contraindications?
2.2 Warnings?
2.3 Additive effects?
2.4 Adverse and additive adverse drug reactions?
2.5 Drug interactions, pharmacodynamic and/or pharmacokinetic?
2.6 Overdosage? Underdosage? Protein binding?
2.7 Possible duplicate prescription, missing prescription?
2.8 Mode of application? Adjustments intravenous, peroral, tubes?
2.9 Temporal aspects of application/incompatibilities?

→ 3. Interdisciplinary networking with pharmacotherapy adaptation

Synoptic internistic/clinical-pharmacologic medication review
(taking into account the Summary of Product Characteristics for each substance, interaction check, guidelines, current PubMed research, if necessary) and communicating the results through interdisciplinary patient rounds or immediate telephone transmission to the attending colleague(s) with consensual decisions on an adapted therapy.

Figure 1. Comprehensive, reproducible IPM based on the hospital's digital patient record. * Comprehensive, digitally based IPM, developed, implemented, and practiced by Ursula Wolf, MD, Head of the Pharmacotherapy Management Department, specialist in internal medicine, and expertised in clinical pharmacology, who performed >52,000 individual medication reviews.

Thus, knowledge gaps about ADRs and DDIs within the outpatient medication lists or in combination with in-hospital drugs have been identified repeatedly from more than 52,000 medication reviews performed regularly as part of IPM and therapeutic drug monitoring (TDM). Their presentation in this selection serves as an informative practice support for therapeutic application. Hospitalized patients with reviewed medication included elderly patients >70 years of age from the traumatology department, all patients in the operative ICUs, in the organ and stem cell transplantation departments, and those in further disciplines within the UKH; the patients' entire ambulatory drug therapy was mostly covered. This additional IPM effort intends to identify measures to improve clinicians' recognition of these DDIs and prevent unwarranted ADRs of antimicrobial agents.

Being familiar with ADRs is essential to prevent drug-induced harm. Within the broad spectrum of antibiotic and antifungal therapy, physicians are aware of most typical ADRs. Thus, explicitly for the assessed relevant pharmacological issues in antimicrobial therapy that are less commonly considered, we provide further practical devices.

Our findings, in terms of relevant complexities in antimicrobial pharmacology that are less frequently considered, are presented in Table 1. They refer to ADRs and DDIs often recognized within a polypharmacy regimen or in patients with pre-existing organ deterioration.

2.1. Anidulafungin

Patients who experience elevated liver enzyme levels during anidulafungin therapy must be monitored for worsening liver function, and continuation of anidulafungin therapy must be subjected to a rigorous benefit–risk assessment. Hypokalemia and hyperglycemia, as well as headache and risk of convulsions, have to be considered [17].

2.2. Carbapenems

The rapid onset of extensive decline in blood levels of valproic acid when co-administered with carbapenem agents results in a 60–100% decrease in valproic acid levels in about two days [15]. This combination therapy is not manageable and must be avoided, and there are enough antiepileptic alternatives.

Neurologic/psychologic symptoms including hallucinations, drowsiness, dizziness, somnolence, and headache are ADRs to be aware of, and more rare symptoms comprise myoclonus, confusional states, seizures, paresthesias, encephalopathia, focal tremor, impaired sense of taste, and hypotension [18]. Seizures have been reported during treatment with carbapenem, including meropenem [18]. Our impression is that older patients are especially more at risk, often because of other concomitant diseases, cerebrovascular concerns, and medications that additionally increase the risk of seizures. Serum magnesium should be in the upper normal range and, if the risk is correspondingly high, such as after a stroke, concomitant medication with a similar seizure potential should be avoided as far as possible. With carbapenem, attention needs to be paid to thrombocytosis and thrombocytopenia. We observed high-grade thrombocytosis with carbapenem in the therapeutic course that started with onset, and if there was no contraindication, but risk related to thrombocytosis as in coronary patients, we temporarily applied acetylsalicylic acid (ASA) in a low dose. A differential diagnosis of thrombocytosis in antibiotic therapy remains, of course, a reactive thrombopoiesis in the recovery phase of an infection or a concomitant use of cortisone or special affections of the spleen.

2.3. Cotrimoxazole and Rifampicin

Not infrequently, hyperkalemia and hyponatremia are induced, necessitating simultaneous serum electrolyte monitoring.

Rifampicin decreases the effect of all currently available direct oral anticoagulants (DOACs) by inducing CYP3A4 and p-glycoprotein (P-gp); therefore, concomitant use is contraindicated. The same applies to rifampicin with calcineurin inhibitors and mammalian target of rapamycin inhibitors (mTORIs), where a dose elevation of the immunosuppressants must be assured under close TDM to prevent transplant rejection.

Table 1. Under-recognized potential ADRs and DDIs of clinical relevance from IPM insights with primary reference to their Summaries of Product Characteristics. This table is intentionally incomplete and excludes the more familiar gastrointestinal, nephrotoxic, several QT prolonging, several cutaneous dermatologic, and all anaphylactic and further risks.

ADR and DDI Antimicrobial	Neurologic/ Psychiatric	Hematologic	Hemostaseologic	Endocrinologic/ Metabolic	Cardiac	Others
Anidulafungin	Convulsions			Hypokalemia Hyperglycemia		Hepatic dysfunction
Azoles			Contraindicated with all currently available DOACs and ticagrelor, severe risk of hemorrhage		QT prolongation and tachyarrhythmia Reduces amiodarone metabolism and vice versa	Risk of rhabdomyolysis with atorvastatin, lovastatin, simvastatin Increase in blood levels and effects of calcineurin inhibitors and mTORIs Increase in bioavailability and effects of several virustatics, e.g., remdesivir * and nirmatrelvir/ritonavir for COVID-19 therapy
Carbapenems	No concomitant use of valproic acid/sodium valproate/valpromide: 60–100% decrease in valproic acid levels in about two days Hallucinations, dizziness, seizures Generalized seizures Imipenem/cilastin and ganciclovir	Thrombocytosis Thrombocytopenia		High-sodium content		Hepatic dysfunction Probenecid (potent inhibitor of organic anion secretion by renal proximal tubule cells) inhibits renal excretion

Table 1. *Cont.*

ADR and DDI Antimicrobial	Neurologic/ Psychiatric	Hematologic	Hemostaseologic	Endocrinologic/ Metabolic	Cardiac	Others
Cotrimoxazole		Leukopenia, neutropenia, thrombocytopenia, agranulocytosis, anemia: megaloblastic, aplastic anemia hemolytic, possible folic acid deficiency anemia Folic acid deficiency		Hyperkalemia Hyponatremia Metabolic acidosis Note thyroid function		Hepatobiliary disorders Life-threatening cutaneous reactions Enhanced effect with probenecid, sulfinpyrazone, indometacin, phenylbutazone, salicylate Reduced serum cyclosporine concentrations with risk of transplant rejection
Daptomycin	Anxiety, insomnia, dizziness, headache	Anemia	False prolongation of prothrombin time (PT) and elevation of international normalized ratio (INR) with certain recombinant thromboplastin reagents			Other medicinal productsassociated with myopathy should be temporarily discontinued, creatine phosphokinase (CPK) levels must be measured at baseline and at regular intervals Elevated transaminase levels NSAIDs and COX-2 inhibitors elevate plasma levels Eosinophilic pneumonia Obesity increases daptomycin mean AUC0-∞ In elderly patients, AUC0-∞ about 58% higher

Table 1. *Cont.*

ADR and DDI Antimicrobial	Neurologic/ Psychiatric	Hematologic	Hemostaseologic	Endocrinologic/ Metabolic	Cardiac	Others
Fluoroquinolones	Psychomotor hyperactivity/ agitation, confusion, depression			Hyperglycemia	Inherent risk of QT prolongation is increased with a broad spectrum of drugs such as macrolides, azoles, antipsychotics, class IA and III antiarrhythmics, tricyclic antidepressants Tachycardia Risk of valvular regurgitation/insufficiency	Ciprofloxacin may increase cyclosporine (CSA) bioavailability and elevate the risk of CSA-induced neurotoxicity and nephrotoxicity Aneurysm, aortic dissection, tendon rupture Myalgia Inhibits via CYP1A2 the metabolism of theophylline, clozapine, olanzapine, ropinirole, tizanidine, duloxetine Contraindicated with tizanidine Contraindicated with methotrexate Hepatic dysfunction

Table 1. *Cont.*

ADR and DDI Antimicrobial	Neurologic/ Psychiatric	Hematologic	Hemostaseologic	Endocrinologic/ Metabolic	Cardiac	Others
Linezolid	Cave: linezolid is a reversible, nonselective inhibitor of monoamine oxidase (MAOI) Serotonin syndrome associated with the coadministration of serotonergic agents Neuropathy (optic and peripheral) Neither with inhibitors of monoamine oxidase A (e.g., phenelzine) or B inhibitors (e.g., phenelzine, isocarboxazid, selegiline, moclobemide), nor with one of these drugs taken within the last 2 weeks Malignant neuroleptic syndrome Taste disorders (metallic taste), dizziness, insomnia	Thrombocytopenia, anemia, pancytopenia		Lactate acidosis Hyponatremia and/or syndrome of inappropriate secretion of antidiuretic hormone (SIADH) Hyperglycemia (not fasting blood sugar)		Not to be used, unless monitoring of blood pressure in patients with uncontrolled hypertension, pheochromocytoma, carcinoid, thyrotoxicosis, bipolar depression, schizoaffective psychosis, acute confusion conditions Not with serotonin reuptake inhibitors, tricyclic antidepressants, serotonin 5HT1 receptor agonists (triptans), direct- or indirect-acting sympathomimetics (including adrenergic bronchodilators, pseudoephedrine and phenylpropanolamine), vasopressor agents (e.g., epinephrine, norepinephrine), dopaminergic agents (e.g., dopamine, dobutamine), pethidine, or buspirone, high-tyramine foods Increased transaminases, lipase, CPK, LDH

Table 1. *Cont.*

ADR and DDI Antimicrobial	Neurologic/ Psychiatric	Hematologic	Hemostaseologic	Endocrinologic/ Metabolic	Cardiac	Others
Macrolides **Cave: three categories to be differentiated in terms of their interaction characteristics**			Contraindicated with all currently available DOACs, ticagrelor, severe risk of hemorrhage		QT prolongation and tachyarrhythmia Reduced amiodarone metabolism and vice versa Reduced azole metabolism and vice versa	Risk of rhabdomyolysis with atorvastatin, lovastatin, simvastatin Increase in blood levels and effects of calcineurin inhibitors and mTORIs Increased bioavailability and effects of several virustatics, e.g., remdesivir * and nirmatrelvir/ritonavir for COVID-19 therapy
Piperacillin/ Tazobactam	Insomnia		Activated partial thromboplastin time and bleeding time prolonged	Hypernatremia: 217 mg of sodium per vial of powder for solution for infusion		In elderly patients, mean half-life for piperacillin and tazobactam was 32% and 55% longer

Table 1. *Cont.*

ADR and DDI Antimicrobial	Neurologic/ Psychiatric	Hematologic	Hemostaseologic	Endocrinologic/ Metabolic	Cardiac	Others
Rifampicin			Reduces plasma level and effects of all currently available DOACs and ticagrelor Vitamin K-dependent coagulation and severe bleeding Supplemental vitamin K administration should be considered when appropriate (vitamin K deficiency, hypoprothrombinemia) Avoid concomitant use of rifampicin with other antibiotics causing vitamin K-dependent coagulopathy such as cefazolin (or other cephalosporins with N-methyl-thiotetrazole side chain) as it may lead to severe coagulation disorder	Reduces bioavailability of oral contraceptives Enzyme induction with enhanced metabolism of endogenous substrates as adrenal hormones, thyroid hormones, and vitamin D	Rifampicin strongly induces CYP2C19, resulting in both an increased level of clopidogrel active metabolite and platelet inhibition, which, in particular, might potentiate the risk of bleeding; avoid combination	Caution in elderly patients with decrease in renal function, especially if there is evidence of liver function impairment Severe cutaneous adverse reactions (SCARs) Reduced blood levels and effects of calcineurin inhibitors and mTORIs with risk of transplant rejection Reduced bioavailability and effects of several virustatics, e.g., remdesivir * and nirmatrelvir / ritonavir for COVID-19 therapy Loss of efficacy of caspofungin, dosage increase in caspofungin necessary Contraindicated with voriconazole (treatment failure), reduced effects of azoles Reduced or loss of effect of a broad spectrum of drugs including opioids, calcium antagonists, etc.

Table 1. *Cont.*

ADR and DDI Antimicrobial	Neurologic/ Psychiatric	Hematologic	Hemostaseologic	Endocrinologic/ Metabolic	Cardiac	Others
Tigecycline			Prolongation of prothrombin time (PT) and activated partial thromboplastin time(aPTT), hypofibrinogenemia			Close monitoring for the development of superinfection Hyperbilirubinemia, hepatic injury, pancreatitis P-gp substrate Cave: inhibitors such as azoles or macrolides enhance bioavailability and effects and P-gp inducers such as rifampicin decrease tigecycline bioavailability and effects Increases serum concentrations of calcineurin inhibitors
Vancomycin	Ototoxicity, periodic testing of auditory function	Thrombocytopenia, neutropenia, agranulocytosis, eosinophilia			Anesthetic-induced myocardial depression may be enhanced by vancomycin	Monitoring of serum concentrations Adapted according to weight, age, and renal function "red man's syndrome" Dyspnea, stridor

* Note: Despite partly confusing published guidance for remdesivir on the management of its potential drug interactions, it is critical to know that the interaction potential and clinical relevance have not been studied yet, although in vitro remdesivir is a substrate for esterases in plasma and tissue, for the metabolic enzymes CYP2C8, CYP2D6, and CYP3A4, and for the organic anion-transporting polypeptide 1B1 (OATP1B1) and the P-glycoprotein transporter. In vitro, remdesivir is an inhibitor of CYP3A4, OATP1B1, and OATP1B3 and remdesivir induces in vitro CYP1A2 and, potentially, CYP3A, which is partially contradictory.

Typically, an inducer effect lasts a few days after discontinuation of the drug, and also manifests itself only maximally after a few days from the start of therapy in contrast to inhibitory metabolic disorders. This always has to be considered, for example, in the timing of TDM of immunosuppressants for contemporary dose adaptation [19].

2.4. Daptomycin

Daptomycin can cause anxiety, insomnia, dizziness, and headache. Use may result in false prolongation of prothrombin time (PT) and elevation of the international normalized ratio (INR) with certain recombinant thromboplastin reagents. Other drugs associated with myopathy, such as statins, should be temporarily discontinued; creatine phosphokinase (CPK) levels must be measured at baseline and at regular intervals. Eosinophilic pneumonia is a potential side effect of daptomycin. Daptomycin is one of the still-rare group of drugs whose bioavailability has been studied in relation to obesity. Obesity increases the area under the curve (AUC0-∞) of daptomycin [20].

2.5. Fluoroquinolones

Simultaneous administration with tube feeds is not recommended due to the minerals, and explicitly, divalent cations. There are interactions with antacids, iron, zinc, magnesium, sucralfate, calcium, didanosine, oral nutritional solutions, and dairy products. Ciprofloxacin should be administered 2 h before or at least 4 h after these products [21].

Systemic and inhaled fluoroquinolones may contribute to an increased risk of aneurysm, aortic dissection, and tendon rupture, but also of valvular regurgitation/insufficiency. A careful risk–benefit assessment is required, and other therapeutic options should be considered in patients at increased risk due to predisposing factors such as existing valvular heart disease, endocarditis, connective tissue diseases, hypertension, rheumatoid arthritis, and infectious arthritis [22].

2.6. Linezolid

Clinical use of linezolid in co-medication with serotonergic agents such as the antidepressant group of selective serotonin reuptake inhibitors (SSRIs) and serotonergic opioids is contraindicated because of the risk of serotonin syndrome. Linezolid is an antibiotic agent with nonspecific MAO-inhibitory activity and a predisposing risk for malignant neuroleptic syndrome. Therefore, it is important to know that excessive amounts of foods and beverages high in tyramine should be avoided, given the potential for significant pressure reactions [Lit]. A risk of myelosuppression (including anemia, leucopenia, pancytopenia, and thrombocytopenia) and lactate acidosis as ADRs needs to be considered with linezolid [23].

2.7. Similarities of Macrolides and Azoles concerning Severe DDIs

The combination of most macrolides or azoles with the widespread use of statins must be viewed very critically [24]. HMG-CoA reductase inhibitor (statin) ADRs such as myositis and rhabdomyolysis are often dose dependent. The bioavailability of the statins atorvastatin, lovastatin, and simvastatin, when co-administered with CYP3A44 inhibiting antimicrobials, increases, as do their ADRs. In this context, pravastatin, rosuvastatin, and fluvastatin are more inert due to their different metabolic and elimination pathways.

With macrolides or azoles, such as erythromycin, ketoconazole, or voriconazole, rhabdomyolysis and severe hepatic dysfunction have been observed to occur suddenly, especially with propofol plus amiodarone, which is also due to the marked, opposing CY3A4 inhibition of metabolism. This is essential to know because amiodarone itself can induce tachyarrhythmias, which, at the same time, evolve from the QT prolongating potential risk of most macrolides and azoles. These combinations aggravate and prolong the dangerous situation of the metabolism and toxicity of any drug involved via CYP3A4 metabolism and/or P-gp transport. Increasing hepatic dysfunction and even organ failure can rapidly escalate under these circumstances as the drugs become increasingly unmetabolized and cumulatively manifest their own hepatotoxic risks. These ADRs can accumulate into multi-organ failure, particularly in the presence of pre-existing renal failure. However, hepatic failure can be reversible with timely targeted drug discontinuation. Withdrawal of amiodarone often results in the timely resolution of the sometimes life-threatening problem. With regard to rhabdomyolysis, to prevent renal failure, we would like to point out that besides drug withdrawal or dose adjustment, the earliest administration of bicarbonate is recommended, always with reference to the simultaneously monitoring of adequate respiratory capacity, the partial pressure of carbon dioxide (pCO2), and the acid–base balance.

Fentanyl and buprenorphine must be withdrawn with the onset of macrolides or azoles because of the risk of, among other consequences, respiratory depressant effects due to decreased metabolism [25].

All currently approved DOACs are contraindicated with most macrolides (notice the metabolic properties of three different groups) and azoles because they increase the risk of hemorrhages due to the decreased DOAC degradation by inhibited P-gp and/or CYP3A4 [26]. The same applies to ticagrelor metabolism [27], which is also a substrate of P-gp and CYP3A4, a fact that is often not taken into account postinterventionally in cardiology patients and is further aggravated by concomitant hepatic dysfunction with reduced metabolic capacity, e.g., related to right-sided heart failure.

Both macrolides and azoles interact with several antiviral agents; in this context, and with respect to its current frequent use in the contemporary COVID-19 pandemic, we refer to remdesivir. Although remdesivir is a substrate of the metabolizing enzymes CYP2C8, CYP2D6, and CYP3A4, as well as a substrate of the transporters for organic anion-transporting polypeptide 1B1 (OATP1B1) and P-gp, no clinical interaction studies have been performed with remdesivir [28]. Concomitant administration of potent inhibitors such as azoles and macrolides may result in intolerably increased remdesivir exposure with all its ADRs. Therefore, the combination has to be avoided. Administration of potent P-gp and CYP3A4 inducers such as rifampicin and cotrimoxazole potentially decrease plasma concentrations of remdesivir and, thus, resulting in a reduction or loss of efficacy; therefore, it cannot be administered in combination. The guidance framework is still imprecise here and it is incomprehensible that a substance approved to be used in COVID-19 therapy worldwide is so understudied for interactions, as polypharmacy is often unavoidable in these patients especially.

It is of great importance to consider the same DDIs of most azoles and most macrolides, or rifampicin and cotrimoxazole, respectively, in analogous terms for nirmatrelvir/ritonavir [29]. Nirmatrelvir/ritonavir should not be started immediately after discontinuation of the antibiotics fusidic acid, rifampicin, and cotrimoxazole. This remains critical because it should be administered as early as possible with the onset of COVID-19 symptoms. Precaution is required with regard to inducer and inhibitor effects with other drugs, as these DDIs may lead to clinically significant ADRs, potentially resulting in severe, life-threatening events from greater exposures of nirmatrelvir/ritonavir due to the concomitant-inhibiting medication and vice versa; with inducers comes a loss of the therapeutic effect of nirmatrelvir/ritonavir and the possible development of viral resistance. Nirmatrelvir/ritonavir is a strong inhibitor of p-GP and CYP3A4. To give an impression of these risks, investigations with rifabutin reveal a 4-fold–2.5-fold increase, whereas the metabolite 25-O-desacetyl rifabutin has a 38-fold–16-fold increase. This large increase in the rifabutin area under the curve (AUC) requires the consequent reduction of the rifabutin dose to 150 mg 3 times per week when co-administered with ritonavir as a strong inhibitor. The ritonavir-induced, elevated AUC change of ketoconazole was 3.4-fold [29].

2.8. Posaconazole

Posaconazole is an azole commonly used in stem cell transplantation. Therefore, it is important to note that its inherent risk of major gastrointestinal side effects [30] with severe diarrhea may mimic or overlap CMV colitis or graft-versus-host disease, or symptomatically mask the success of their therapies and, therefore, must always be considered in differential diagnosis, especially in this accordingly most critical patient population.

2.9. Rifampicin

As a CYP3A4 and P-gp inducer, rifampicin, conversely, can significantly reduce the availability of CYP3A4 and P-gp substrates and, thus, their efficacy, such as in calcineurin inhibitors and mTORIs; therefore TDM and dose adjustments are required [31]. For these inducing effects, the combination with DOACs is contraindicated. There is also a wide range of DDIs resulting from inducing effects of rifampicin, e.g., with virustatics, antiepileptic drugs, and opioids, that reduce or lose their effects.

Similarly, for *metamizole*, a very common postoperative analgesic in Germany, a potential 2.9-fold hepatocyte CYP3A4 induction has been described [32], whereby a potential reduction, e.g., in the efficacy of various macrolides or azoles, cannot be excluded. We also regularly have to counteract it through dosage elevation in co-administered calcineurin inhibitors and mTORIs. Rifampicin is contraindicated in the presence of jaundice and hepatic dysfunction. Its enzyme induction can enhance the metabolism of endogenous substrates including adrenal hormones, thyroid hormones, and vitamin D. Additionally, the reported incidence of a paradoxical drug reaction ranges between 9.2 and 25%. In these cases, after the initial improvement of the tuberculosis during therapy with rifampicin, the symptoms may worsen again. An excessive immune reaction is suspected as a possible cause, which can be treated symptomatically while continuing antituberculostatic therapy [33].

2.10. Tigecycline

Tigecycline is known to possibly induce hyperbilirubinemia and prolonged PTT with the elevated risk of bleeding [34]. Especially in the case of concomitant continuous intravenous heparin administration, the latter often has to be adjusted according to the PTT values that need to be monitored. Based on an in vitro study, tigecycline is a P-gp substrate [34]. Co-administration of P-gp inhibitors (e.g., ketoconazole or cyclosporine) affect the pharmacokinetics of tigecycline and increase the bioavailability, including its ADRs. Concomitant P-gp inducers (e.g., rifampicin) may reduce the antibiotic effect [31]. The drug may increase serum concentrations of calcineurin inhibitors, necessitating close TDM. Tigecycline requires close monitoring for the development of superinfection, hyperbilirubinemia, hepatic injury, and pancreatitis [34].

2.11. Vancomycin

Thrombocytopenia is not infrequently seen with vancomycin. It is worth mentioning a reference to obese patients, whom we increasingly have to consider. Since an adjustment of the usual daily doses of vancomycin may be necessary in obese patients [35], TDM is appropriately helpful in this situation with regular level measurements because of its nephrotoxicity and further risks. Rapid bolus administration (i.e., over several minutes) may be associated with severe hypotension (including shock and rare cardiac arrest), histamine-like responses and maculopapular or erythematous rash ("red man's syndrome" or "red neck syndrome"), and pain and spasm of the chest and spine muscles [35].

2.12. Voriconazole

The preclinical repeated-dose toxicity studies of the intravenous vehicle sodium betacyclodextrin sulfobutyl ether (SBECD) revealed vacuolization of the urinary tract epithelium and activation of macrophages in the liver and lungs as the main effects [36]. Carcinogenicity studies have not been performed with SBECD, although an SBECD-containing impurity has been shown to be a mutagenic substance with demonstrated carcinogenic potential in rodents; thus, this compound must also be considered to have carcinogenic potential in humans [36].

2.13. Probenecid

When used concomitantly with probenecid due to its slower excretion, the plasma levels, effect, and ADRs of several antimicrobials may be increased, and dose adjustments may be necessary in penicillin, cephalosporin, quinolone (e.g., ciprofloxacin, norfloxacin), dapsone, sulfonamides, nitrofurantoin, nalidixic acid, and rifampicin [34]. Probenecid is contraindicated in simultaneous treatment with β-lactam-antibiotics and with pre-existing renal impairment [37].

2.14. Proton Pump Inhibitors

Proton pump inhibitors are known to be associated with an elevated risk of bacterial infections [38]. They should be applied more restrictively, and for stress ulcer prevention, be reduced to the adequate prophylactic dosage of, e.g., daily 20 mg of pantoprazole only.

2.15. Benefits and Preliminary Effects of the Applied IPM Strategy

We have elaborated the domain aspects extracted and tabulated for antimicrobial ADRs and DDIs and are continuously focusing on their application in the IPM.

The unavailability despite physicians' requests for such compilations attests to the clinical weight of this overview of risks. Studying the innumerable ADRs and DDIs from SmPCs and DDI checks takes time, which is not available in everyday clinical practice. Additionally, since countless ADRs and DDIs have emerged, especially within the broad medication list for the critically ill patients in the ICU and for polypharmacy in elderly traumatology patients who require additional antimicrobial therapy, which are impossible to comprehensively research in everyday life, we provide our insights resulting from a synoptic point of view of internal medicine and clinical pharmacology based on the daily performance of 35 to 52 IPM in an everyday real life condition by this tabulated overview. Applying this focus as one elementary building block among others of the comprehensive IPM enables us to perform a fast medication analysis in 6.5 min, whereas the extensive procedure otherwise is almost unmanageable due to the particularly broad medication lists for each patient within these settings and may take 1 to 2 h.

In addition to this benefit, the strategy employed, including maintained awareness of these knowledgeable, selective ADRs and DDIs and applying this information to each individual patient, also may contribute to the apparently highly effective IPM measures. There are compelling preliminary data suggesting that this preventive IPM strategy outlined here, documenting our IPM attention and strategic aspects, including the broad component of antimicrobials, contributes to risk reduction and optimized medication therapy in patients most vulnerable, such as those with unavoidable polypharmacy who suffer from concomitant organ dysfunction already.

Based on the annually documented reports of the controlling department of the UKH, the first preliminary descriptive results, which have to be statistically analyzed further on for corresponding publications, document that coinciding with the onset of IPM implementation for the operative ICU patients most critically ill, and for whom the ICU was the treatment ward in which thy had the longest length of stay at the UKH, the average length of stay decreased from 29.1 to 23.2 days from 2015 to 2018. In parallel, the number of patients treated even increased, e.g., by 32.7% in the ICU ward, where critically ill patients are often mechanically ventilated and dialyzed. Despite higher numbers of patients treated, medication costs concurrently decreased by 27.9% for the operative ICU wards. These results will be published separately for the entire evaluation of the IPM implementation after further analytical statistics are performed.

Regarding the department of geriatric traumatology, we retrospectively analyzed the association of the overall implemented IPM elements, including the exclusively presented, specific antimicrobial strategies with an ADR and DDI focus on prevention of complicating delirium, in-hospital fall events, and impaired renal function within a subset of 404 patients in a group-matched study. There was an IPM-associated 90.2% relative reduction in complicating delirium [16] and almost analogous results for the reduction in in-hospital fall events and in renal function impairment (for separate publication).

3. Discussion

IPM in the context of antimicrobial therapy reveals several clinical important aspects to consider. Most of them should be familiar as more or less typical ADRs and, thus, have not been presented here repeatedly. The selected insights into lesser-known ADRs and essential DDIs result from clinical IPM practice and observations in our patient treatments, and here, represent only excerpts from the respective ADRs and DDIs. A further number of unidentified ADRs must be assumed, since they do not always manifest in their fulminant form or a complete spectrum, but also in a crytogenic or undulating manner. This incompleteness may also apply, in part, to the identified and listed ADRs referred to in the drug product information.

Therefore, there is almost no way to definitively determine a uniform general rate of ADRs for a given drug, but the more vulnerable the patient is due to his pre-existing conditions and illnesses, organ dysfunction, and polypharmacy, the higher the risk of ADR and DDI manifestation. For this reason, post-marketing evaluations, documentations, and studies are very important because the pre-marketing trial phase of the approval studies almost never include these particularly susceptible patients.

Drugs and, especially, antibiotics can rapidly worsen an already developing single- or multi-organ failure, e.g., as a result of pre-existing hypoxia, and require special therapeutic attention with fine-tuning of dose adjustments. The essential knowledge and consideration of drug-specific ADRs, particularly in frequent antimicrobial combination therapy, is a preventive tool in patient and drug safety, both of which are continuously serious concerns as repeatedly claimed by the WHO for over a half century already [1]. A decline in renal function, and hepatic and further organ deterioration up to multi-organ failure always must exclude pharmacological iatrogenic drug-associated causation. This presentation is a sub-aspect of extensive studies of our IPM demonstrating IPM-associated optimization of medication, e.g., effectively preventing delirium by 90.2% in elderly trauma patients [16].

As ADRs are categorized inconsistently and DDIs are subsumed, this makes the important differentiation of these two risky issues less clear, and thus, the preventative, necessary countermeasures might be less appropriate or even missed.

In addition, we must take into account the most frequently prevalent hypoalbuminemia, especially in our demographically increasing elderly patient population, as well as in critically ill intensive care patients. This aspect remains almost neglected and should be focused on more intensely because, besides the predominant, clinically significant pharmacokinetic interactions observed at the metabolic level, another large proportion of competing interactions and ADR manifestations must be considered at the serum protein-binding level [39]. This further concern has not been addressed within the presented aspects. Nevertheless, we must pay additional attention to it, especially in view of the extremely common manifestation of hypoalbuminemia in our elderly and ICU patients. This can lead to severe and more rapid overexposures of high protein-binding drugs, including their ADRs. The dilemma is further exacerbated by simultaneously competing efforts of different drugs to bind to the remaining serum protein residual capacity.

It is worthwhile to depict that DDIs are of such high clinical relevance, both on the metabolic and protein-binding level, that they are even targeted in drug design and development by pharmaceutical companies. This is found with the combined therapeutic use of nirmatrelvir/ritonavir [29], and also with imipenem plus cilastin [40] and with taxanes; here, the albumin-bound nanoparticle formulation of paclitaxel and nab-paclitaxel exhibits enhanced paclitaxel tissue distribution and enhanced tumor penetration through additional active, selective transport into tumor tissue via target proteins [41,42].

Further IPM efforts are intended for communicating interventions that improve the physicians' recognition of these DDIs and ADRs of antimicrobials. With this in mind, we have outlined the detailed process of effective IPM enabling the early detection of ADRs and

DDIs and provide the critical basis for causal, preventive medical countermeasures aimed at successful and safe antimicrobial therapy, even in the most vulnerable risk patients.

Knowledge of documented ADRs and DDIs enable early differentiation from further disease progression, which, if unrecognized, would typically and inevitably lead to additive and accumulative drug therapy to treat these unconscious iatrogenic symptoms, and thus, might be the beginning of a fatal course for the patient. The implementation of the comprehensive IPM has been found to be highly preventive of several complications, as was just published for delirium [16], and thus, IPM guarantees an important step forward in eliminating severe drug and patient safety concerns. In this context, antimicrobials are also known to potentially directly induce iatrogenic delirium, e.g., from anticholinergic properties in ampicillin, clindamycin, and gentamycin or from the dopaminergic and serotonergic agent linezolid and hyponatremia-inducing antibiotics such as cotrimoxazole [43].

In 1975, in their critical review on ADRs, Karch and Lasagna already stated, "The data on adverse drug reactions (ADRs) are incomplete, unrepresentative, uncontrolled, and lacking in operational criteria for identifying ADRs. No quantitative conclusions can be drawn from the reported data as for morbidity, mortality, or the underlying causes of ADRs, and attempts to extrapolate the available data to the general population would be invalid and perhaps misleading" [44]. There have been no sufficiently successful attempts to overcome this serious grievance for nearly 50 years now, although it is becoming an increasingly threatening problem worldwide [45] with an inherent, extreme socioeconomic healthcare burden. Since 1998, the U.S. Food and Drug Administration has been operating the Adverse Event Reporting System, collecting all voluntary reports of adverse drug events submitted directly to the agency or through drug manufacturers. These data show a significant increase in reported deaths and serious injuries related to drug therapy during the observation period from 1998 to 2005. Within a 7-year period, the reported number of serious adverse drug events increased 2.6-fold, from 34,966 to 89,842, and the number of fatal adverse drug events increased 2.7-fold, from 5519 to 15,107 [46]. Davies et al. documented patients with adverse drug events in the hospital that had a longer length of stay and were more likely to be older, female, and taking a higher number of medications. The latter was the only significant predictor of an ADR in the multivariable analysis, with each additional medication increasing the risk of an episode of ADR by 1.14 (95% CI 1.09, 1.20) [47]. A causality and preventability assessment of adverse drug events of antibiotics among inpatients by Saqib et al. revealed 59.3% were preventable and caused by medication errors due to the nonadherence of policies (38.4%) and lack of information about antibiotics (32%) [12].

The inclusion of ADRs and DDIs as a subset of the International Classification of Diseases (ICD) is an important step to focus on these often serious to life-threatening concerns and weighting them on the level of diseases. This should be addressed more progressively. With 28.4%, Joshua et al. noted a high incidence of serious, even lethal, multiple ADRs in patients admitted to the medical ICU. Similar to our observations, they reported a broad clinical spectrum of ADRs and found infrequently documented ADRs from newer drugs. Antimicrobials (27%) were the commonly involved drugs in these ADR-susceptible patients with pre-existing multi-organ dysfunction and altered pharmacokinetic parameters [13]. In line with our IPM observations, ADRs were significantly associated with comorbidity, polypharmacy, and length of stay [13]. Severe coagulation problems and hemorrhagia, as they may occur from consumptive coagulopathy and sepsis, may also result from the ADRs and DDIs partly listed in our table. The very comprehensive IPM is a helpful tool to differentiate them more easily, particularly through its daily performance and through close patient follow-ups at the most acute, individual level.

The frequent QT prolongation risk of numerous drugs, which is often aggravated by concomitant medication in a cumulative manner, has to be regarded and is not always avoidable. Because most of the risks should be familiar, we left them out of consideration in this presentation, although the clinical consequences may become most critical. As a preventive measure, serum potassium and magnesium should be kept in an upper normal range and monitored closely, and an acidotic metabolic state should be avoided. Macrolides and azoles, most of which are highly potent inhibitors and pharmacokinetic enhancers, are antimicrobials that are important to a keep in mind, as is fluorchinolone in this context. Due to their transporter- and metabolism-inhibiting capacities, azoles and macrolides even cumulate the tachyarrhythmia risks to a higher degree than merely additively with co-administered QT prolonging drugs, which are respective substrates to the inhibited enzymes. The classification of macrolides into three different groups according to their affinity for CYP3A4, and thus, their propensity to cause pharmacokinetic drug interactions, may help to clinically estimate the severity of potentially resulting DDIs. Group 1 with troleandomycin, erythromycin, and its prodrugs decrease drug metabolism and may engender drug interactions, whereas group 2 with clarithromycin, flurithromycin, midecamycin, midecamycin acetate, josamycin, and roxithromycin should cause less frequent intense interactions [48], although according to our TDM routine data, clarithromycin is associated with a significant elevation in cyclosporine when co-administered, necessitating TDM and the timely dose adjustment of cyclosporine [49]. Additionally, the study of Hill et al. indicates that in elderly patients taking a DOAC, concomitant clarithromycin was associated with a small but statistically significant greater 30-day risk of hospital admission with major hemorrhage compared to azithromycin, a group 3 macrolide [50].

Since CYP450 enzyme generation is downregulated by elevated cytokine levels, such as from IL-1 in chronic inflammation, this would be expected to be renormalized by an IL-1 receptor antagonist such as anakinra. The clinical relevance for CYP450 substrates with a narrow therapeutic range that may be affected currently remains unknown [51]. A further aspect of interindividual varying metabolization and drug effects from mitochondrial genetics addressing personalized medicine and drug toxicity has been reviewed by Penman et al. To date, the implications remain unclear, although clinical studies have reported associations between mitochondrial haplogroup and antiretroviral therapy, chemotherapy, and antibiotic-induced toxicity. Mitochondrial DNA differences may reveal differential functions as a factor in idiosyncrasies, leading to unpredictable adverse effects and drug-induced toxicities [52]. The effect of polymorphism of the ABCB1 gene encoding P-gp was assessed by a French group for the substrates rivaroxaban and dabigatran by studying the contribution of ABCB1 genetic polymorphisms, as well as the interaction with clarithromycin, to interindividual variability under dabigatran and rivaroxaban exposure [53]. Whereas the ABCB1 genotype turned out not to be a clinically relevant determinant of both drugs' pharmacokinetics, the co-administration of the P-gp inhibitor clarithromycin with dabigatran or rivaroxaban resulted in a clinically relevant two-fold increase in both drugs' AUC, irrespective of the ABCB1 genotype clearly indicating the risk of DDIs [53]. With regard to almost extremely frequently administered drugs such as atorvastatin, we even find gaps in the knowledge and unawareness of significant DDIs resulting from antimicrobials. E.g., even clarithromycin (a macrolide group 2) induces a 4,5-fold increase in atorvastatin's AUC, and for itraconazole, a 3.3-fold elevated AUC with standard dosages [54]. Given that the dual-interaction mechanism of rifampicin as an inducer of CYP3A4 and an inhibitor of the hepatocellular uptake transporter OATP1B1 is difficult to sum up already, there is even a further varying effect depending on the time of application, with a 5-fold decrease in atorvastatin when it is co-administered with rifampicin simultaneously [54].

ADRs and DDIs, as documented in antimicrobial therapy from our IPM insights, are an increasing challenge in today's healthcare worldwide, especially given the growing complexity of chronic and acute combination therapy interventions, interdisciplinary treatments by different specialists, an aging population with increasing multimorbidity, and additionally, disease interactions from renal impairment or hepatic dysfunction. An alarming deficit of ADR and DDI awareness is burdening the patients, the hospitals, the entire healthcare sector, and the public budget. Urgent intervention strategies for prevention are to be promoted and anchored politically. This may be achieved via the development of increasingly more effective digital medication tools and making institutions of specialized drug safety managers mandatory in hospitals and health insurance companies to cover the inpatient and outpatient care sectors.

4. Methods

This study from ongoing, real-life observations was designed to summarize and tabulate the under-recognized, clinically most critical risks of ADRs and DDIs in the context of current medication regimens and in the setting of elderly, multimorbid, or even seriously ill hospitalized patients. Critical, but less recognized, ADR and DDI risks from daily 35–52 medication reviews were selected through comprehensive individual pharmacotherapy management (IPM) in these vulnerable operative intensive care, elderly traumatology, and organ and stem cell transplant patients. The underlying observation period was from 12/2014 to 8/2022. We examined the drug combinations found in current, real-life practice based on medication reviews of >52,000 IPMs. As a consequent part of the study, we also provide examples of our own practice-proven advice on control and countermeasures.

IPM was designed and implemented by a University Hospital Halle (UKH) physician with expertise in internal medicine and clinical pharmacology, and who is responsible for the pharmacotherapy management at the UKH in continuous cooperation with senior physicians of the operative intensive care units (ICUs) and geriatric traumatology. The focus is on patient and drug safety, especially in these most vulnerable groups with polypharmacy, often additionally associated with impaired organ function, as in interdisciplinary intensive care patients, organ or stem cell transplanted patients, and elderly trauma patients. IPM considers the entire digital patient record from the internal clinical information systems for data management, the Integrated Care Manager (ICM) and Orbis software for advice on clinical documentation. Always in synopsis internal medicine/clinical pharmacology, the comprehensive medication review has been performed by the same study physician for each patient daily in the operative ICUs and fortnightly for interdisciplinary patient visitations in geriatric traumatology. The review was conducted primarily based on the Summary of Product Characteristics (SmPC), particularly considering pharmacokinetics, pharmacodynamics, dosage, contraindications, interactions, and all ADRs. In addition, the literature research was carried out to answer extended or unresolved questions. All diagnoses and indications were recorded as the basis for each individual medication review. Further on, current medical guidelines were taken into account.

Recommendations result from the entire consideration of the individual patient's situation, both regarding his chronic and current clinically relevant disease process, and the acute pre- and postoperative situation, considering all available laboratory medical data on organ functions; albumin; lactate and inflammation parameters; electrocardiography and imaging examination results; the continuous course of vital parameters (blood pressure, heart rate, respiratory function, and acid–base balance); body weight; BMI; cognitive disorders; other subjective complaints; pain intensity/profile; the results of assessments and risk scores; demographic data such as age, gender, and living situation; anamnesis/external anamnesis; and the course of clinical examination findings. There is always an additional focus on anticholinergic components and electrolyte disturbances, glucose metabolism, vitamin D, parathormone balance and endocrinological thyroid functions, coagulation parameters, transient organ replacement therapies, and manifest/potential hematologic disorders, in particular, on all forms of anemia. This comprehensive digital overall view of a patient represents the basis for the internistic/clinical pharmacologic intervention with the simultaneous synoptic and adaptive examination of all medications.

By the consequent implementation of this regularly systematic individual pharmacotherapy management, the focus is on prevention and elimination of iatrogenic medication-associated: 1. severe hypotension from cumulative drug effects, 2. renal injury, 3. single and multiple organ failure, 4. neurologic/psychiatric disorders including delirium, 5. in-hospital fall events, 6. arrhythmias, 7. hemorrhages from hemostaseologic disorders, 8. cerebrocardiovascular events, 9. hematologic/myeloproliferative disorders, 10. all types of differentiated anemias, 11. venous thrombosis, 12. acid–base and electrolyte imbalance, 13. metabolic/endocrinologic disorders, 14. gastrointestinal effects such as paralytic ileus and ulcer of the upper intestine, 15. multidrug-resistant nosocomial infections, and 16. oropharyngeal dysphagia.

In this context, during the 8 year follow-up of IPM we recognized risks of ADRs and DDIs resulting from antimicrobials, which are of predominant clinical relevance concerning the regarded aspects. As we obviously work with a similar time pressure as most physicians nowadays do, we, exactly for this reason, sum up and compile into a table for overview the most critical risks resulting from our clinical insights of the comprehensive IPM, that always adapts drug information to the individual and acute patient clinical condition.

The findings presented here are a specific sub-focus of lesser referenced ADRs and relevant DDIs associated with the concomitant administration of antimicrobials in the context of our IPM. We explicitly exclude the more familiar nephrotoxic risks, QT prolongation, and myelosuppression as the most typical ADRs of several antimicrobial agents. We covered the relevant pharmacological aspects with which we are not so familiar to provide additionally corresponding further practical guidance through IPM.

5. Strengths and Weaknesses

The analysis refers to continuous observations on selected wards of the UKH. Nevertheless, the aspects addressed may be representative for the entire patient population. IPM is an individual medication review that is conducted as intensively as possible. Genetic pharmacological aspects are not considered additionally. We did not measure blood values that attribute adverse effects to presumably iatrogenic medical agents due to elevated levels. Only from the time of occurrence during application and reversal of the critical situation by selective drug withdrawal could a possible causal relationship be postulated. However, the causal association remains difficult to definitively prove due to numerous pathophysiological disturbances often occurring in parallel in our severely ill patients. The more than 8 years of daily IPM experience with a critically ill or multimorbid elderly interdisciplinary patient population covers a particularly vulnerable group with polypharmacy and, probably, already reduced organ function. There is almost no possibility to clearly identify the entire frequency of ADR manifestation, but the more vulnerable the patient is as a result of pre-existing illnesses, organ deterioration, and polypharmacy, the higher the ADR risk. The spectrum of ADRs and DDIs addressed is based primarily on the SmPCs, whose completeness and, in rare cases, even the presentation of clearly defined metabolic pathways of active substances and risks from inactive metabolites are not always fulfilled, thus leaving out still unknown grey areas, which may be additionally dangerous for the treated patients.

6. Conclusions and Outlook

IPM-based awareness of serious but lesser-known ADRs and DDIs of antimicrobial agents can counteract or reverse single- or multi-organ failure, cardiac arrhythmias, life-threatening hemorrhage or thrombosis, hematologic/myeloproliferative disorders, electrolyte imbalances, neurologic/psychiatric disorders, and drug escalations by enabling timely and targeted correction of the underlying causation individually. The conceptionalized IPM strategy applied to optimize drug treatment and prevent ADRs and DDIs, both common risks of polypharmacy, is accompanied by a reduction in length of hospital stay and associated costs. We plan to implement IPM's longstanding experience and lessons learned for wider use on a digitized platform to share a comprehensive compass, in addition to the documented findings on antimicrobials, as an important contribution to improving patient and drug safety collectively when facing the high-risk issue of under-recognized ADRs and DDIs worldwide.

Author Contributions: Conceptualization, U.W.; formal analysis, U.W.; investigation, U.W., H.B., R.N. and T.S.; methodology, U.W., H.B., R.N. and T.S.; project administration, U.W., H.B., R.N. and T.S.; validation, U.W.; writing—original draft, U.W.; writing—review and editing, U.W., H.B., R.N. and T.S. Additionally, U.W. conceptionalized and performed the IPM. T.S., H.B. and R.N. put forward the implementation of digital records for patient management. T.S, H.B. and R.N. introduced IPM into interdisciplinary patient care within their departments and have provided continuous and intense collaboration ever since. All authors have read and agreed to the published version of the manuscript.

Funding: We acknowledge the financial support from the funding program, Open Access Publishing by the German Research Foundation (DFG).

Institutional Review Board Statement: We confirm that all methods were performed in accordance with the relevant guidelines and regulations, such as the ethical standards of the institutional ethics committee and with the 1964 Helsinki Declaration and its subsequent amendments. Ethical review and approval were waived for this study, as no ethical concerns were raised with regard to the anonymized data collection and nonperson presentation. Patient interests worthy of protection are not affected by the completely anonymized data obtained for clinical purposes from routine care.

Informed Consent Statement: Requirement for informed consent was waived for the entirely nonperson-related anonymized evaluation and presentation.

Data Availability Statement: There was no data collection for the purpose of this article, only observational results and recommendations.

Acknowledgments: The authors gratefully acknowledge all team physicians who reliably collaborate with IPM in trauma, transplant, and intensive care medicine, whether during patient rounds or daily telemedicine recommendations. The personal commitment and dedication of every individual physician is what contributes to the highly effective IPM, and the pleasant collegial cooperation on an equal footing promotes patient and drug safety at UKH.

Conflicts of Interest: H.B., R.N., and T.S. declare no conflict of interest. U.W. received honoraria for scientific lectures on risks of polypharmacy from Bristol Myers Squibb and Pfizer.

References

1. WHO Meeting on International Drug Monitoring: The Role of National Centres (1971: Geneva, Switzerland) & World Health Organization. (1972). International Drug Monitoring: The Role of National Centres, Report of a WHO Meeting [Held in Geneva from 20 to 25 September 1971]. World Health Organization. Available online: https://apps.who.int/iris/handle/10665/40968 (accessed on 12 July 2022).
2. Cacabelos, R.; Naidoo, V.; Corzo, L.; Cacabelos, N.; Carril, J.C. Genophenotypic Factors and Pharmacogenomics in Adverse Drug Reactions. *Int. J. Mol. Sci.* **2021**, *22*, 13302. [CrossRef]
3. European Medicines Agency and Heads of Medicines Agencies, 2017: Guideline on Good Pharmacovigilance Practices (GVP)—Annex I (Rev 4) EMA/876333/2011 Rev 4, page 8/33. Available online: https://www.ema.europa.eu/en/documents/scientific-guideline/guideline-good-pharmacovigilance-practices-annex-i-definitions-rev-4_en.pdf. (accessed on 18 August 2022).
4. Edwards, I.R.; Aronson, J.K. Adverse drug reactions: Definitions, diagnosis, and management. *Lancet* **2000**, *356*, 1255–1259. [CrossRef]
5. Vervloet, D.; Durham, S. Adverse reactions to drugs. *BMJ* **1998**, *316*, 1511–1514. [CrossRef] [PubMed]
6. Coleman, J.J.; Pontefract, S.K. Adverse drug reactions. *Clin. Med.* **2016**, *16*, 481–485. [CrossRef]
7. Osanlou, O.; Pirmohamed, M.; Daly, A.K. Pharmacogenetics of Adverse Drug Reactions. *Adv. Pharmacol.* **2018**, *83*, 155–190. [CrossRef] [PubMed]
8. Davies, E.C.; Green, C.F.; Mottram, D.R.; Pirmohamed, M. Adverse drug reactions in hospitals: A narrative review. *Curr. Drug Saf.* **2007**, *2*, 79–87. [CrossRef]
9. Chan, S.L.; Ang, X.; Sani, L.L.; Ng, H.Y.; Winther, M.D.; Liu, J.J.; Brunham, L.R.; Chan, A. Prevalence and characteristics of adverse drug reactions at admission to hospital: A prospective observational study. *Br. J. Clin. Pharmacol.* **2016**, *82*, 1636–1646. [CrossRef]
10. Eissenberg, J.C.; Aurora, R. Pharmacogenomics: What the Doctor Ordered? *Mo. Med.* **2019**, *16*, 217–225.
11. Ross, C.J.; Carleton, B.; Warn, D.G.; Stenton, S.B.; Rassekh, S.R.; Hayden, M.R. Genotypic approaches to therapy in children: A national active surveillance network (GATC) to study the pharmacogenomics of severe adverse drug reactions in children. *Ann. N. Y. Acad. Sci.* **2007**, *1110*, 177–192. [CrossRef]
12. Saqib, A.; Sarwar, M.R.; Sarfraz, M.; Iftikhar, S. Causality and preventability assessment of adverse drug events of antibiotics among inpatients having different lengths of hospital stay: A multicenter, cross-sectional study in Lahore, Pakistan. *BMC Pharmacol. Toxicol.* **2018**, *19*, 34. [CrossRef] [PubMed]

13. Joshua, L.; Devi, P.; Guido, S. Adverse drug reactions in medical intensive care unit of a tertiary care hospital. *Pharmacoepidemiol. Drug Saf.* **2009**, *18*, 639–645. [CrossRef] [PubMed]

14. Rochon, P.A.; Gurwitz, J.H. Optimising drug treatment for elderly people: The prescribing cascade. *BMJ* **1997**, *315*, 1096–1099. [CrossRef]

15. McCarthy, L.M.; Visentin, J.D.; Rochon, P.A. Assessing the Scope and Appropriateness of Prescribing Cascades. *J. Am. Geriatr. Soc.* **2019**, *67*, 1023–1026. [CrossRef] [PubMed]

16. Drewas, L.; Ghadir, H.; Neef, R.; Delank, K.S.; Wolf, U. Individual Pharmacotherapy Management (IPM)—I: A group-matched retrospective controlled clinical study on prevention of complicating delirium in the elderly trauma patients and identification of associated factors. *BMC Geriatr.* **2022**, *22*, 29. [CrossRef]

17. Ecalta 100 mg. Summary of Product Characteristics. Available online: https://www.ema.europa.eu/en/documents/product-information/ecalta-epar-product-information_de.pdf (accessed on 12 June 2022).

18. Annmer 500 mg (Meropenem for Injection USP 500 mg). Summary of Product Characteristics. Available online: https://www.nafdac.gov.ng/wp-content/uploads/Files/SMPC/a_v4/Summary-of-Product-Characteristics-Annmer-500mg.pdf (accessed on 11 June 2022).

19. Co-Trimoxazole Tablets 80/400 mg. Summary of Product Characteristics. Available online: https://www.medicines.org.uk/emc/product/5752/smpc (accessed on 13 June 2022).

20. Cubicin 350 mg Powder for Solution for Injection or Infusion, Cubicin 500 mg Powder for Solution for Injection or Infusion. Summary of Product Characteristics. Available online: https://www.ema.europa.eu/en/documents/product-information/cubicin-epar-product-information_en.pdf (accessed on 11 July 2022).

21. Fragen zur Sondenapplikation von Cipro—1 A Pharma 100 mg/-250 mg/-500 mg/-750 mg. 1 A Pharma. Available online: http://www.1a-files.de/pdf/sondenboegen/Cipro_alle.pdf (accessed on 13 June 2022).

22. BfArM: Rote-Hand-Brief zu Systemisch und Inhalativ Angewendeten Fluorchinolonen: Risiko Einer Herzklappenregurgitation/-Insuffizienz. 29 Oktober 2020. Available online: https://www.bfarm.de/SharedDocs/Risikoinformationen/Pharmakovigilanz/DE/RHB/2020/rhb-fluorchinolone.html (accessed on 12 June 2022).

23. Linezolid 600 mg Tablets. Summary of Product Characteristics. WHOPAR Part 4, February 2017, Section 6 Updated: June 2017. Available online: https://extranet.who.int/pqweb/sites/default/files/TB299part4v1_0.pdf (accessed on 11 July 2022).

24. Hylton Gravatt, L.A.; Flurie, R.W.; Lajthia, E.; Dixon, D.L. Clinical Guidance for Managing Statin and Antimicrobial Drug-Drug Interactions. *Curr. Atheroscler. Rep.* **2017**, *19*, 46. [CrossRef]

25. Fentanyl Beta 12/-25/-50/-75/-100 Mikrogramm/Stunde Matrixpflaster. Fachinformation. Available online: https://s3.eu-central-1.amazonaws.com/prod-cerebro-ifap/media_all/77630.pdf (accessed on 9 May 2022).

26. Ferri, N.; Colombo, E.; Tenconi, M.; Baldessin, L.; Corsini, A. Drug-Drug Interactions of Direct Oral Anticoagulants (DOACs): From Pharmacological to Clinical Practice. *Pharmaceutics* **2022**, *14*, 1120. [CrossRef]

27. Ticagrelor. Brilique 60 mg Film-Coated Tablets. Summary of Product Characteristics. Available online: https://www.ema.europa.eu/en/documents/product-information/brilique-epar-product-information_en.pdf (accessed on 28 July 2022).

28. Remdesivir. Veklury 100 mg Powder for Concentrate for Solution for Infusion. Summary of Product Characteristics. Available online: https://www.ema.europa.eu/en/documents/product-information/veklury-epar-product-information_en.pdf (accessed on 29 July 2022).

29. Paxlovid 150 mg + 100 mg Film-Coated Tablets. Nirmatrelvir/Ritonavir. Summary of Product Characteristics. Available online: https://www.ema.europa.eu/en/documents/product-information/paxlovid-epar-product-information_en.pdf (accessed on 12 July 2022).

30. Posaconazol. Noxafil 40 mg/mL Oral Suspension. Summary of Product Characteristics. Available online: https://www.ema.europa.eu/en/documents/product-information/noxafil-epar-product-information_en.pdf (accessed on 11 July 2022).

31. Rifampicin 300 mg Capsules. Summary of Product Characteristics. Available online: https://www.medicines.org.uk/emc/product/8789/smpc#gref (accessed on 12 July 2022).

32. Saussele, T.; Burk, O.; Blievernicht, J.K.; Klein, K.; Nussler, A.; Nussler, N.; Hengstler, J.G.; Eichelbaum, M.; Schwab, M.; Zanger, U.M. Selective induction of human hepatic cytochromes P450 2B6 and 3A4 by metamizole. *Clin. Pharmacol. Ther.* **2007**, *82*, 265–274. [CrossRef]

33. Rifampicin-Natrium. Eremfat i.v. 300 mg/600 mg. Fachinformation. Available online: https://s3.eu-central-1.amazonaws.com/prod-cerebro-ifap/media_all/69316.pdf (accessed on 12 July 2022).

34. Tigecycline. Tygacil 50 mg Powder for Solution for Infusion. Summary of Product Characteristics. Available online: https://www.ema.europa.eu/en/documents/product-information/tygacil-epar-product-information_en.pdf (accessed on 10 July 2022).

35. Vancomycin 500 mg, Powder for Concentrate for Solution for Infusion. Summary of Product Characteristics. Available online: http://bowmed.com/product-info-downloads/spc_vanco500mg_11_04_2016.pdf (accessed on 11 July 2022).

36. Voriconazol. VFEND 200 mg Pulver zur Herstellung einer Infusionslösung. Summary of Product Characteristics. Available online: https://www.pfizer.de/sites/default/files/FI-15292.pdf (accessed on 11 July 2022).

37. Probenecid. Probenecid Biokanol Tabletten. Summary of Product Characteristics. Available online: https://s3.eu-central-1.amazonaws.com/prod-cerebro-ifap/media_all/60828.pdf (accessed on 12 July 2022).

38. Bavishi, C.; Dupont, H.L. Systematic review: The use of proton pump inhibitors and increased susceptibility to enteric infection. *Aliment Pharmacol Ther.* **2011**, *34*, 1269–1281. [CrossRef]

39. Patsalos, P.N.; Duncan, J.S. Antiepileptic drugs. A review of clinically significant drug interactions. *Drug. Saf.* **1993**, *9*, 156–184. [CrossRef]

40. Imipenem/Cilastatin-Actavis 500 mg/500 mg Pulver zur Herstellung einer Infusionslösung. Fachinformation. Available online: https://s3.eu-central-1.amazonaws.com/prod-cerebro-ifap/media_all/43955.pdf (accessed on 12 July 2022).

41. Gradishar, W.J. Albumin-bound paclitaxel: A next-generation taxane. *Expert Opin. Pharmacother.* **2006**, *7*, 1041–1053. [CrossRef] [PubMed]

42. Chen, N.; Brachmann, C.; Liu, X.; Pierce, D.W.; Dey, J.; Kerwin, W.S.; Li, Y.; Zhou, S.; Hou, S.; Carleton, M.; et al. Albumin-bound nanoparticle (nab) paclitaxel exhibits enhanced paclitaxel tissue distribution and tumor penetration. *Cancer Chemother. Pharmacol.* **2015**, *76*, 699–712. [CrossRef]

43. Baum, S. Arzneimitteltherapiesicherheit. Gefürchtetes Delir. *DAZ* **2020**, *25*, 66. Available online: https://www.deutsche-apotheker-zeitung.de/daz-az/2020/daz-25-2020/gefuerchtetes-delir (accessed on 11 July 2022).

44. Karch, F.E.; Lasagna, L. Adverse drug reactions: A critical review. *JAMA* **1975**, *234*, 1236–1241. Available online: https://samizdathealth.org/wp-content/uploads/2020/12/1975-Karch-and-Lasagna-1975.pdf (accessed on 12 July 2022). [CrossRef] [PubMed]

45. Rodrigues, M.C.; Oliveira, C.D. Drug-drug interactions and adverse drug reactions in polypharmacy among older adults: An integrative review. *Rev. Lat. Am. Enfermagem.* **2016**, *24*, e2800. [CrossRef] [PubMed]

46. Moore, T.J.; Cohen, M.R.; Furberg, C.D. Serious adverse drug events reported to the Food and Drug Administration, 1998–2005. *Arch. Intern. Med.* **2007**, *167*, 1752–1759. [CrossRef]

47. Davies, E.C.; Green, C.F.; Taylor, S.; Williamson, P.R.; Mottram, D.R.; Pirmohamed, M. Adverse drug reactions in hospital in-patients: A prospective analysis of 3695 patient-episodes. *PLoS ONE* **2009**, *4*, e4439. [CrossRef]

48. von Rosensteil, N.A.; Adam, D. Macrolide antibacterials. Drug interactions of clinical significance. *Drug Saf.* **1995**, *13*, 105–122. [CrossRef] [PubMed]

49. Azithromycin 500 mg Tablets. Summary of Product Characteristics. Available online: https://www.medicines.org.uk/emc/product/6541/smpc#gref (accessed on 11 July 2022).

50. Hill, K.; Sucha, E.; Rhodes, E.; Carrier, M.; Garg, A.X.; Harel, Z.; Hundemer, G.L.; Clark, E.G.; Knoll, G.; McArthur, E.; et al. Risk of Hospitalization With Hemorrhage Among Older Adults Taking Clarithromycin vs Azithromycin and Direct Oral Anticoagulants. *JAMA Intern. Med.* **2020**, *180*, 1052–1060. [CrossRef] [PubMed]

51. Anakinra. Kineret 100 mg/0.67 mL Injektionslösung in einer Fertigspritze. Zusammenfassung der Merkmale des Arzneimittels. Available online: https://www.ema.europa.eu/en/documents/product-information/kineret-epar-product-information_de.pdf (accessed on 12 July 2022).

52. Penman, S.L.; Carter, A.S.; Chadwick, A.E. Investigating the importance of individual mitochondrial genotype in susceptibility to drug-induced toxicity. *Biochem. Soc. Trans.* **2020**, *48*, 787–797. [CrossRef] [PubMed]

53. Gouin-Thibault, I.; Delavenne, X.; Blanchard, A.; Siguret, V.; Salem, J.E.; Narjoz, C.; Gaussem, P.; Beaune, P.; Funck-Brentano, C.; Azizi, M.; et al. Interindividual variability in dabigatran and rivaroxaban exposure: Contribution of ABCB1 genetic polymorphisms and interaction with clarithromycin. *J. Thromb. Haemost.* **2017**, *15*, 273–283. [CrossRef]

54. Atorvastatin-ratiopharm 10 mg/20 mg/40 mg/80 mg Filmtabletten. Fachinformation. Available online: https://www.ratiopharm.de/assets/products/de/label/Atorvastatin-ratiopharm%2010%20mg20%20mg40%20mg80%20mg%20Filmtabletten%20-%20 5.pdf?pzn=9292760 (accessed on 11 July 2022).

Article

Do Different Sutures with Triclosan Have Different Antimicrobial Activities? A Pharmacodynamic Approach

Frederic C. Daoud [1,*], Fatima M'Zali [2], Arnaud Zabala [2], Nicholas Moore [1] and Anne-Marie Rogues [1,3]

[1] INSERM, BPH, U1219, Université de Bordeaux, 33000 Bordeaux, France
[2] UMR 5234 CNRS, Laboratoire de Microbiologie Fondamentale et Pathogénicité, Université de Bordeaux, 33000 Bordeaux, France
[3] Pôle de Santé Publique, Service d'Hygiène Hospitalière, CHU Bordeaux, 33000 Bordeaux, France
* Correspondence: frederic.daoud-pineau@u-bordeaux.fr or fcdaoud@gmail.com; Tel.: +33-06-03-00-68-98

Abstract: (1) Background: Three antimicrobial absorbable sutures have different triclosan (TS) loads, triclosan release kinetics and hydrolysis times. This in vitro study aims to analyse and compare their antimicrobial pharmacodynamics. (2) Methods: Time-kill assays were performed with eight triclosan-susceptible microorganisms common in surgical site infections (SSIs) and a segment of each TS. Microbial concentrations were measured at T0, T4, T8 and T24 h. Similar non-triclosan sutures (NTS) were used as controls. Microbial concentrations were plotted and analysed with panel analysis. They were predicted over time with a double-exponential model and four parameters fitted to each TS × microorganism combination. (3) Results: The microbial concentration was associated with the triclosan presence, timeslot and microorganism. It was not associated with the suture material. All combinations shared a common pattern with an early steep concentration reduction from baseline to 4–8 h, followed by a concentration up to a 24-h plateau in most cases with a mild concentration increase. (4) Conclusions: Microorganisms seem to be predominantly killed by contact or near-contact killing with the suture rather than the triclosan concentration in the culture medium. No significant in vitro antimicrobial pharmacodynamic difference between the three TS is identified. Triclosan can reduce the suture microbial colonisation and SSI risk.

Keywords: suture; antimicrobial; pharmacodynamics; triclosan; surgical site infection; time-kill; contact killing; translational modelling

Citation: Daoud, F.C.; M'Zali, F.; Zabala, A.; Moore, N.; Rogues, A.-M. Do Different Sutures with Triclosan Have Different Antimicrobial Activities? A Pharmacodynamic Approach. *Antibiotics* **2022**, *11*, 1195. https://doi.org/10.3390/antibiotics11091195

Academic Editor: Dóra Kovács

Received: 26 July 2022
Accepted: 30 August 2022
Published: 3 September 2022

Publisher's Note: MDPI stays neutral with regard to jurisdictional claims in published maps and institutional affiliations.

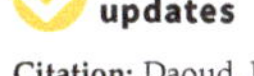

1. Introduction

There is a broad array of surgical wound closure methods, including thousands of suture types, staples and surgical adhesives [1]. Sutureless surgery is also being developed in various fields, including maxillofacial and cardiac surgery [2–5]. Minimising the risk of surgical site infection (SSI) is an important consideration when developing surgical wound closure techniques.

Triclosan is a synthetic, hydrophobic bisphenol (5-Chloro-2-(2,4-dichlorophenoxy) phenol) [6]. It is solid below 54 °C and displays low solubility in water (10 µg/mL in pure water at 20 °C) compared to nonpolar solvents such as olive oil (approximately 600,000 µg/mL) or ethanol (>1 million µg/mL) [7]. Hydrolysis and photodegradation are the two main triclosan degradation pathways. Both are too slow to be measurable over 24 h, providing the assays are protected from intense light. Triclosan has several properties that make it a broad-spectrum antimicrobial, especially its nonpolarity, which brings triclosan molecules together, among other nonpolar substances such as phospholipids influences bacterial cell membranes [8–11]. Triclosan lipid-membranotropism facilitates its concentration inside cell phospholipids and the membranes of gram-positive cocci (mix of peptidoglycan and phospholipids) and gram-negative bacteria (predominantly phospholipids). It also partially explains triclosan's lower ability to penetrate the outer walls of *C. albicans*, predominantly consisting of polysaccharides [12]. Triclosan's antimicrobial activity has multiple targets, but the main one is reported to be NADH-dependent enoyl-[acyl carrier protein] reductase (FabI). This inhibits cell membrane fatty acid synthesis, thus disrupting membranes [13,14].

Cell membrane lipidic composition is not enough to explain triclosan susceptibility. For example, *Pseudomonas aeruginosa* cell membrane is predominantly lipidic but has an enzymatic membrane efflux pump sufficient to expel triclosan, thus reducing its concentrations and antimicrobial effects [15,16].

Triclosan is added to absorbable sutures to inhibit microbial colonisation and thus reduce the risk of SSI [17–22]. Braided polyglactin-910 sutures are available with a maximum triclosan load of 472 µg/m (V+) [23]. Monofilament polydioxanone (P+) and monofilament poliglecaprone 25 sutures with a maximum of 2360 µg/m (M+) [23–25]. These sutures are also available without triclosan (V, P and M).

A previous study analysed the triclosan release kinetics of V+, P+ and M+ in pure static water and accelerated conditions calibrated to reproduce subcutaneous and intramuscular release in operated large animals [26]. It established the relation between the triclosan release rate and the antistaphylococcal activity in V+. However, the pharmacodynamics of V+, P+ and M+ once implanted in live operated tissues are not documented.

Several surface static agar cultures have measured zones of inhibition (ZOI) of sutures with triclosan (TS) vs. non-triclosan controls (NTS). One assay showed that V+ inhibited the growth of *Staphylococcus aureus* and *Staphylococcus epidermidis* after 24-h exposure, while V caused no inhibition [27]. Others showed similar results with M+ vs. M and P+ vs. P. *Escherichia coli* was inhibited up to approximately 1 cm from P+ [28–30]. Those in vitro experiments also showed the inhibition of methicillin-resistant strains of *S. epidermidis*, *S. aureus* and *Klebsiella pneumoniae*. In vivo experiments in one study showed that subcutaneous TS segments in mice with 7×10^6 *E. coli* colony-forming units (CFU) inoculum displayed, when removed after 48 h, a 90% microbial reduction, while NTS controls were colonised [28]. The same study showed in guinea pigs with a 4×10^5 *S. aureus* CFU inoculum a 99.9% microbial reduction.

While these studies confirm triclosan's antibacterial activity, they do not demonstrate the translation of the results to operated human tissues regarding triclosan bioavailability and TS antimicrobial activity. The level, duration and volume of antimicrobial activity around TS are uncertain.

The fact that 88% (22/25) (11,957 patients) of parallel-arm prospective randomised controlled clinical trials (RCT), which are the most comprehensive meta-analyses published to date, are non-significant supports translational uncertainty, although the pooled relative risk (RR) was 0.73 [0.65, 0.82] [31]. The WHO has published a conditional guideline recommending the use of TS to reduce the risk of SSI, stating that the quality of the evidence is moderate [32].

Many factors influence SSI risk, including the suture material, microbial concentration, infected volume, microbial multiplication, susceptibility to triclosan, exposure duration, surgical site characteristics and patient's natural defences [33].

The objective of this study (AD16-174/AST2016-181/IIS15-216/2016-11-09) was to analyse the in vitro pharmacodynamics of the three TS, understand their translational antimicrobial characteristics and compare them (Table S1).

2. Materials and Methods

2.1. Microbiology: Time-Kill Assays

Nine microorganisms common in SSIs were selected, representing a range of triclosan minimum inhibitory concentrations (MIC). Those were *E. coli* ATCC 25922 (MIC 0.03 µg/mL), *E. coli* ESBL producer collection clinical strain (MIC 0.03 µg/mL), *S. epidermidis* CIP 8155T (MIC 0.03 µg/mL), *S. aureus* ATCC 29213 (MIC 0.03 µg/mL), Methicillin-resistant *S. aureus* (MRSA) ATCC 33592 (MIC 0.03 µg/mL), MRSA collection clinical strain, *Candida albicans* ATCC 10231 (MIC 4 µg/mL), *C. albicans* collection clinical strain and *P. aeruginosa* collection clinical strain (MIC 256 µg/mL). V, V+, P, P+, M and M+ challenged all the microorganisms. All sutures had USP 2-0 calibres, i.e., a 0.35 to 0.399 mm diameter, a 35 cm length and, thus, a 0.03 to 0.04 cm^3 volume.

A time-kill assay protocol was specified according to CLSI standards [34]. Sutures and microbial cultures were handled in a safety cabinet. Sutures were unpackaged and immediately cut into various lengths, including 35 cm-long segments. Suture incubation on agar plates for 18 h at 37 °C checked sterility.

Agar plate cultures each microorganism's purity. A single colony was inoculated in a 10 mL sterile tryptic soy culture broth (TSB; Difco, BD Diagnostics, Sparks, MD, USA) tube. The suspension cultures were incubated for 4 h at 37 °C with continuous shaking until exponential growth was reached with 0.5 McFarland turbidity (approximately 1.5×10^8 colony-forming units CFU/mL). A densitometric controlled inoculum of the culture was extracted. It was added to 10 mL of the culture medium in a sterile tube setting at an approximately 10^6 CFU/mL baseline microbial concentration. After testing different segment lengths, 35 cm was selected because it enabled the distinguishing of concentration differences between each timeslot. Each culture had one immersed segment incubated for 24 h at 37 °C. Microbial concentrations were determined at four timeslots: baseline, 4, 8 and 24 h (T0, T4, T8, T24).

Culture tubes underwent 4 min of 48 Mhz sonication to detach viable microorganisms from the tube walls and suture. A 100 µL sample was drawn and then underwent five serial $1/10$ v/v dilutions. The original sample and dilutions were spread on separate agar plates with Mueller Hinton (MH) medium. The plates were incubated for 24 h at 37 °C and used if colonies were countable. Plate colony count multiplied by the dilution factor was the source culture microbial concentration (CFU/mL). Six copies of the 54 sutures × microorganism combinations were performed.

2.2. Data Analysis

2.2.1. Plots of Time-Kill Assays

For each combination and each timeslot (T0 through T24), the viable microbial concentration was converted to $\log_{10}$(CFU/mL). Repeated time-kill assays were plotted jointly.

2.2.2. Statistical Analysis

A comprehensive panel analysis tested the association of microbial concentration with four factors (1). This model included TS, NTS and *P. aeruginosa*.

$$C(t) = b_0 + b_1 x_{triclosan.i} + b_2 x_{material.i} + b_3 x_{microorganism.i} + b_4 x_{timeslot.i} \tag{1}$$

A second-panel analysis focused only on TS and triclosan-sensitive microorganisms (2).

$$C(t) = b_0 + b_1 x_{material.i} + b_2 x_{microorganisms.i} + b_3 x_{timeslot.i} \tag{2}$$

The $.i$ indices refer to the dummy variables defined for each modality of each independent factor. The models were calculated using random effects (RE) panel regression if the Hausman test for random effects was non-significant and in the presence of the heteroskedasticity of microbial concentration residues, confirmed by a significant Breusch–Pagan Lagrange multiplier test (LM). If those conditions were not met, the model would be calculated using the pooled ordinary least squares (OLS) method [35].

Specific findings underwent exploratory post hoc tests.

2.2.3. Pharmacodynamic Fitting of Microbial Concentration

Triclosan bactericidal and antifungal activities and microbial multiplication are two competing dynamics. Therefore, a predictive two-term model was used to predict observations (3).

$$C(t) = C_1 e^{-tL_1} + C_2 e^{tL_2} \tag{3}$$

$C(t)$ is the microbial concentration in the tube at time t from baseline. The first term describes the microbial decrease over 24 h, with C_1 solved using $C(0)/2$, i.e., the half of the mean microbial concentration among the six repetitions at baseline. It is equal to the inoculum divided by the culture medium volume. $Ln2/L_1$ is the half-life $(T1_{1/2})$ of microbial killing over 24 h. The negative exponent defines an exponential decay.

The second term describes microbial growth, with C_2 solved using $C(0)/2$. $Ln2/L_2$ is the half-life $(T1_{1/2})$ over 24 h. The implicit positive exponent defines the exponential growth.

The four unknown parameters were solved jointly given the observed datasets consisting of (time, microbial concentration) couples for each TS × strain combination and each repetition. The solution targeted minimising the least-squares of observations vs. the predicted values with the following constraints:

(1) The targeted adjusted coefficient of determination $R^2_{adjusted}$ (4) was within [0.9, 1] and

(2) The predicted concentration at T24 was equal to the observed mean concentration at T24 among the six repetitions. n was the number of observations per combination (n = 4 measurements × 6 repetitions = 24 per combination), and k was the number of estimated explanatory parameters in the function using the n observations (k = 4, per combination).

$$R^2_{adjusted} = 1 - \left(1 - \frac{\sum_{i=1}^{n} (y_i - \hat{y}_i)^2}{\sum_{i=1}^{n} (y_i - \bar{y})^2}\right) \frac{n-1}{n-(k+1)} \tag{4}$$

The $R^2_{adjusted}$ measures the goodness-of-fit (GoF) of the model and thus the degree of prediction of observations. An $R^2_{adjusted}$ equal to 1 indicates the perfect prediction of the observed concentrations, while an $R^2_{adjusted}$ equal to 0 shows that the model does not predict any better than random guesses.

Two required initial inputs (the partial concentrations at T0: C_1 and C_2) were defined such that $C_1 e^0 + C_2 e^0 = C(0)$, given $e^0 = 1$.

Data management and statistical analyses were performed with the xt module in Stata 17, StataCorps LLC, College Station, TX, USA.

The fitting of the pharmacodynamic model to the data was performed in Microsoft Excel 2019 version 2205, Microsoft Corporation, Redmond, WA, USA, using the Solver function. Parameter resolutions were checked with Maple 2021.1, Maple Inc., Waterloo, ON, Canada.

3. Results

3.1. Time-Kill Assays

The plots show growth with the eight triclosan-sensitive strains with NTS control sutures (Figure 1a–c, Figure 2a–c, and Figure 3a,b) and *P. aeruginosa*, which was used as the triclosan-resistant control strain (Figure 3c).

The plots with TS and triclosan-sensitive strains show an initial rapid microbial concentration reduction between baseline and T4 or T8. The reduction magnitude ranges between about 1.5 $\log_{10}$ and 3 $\log_{10}$. It is followed by a microbial concentration plateau between T8 and T24. The plateau ranges between a mild decline, a steady concentration and a mild increase.

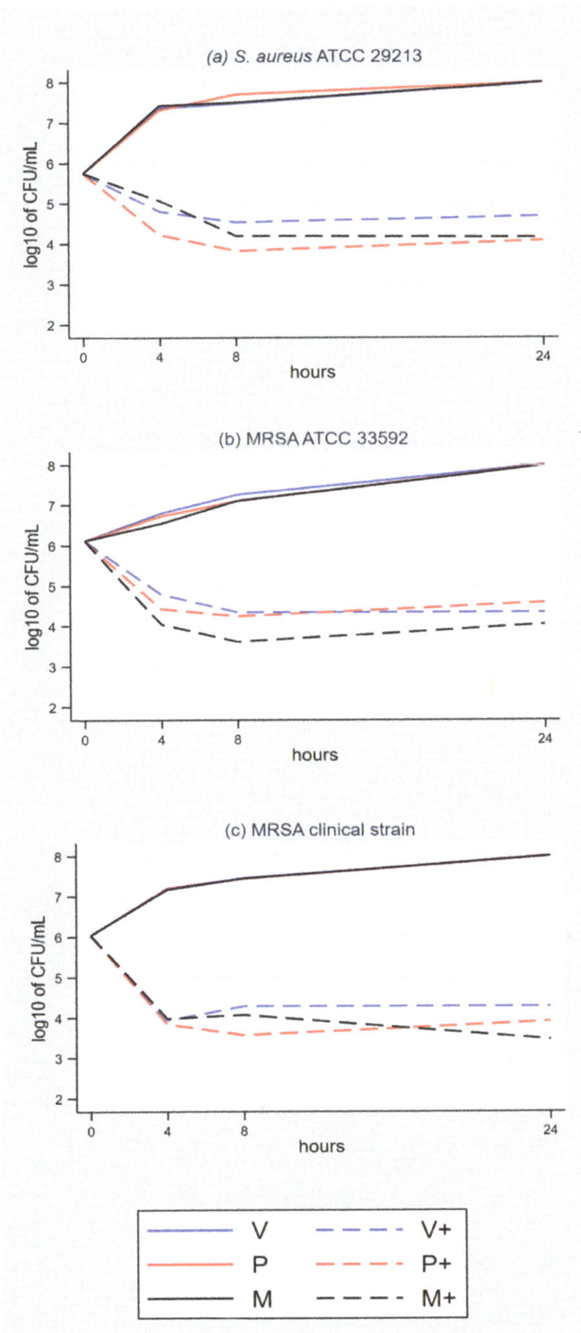

Figure 1. (**a–c**) Time-kill analyses *S. aureus* and MRSA with the three TS.

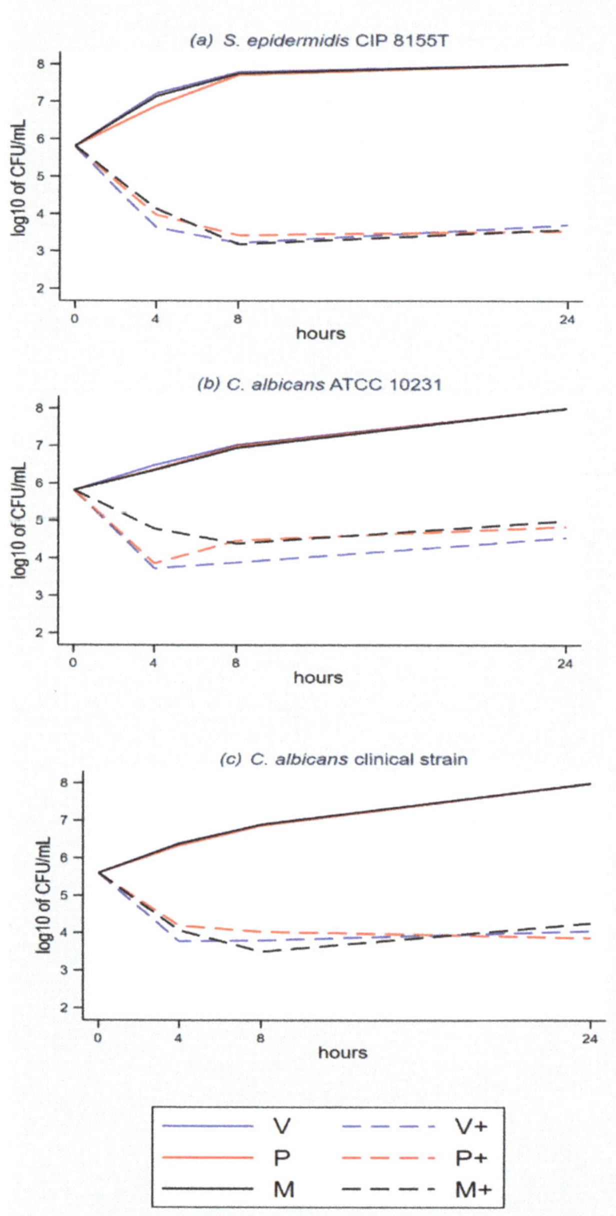

Figure 2. (**a–c**) Time-kill analyses *C. albicans* and *S. epidermidis* with the three TS.

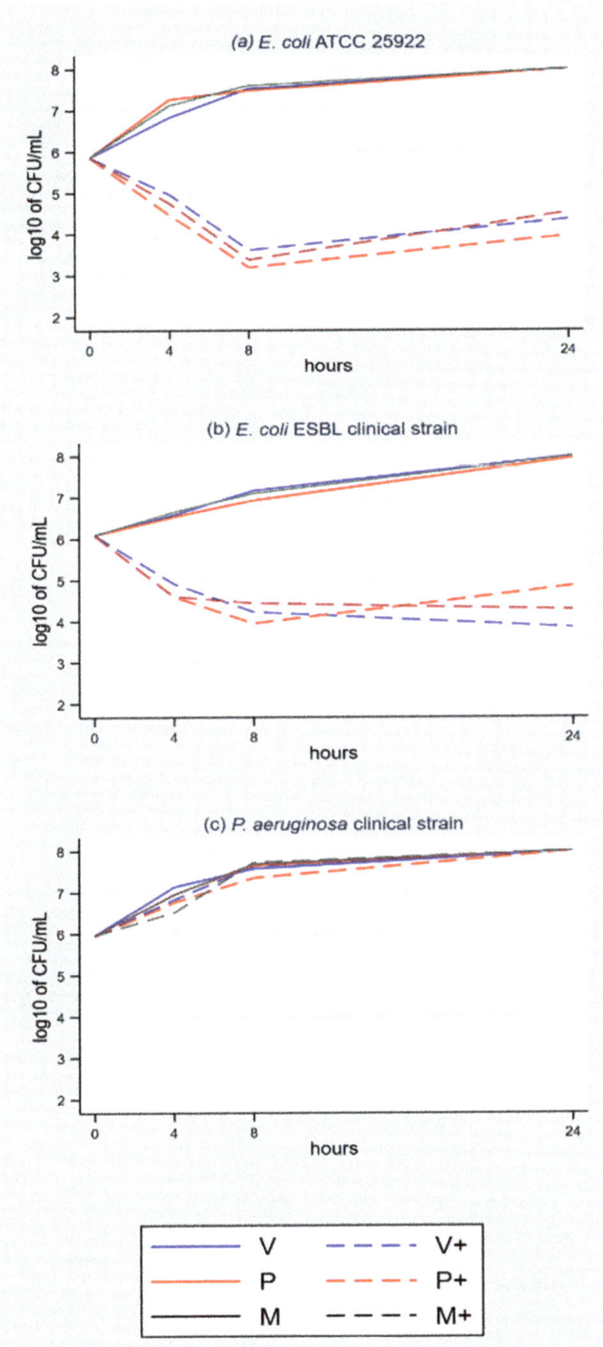

Figure 3. (**a–c**) Time-kill analyses *E. coli* and *P. aeruginosa* with the three TS.

3.2. Statistical Analysis–Comprehensive Panel Model

The test eligibility criteria applied to the comprehensive model indicate the applicability of OLS regression (Table 1). The adjustment of the model is moderate, given R^2 = 0.578. The regression coefficients of the model are large and significant when comparing TS to the NTS controls, when comparing timeslots to baseline T0 and when comparing the *P. aeruginosa* triclosan-resistant control to triclosan-sensitive *S. epidermidis*, the reference level in the model. All other coefficients have a small magnitude and are non-significant. This shows that the microbial concentration is significantly associated with the presence or absence of triclosan, timeslots and *P. aeruginosa*. Suture materials and other microbial strains are not associated with microbial concentration in this comprehensive model.

Table 1. Comprehensive OLS regression of $\log_{10}$ microbial concentration depending on triclosan, material, strain and timeslot.

log(CFU)	Coef.	St.Err.	*t*-Value	*p*-Value	95% Conf	Interval	Sig.
S. epidermidis CIP 8155T	0		Reference level for microorganisms				
C. albicans ATCC 10231	−0.097	0.124	−0.78	0.436	−0.339	0.147	
C. albicans clinical	−0.205	0.124	−1.66	0.098	−0.448	0.038	
E. coli ATCC 25929	0.052	0.124	0.42	0.674	−0.191	0.295	
E. coli ESBL clinical	0.183	0.124	1.48	0.140	−0.060	0.426	
MRSA ATCC 33592	0.152	0.124	1.23	0.221	−0.195	0.395	
MRSA clinical strain	0.048	0.124	0.38	0.700	−0.205	0.291	
P. aeruginosa clinical	1.306	0.124	10.55	0.000	1.063	1.549	*
S. aureus ATCC 29213	0.170	0.124	1.37	0.170	−0.073	0.413	
Polyglecaprone 25	0		Reference level for suture materials				
Polyglactin 910	0.035	0.072	0.47	0.637	−0.105	0.175	
Polydioxanone	−0.015	0.072	−0.21	0.835	−0.156	0.125	
No triclosan	0		Reference level for use of triclosan				
Triclosan	−2.256	0.058	−38.64	0.000	−2.370	−2.141	*
0 h	0		Reference level for hours				
4	−0.379	0.083	−4.66	0.000	−0.541	−0.217	*
8	−0.237	0.083	−2.90	0.005	−0.399	−0.075	**
24	0.416	0.083	5.11	0.000	0.254	0.578	*
Overall model intercept	4.400	0.113	38.92	0.000	4.179	4.622	*
Number of obs		1296		Number of groups		324	
F-test		df = 141,281		*p*-value		<0.0001	
R-squared		0.5847		Adjusted R-squared		0.5802	

* $p < 0.001$, ** $p < 0.005$, Sig.: significant, df: degrees of freedom, St. Err.: standard error. Hausman test: significance of: fails to meet the asymptotic assumptions. Breusch and Pagan Lagrangian multiplier test (LM) for random effects: *p*-value = 0.2085.

3.3. Statistical Analysis–Focused Panel Model

The test eligibility criteria applied to the focused model indicated the use of random-effects panel regression (Table 2). The adjustment of the model is improved with R^2 = 0.687. The regression coefficients of the model are large and significant when comparing the timeslots to the T0 baseline and are significant but mild when comparing the 5 triclosan-sensitive microbial strains to *S. epidermidis* as a reference level for microorganisms in the model. The coefficients for suture materials have a small magnitude and are non-significant. This model shows that the microbial concentration is significantly associated with timeslots, mildly associated with some microbial strains and not associated with suture materials.

Table 2. Focused random effects panel regression of $\log_{10}$ microbial concentration depending on material, strain and timeslot.

log(CFU)	Coef.	St.Err.	*t*-Value	*p*-Value	95% Conf	Interval	Sig.
S. epidermidis CIP 8155T	0		Reference level for microorganisms				
C. albicans ATCC 10231	0.341	0.128	2.65	0.008	0.089	0.592	
C. albicans clinical	0.098	0.128	0.77	0.444	−0.153	0.350	
E. coli ATCC 25929	0.265	0.128	2.07	0.039	0.014	0.517	
E. coli ESBL clinical	0.551	0.128	4.30	0.000	0.210	0.803	*
MRSA ATCC 33592	0.537	0.128	4.19	0.000	0.286	0.789	*
MRSA clinical strain	0.221	0.128	1.72	0.085	−0.030	0.472	
S. aureus ATCC 29213	0.545	0.128	4.25	0.000	0.294	0.796	*
Polyglecaprone 25	0		Reference level for suture materials				
Polyglactin 910	0.071	0.079	0.90	0.369	−0.083	0.225	
Polydioxanone	−0.008	0.079	−0.11	0.915	−0.162	0.146	
0 h	0		Reference level for hours				
4	−1.795	0.063	−28.28	0.000	−1.919	−1.670	*
8	−2.148	0.063	−33.85	0.000	−2.273	−2.023	*
24	−1.936	0.063	−30.50	0.000	−2.060	−1.811	*
Overall model intercept	5.392	0.109	49.64	0.000	5.179	5.605	**
Number of obs		576		Number of groups		144	
Wald Chi-squared		df = 12		*p*-value		<0.0001	
R-squared		0.687					

P. aeruginosa and NTS not included in the focused model; * $p < 0.001$, ** $p < 0.005$, Sig.: significant, df: degrees of freedom, St. Err.: standard error. Hausman test: $p = 1$. Breusch and Pagan Lagrangian multiplier test (LM) for random effects: p-value < 0.0001.

3.4. Statistical Analysis–Post Hoc Paired t-Test

The mean microbial reduction among triclosan-susceptible microorganisms, from baseline (T0) to trough (T4 or T8), was $-2.29 \log_{10}$ of CFU/mL [–2.40, –2.19].

The post hoc paired t-test showed a mild but significant mean concentration increase from trough to T24 (+0.36 $\log_{10}$ of CFU/mL [+0.26; +0.46], $p < 0.0001$).

3.5. Pharmacodynamic Fitting of Microbial Concentration

Each of the 24 combinations was fitted with a predictive pharmacodynamic function. Examples with *S. epidermidis* are shown in Figure 4, *S. aureus* in Figure 5, *E. coli* in Figure 6, and examples with ESBL-producing *E. coli* in Figure 7. Subfigures (a), (b) and (c) are fittings with V+, P+ and M+, respectively.

The predictive pharmacodynamic functions of the 24 combinations, based on a common algebraic function and fitted parameters, are listed in Table 3. The GoF, estimated by the adjusted $R^2_{adjusted}$, is between 0.61 and 0.89 in 22 fitted functions. There is a poor fit in the two other functions ($R^2_{adjusted}$ between 0.38 and 0.49) with *C. albicans* ATCC 33529 and the monofilament sutures P+ and M+ (Figures S1–S12).

All functions show an early fast microbial concentration decline from baseline to T4 or T8 h, followed by a plateau until T24. The plateau has a steady concentration in 2 cases (8.33%), a mild concentration decrease in 4 cases (16.7%) and a mild concentration increase in 18 cases (75%).

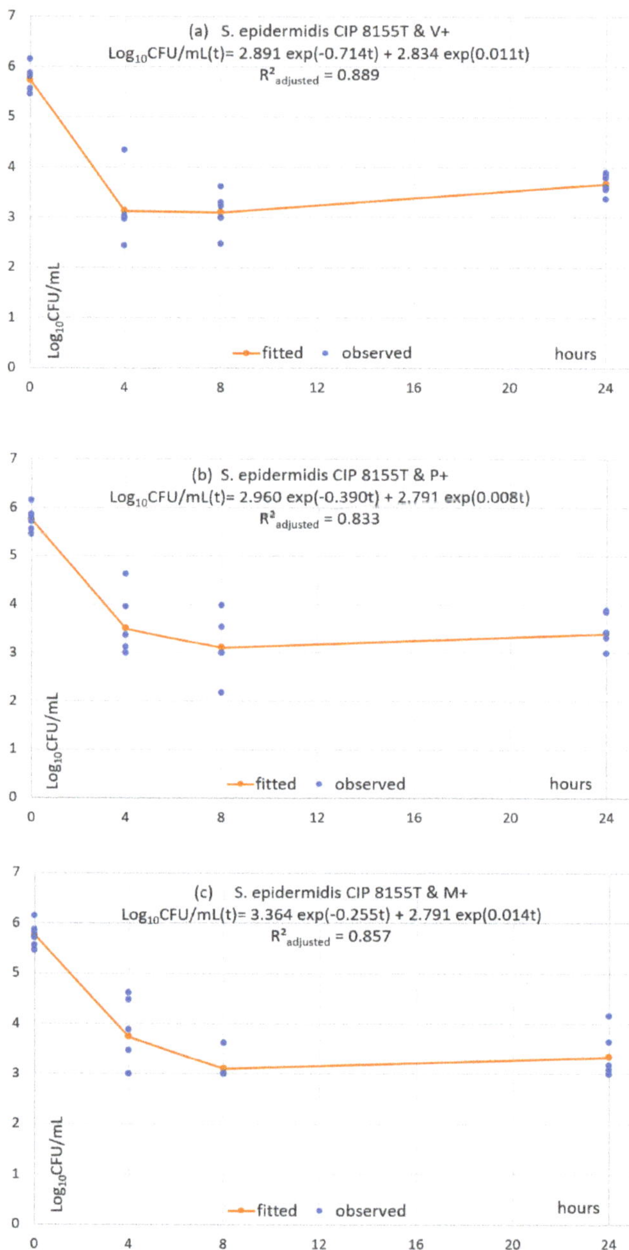

Figure 4. (**a–c**) Fitted predictive pharmacodynamic functions *S. epidermidis* with the three TS.

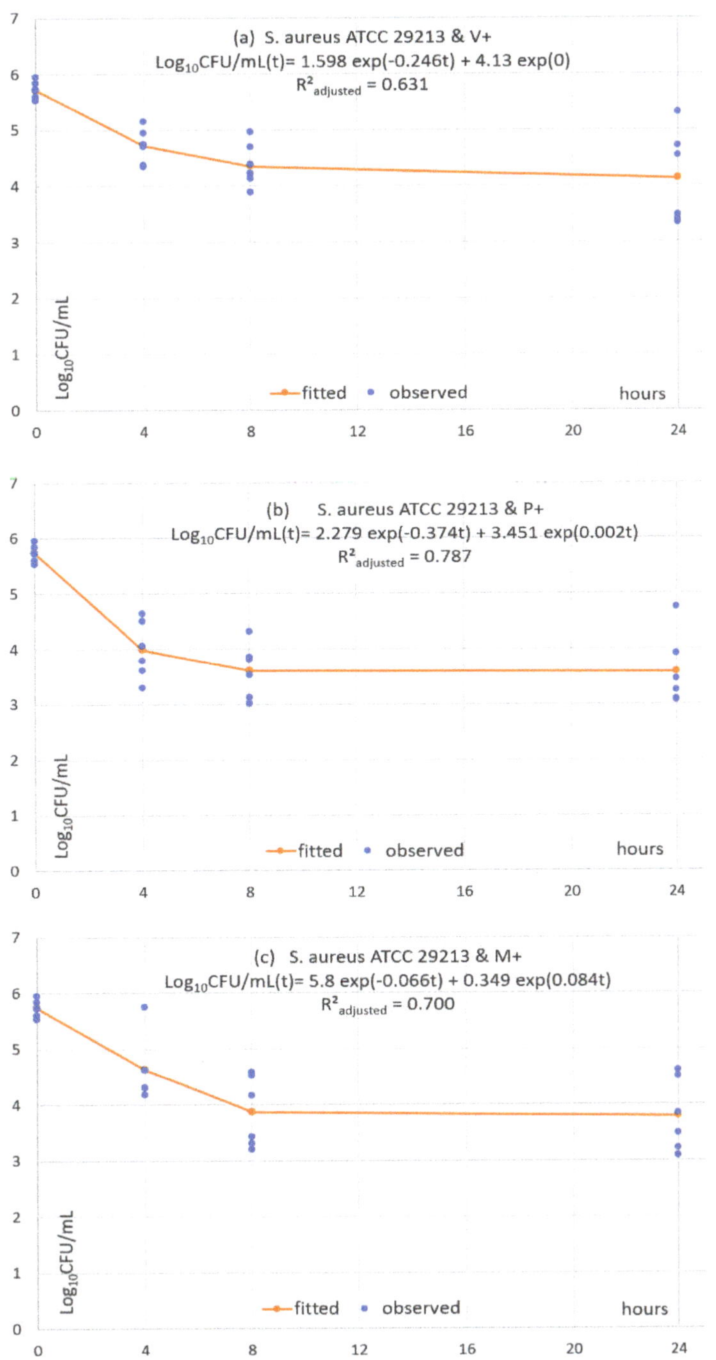

Figure 5. (**a**–**c**) Fitted predictive pharmacodynamic functions *S. aureus* with the three TS.

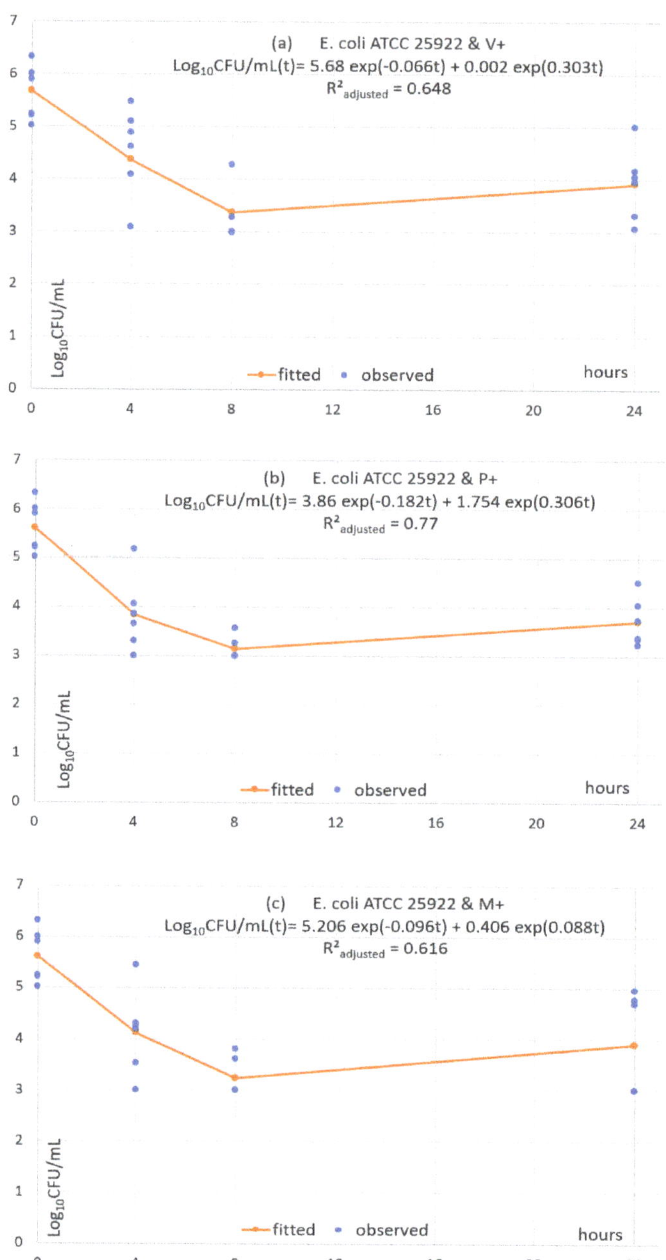

Figure 6. (a–c) Fitted predictive pharmacodynamic functions *E. Coli* with the three TS.

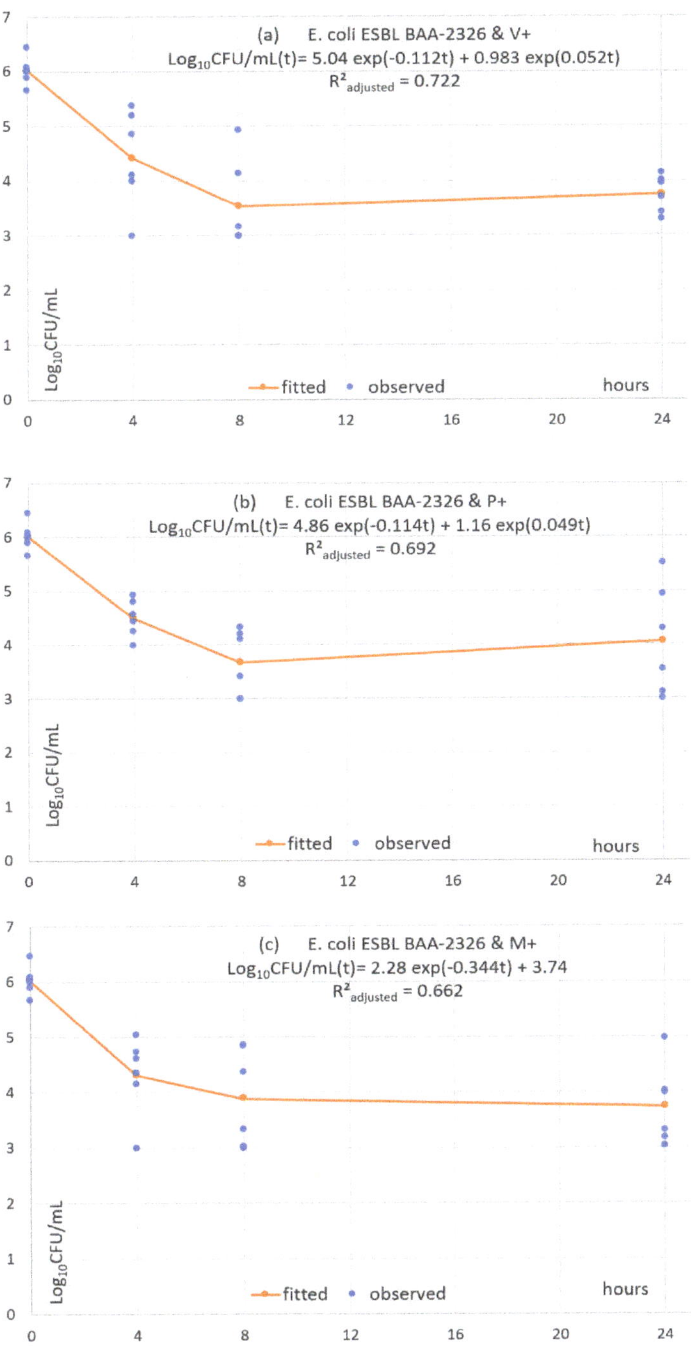

Figure 7. (**a–c**) Fitted predictive pharmacodynamic functions ESBL *E. Coli* with the three TS.

Table 3. Fitting pharmacodynamic model (3) microbial concentrations from T0 to T24, excluding NTS and *P. aeruginosa*.

Microbial Strain	Suture	$R^2_{adjusted}$	C1 µg/mL	HF1 Hours	C2 µg/mL	HL2 Hours	Plot Shape
S. aureus ATCC 29213	V+	0.631	1.600	2.820	4.129	NA.	T0–T8: approx. −1.5log$_{10}$ T8–T24: plateau & mild decrease
S. aureus ATCC 29213	P+	0.787	2.279	1.852	3.451	451	T0–T8: approx. −2log$_{10}$ T8–T24: steady plateau
S. aureus ATCC 29213	M+	0.700	5.382	10.558	0.359	8.171	T0–T8: approx. −2log$_{10}$ T8–T24: plateau & mild decrease
MRSA ATCC 33592	V+	0.676	2.542	2.633	3.445	98.943	T0–T8: approx. −2log$_{10}$ T8–T24: plateau & mild increase
MRSA ATCC 33592	P+	0.710	2.696	1.974	3.691	210.724	T0–T8: approx. T8–T24: plateau & mild increase
MRSA ATCC 33592	M+	0.858	3.017	2.000	2.9700	64.903	T0–T8: approx. −2.5log$_{10}$ T8–T24: plateau & mild increase
MRSA clinical	V+	0.700	2.070	0.633	3.776	NA.	T0–T4: approx. −2log$_{10}$ T4–T24: steady plateau
MRSA clinical	P+	0.836	2.680	1.613	3.166	152.646	T0–T8: approx. −2.5log$_{10}$ T8–T24: plateau & mild increase
MRSA clinical	M+	0.818	2.547	1.677	3.297	NA.	T0–T8: approx. −2.5log$_{10}$ T8–T24: plateau & mild decrease
S. epidermidis CIP 8155T	V+	0.889	2.891	0.971	2.834	64.203	T0–T8: approx. −2.5log$_{10}$ T8–T24: plateau & mild increase
S. epidermidis CIP 8155T	P+	0.833	2.960	1.778	2.790	83.821	T0–T8: approx. −2.5log$_{10}$ T8–T24: plateau & mild increase
S. epidermidis CIP 8155T	M+	0.857	3.364	2.714	2.389	50.196	T0–T8: approx. −2.5log$_{10}$ T8–T24: plateau & mild increase
E. coli ATCC 25922	V+	0.648	5.682	10.528	0.0019	2.287	T0–T8: approx. −2log$_{10}$ T8–T24: plateau & mild increase
E. coli ATCC 25922	P+	0.771	3.858	3.791	1.754	22.728	T0–T8: approx. −2.5log$_{10}$ T8–T24: plateau & mild increase
E. coli ATCC 25922	M+	0.616	5.206	7.196	0.406	7.884	T0–T8: approx. −2.5log$_{10}$ T8–T24: plateau & mild increase
E. coli ESBL BAA-2326	V+	0.721	5.037	6.162	0.983	13.36	T0–T8: approx. −2.5log$_{10}$ T8–T24: plateau & mild increase
E. coli ESBL BAA-2326	P+	0.692	4.857	6.091	1.162	14.216	T0–T8: approx. −2.5log$_{10}$ T8–T24: plateau & mild increase
E. coli ESBL BAA-2326	M+	0.662	2.78	2.014	2.015	NA.	T0–T8: approx. −2log$_{10}$ T8–T24: plateau & mild decrease
C. albicans ATCC 10231	V+	0.629	2.104	0.185	3.371	93.158	T0–T4: approx. −2log$_{10}$ T4–T24: plateau & mild increase
C. albicans ATCC 10231	P+	0.494	1.864	0.166	3.611	86.005	T0–T4: approx. −1.7log$_{10}$ T4–T24: plateau & mild increase
C. albicans ATCC 10231	M+	0.381	1.824	0.184	3.651	98.992	T0–T4: approx. −1.7log$_{10}$ T4–T24: plateau & mild increase
C. albicans clinical	V+	0.647	5.673	10.400	0.006	2.723	T0–T8: approx. −2log$_{10}$ T8–T24: plateau & mild increase
C. albicans clinical	P+	0.648	5.675	10.568	0.001	2.070	T0–T8: approx. −2log$_{10}$ T8–T24: plateau & mild increase
C. albicans clinical	M+	0.648	5.680	10.534	0.001	2.157	T0–T8: approx. −2log$_{10}$ T8–T24: plateau & mild increase

Note: "mild" means a variation of less than 1log$_{10}$.

4. Discussion

4.1. Protocol Specifications and Interpretation

This study performed the first in vitro pharmacodynamics analysis of TS antimicrobial activity using time-kill assays. V, P, M and triclosan-resistant *P. aeruginosa* were the controls. The experimental settings were the same as those used in static water release kinetics [26].

The suspension cultures had enough volume and nutrients to sustain microbial growth beyond the T24 timeslot, as confirmed by the growth in the NTS cultures (Figure 1 to Figure 3). Microbial colony count was proportional to the microbial concentration in the cultures, with a degree of random error between repetitions.

Microbial concentrations exceeding the 10^8 CFU/mL upper boundary could not be accurately estimated, but that had no impact on the analysis. None of them reached the lower detection boundary below 10^2 CFU/mL.

4.2. Key Results

The plots and statistical analyses showed that the microbial concentration is significantly associated with triclosan, the timeslot and the microorganism. It is not associated with the suture material.

The plots with all TSs and triclosan-susceptible microorganisms consisted of an initially rapid microbial reduction from T0 to T4 or T8, with a mean reduction of $-2.29 \log_{10}$ of CFU/mL followed by a plateau without a microbial concentration change, mild decrease or mild increase. The mean change between the T8 and T24 was a mild but significant increase.

The underlying mechanisms of the predictive pharmacodynamic models were assumed to be microbial kill and multiplication. The fitting was good in all combinations except for two. The models differed little between TS types for a given microorganism. Most differences were between microorganisms. The functions reproduced the initial microbial concentration rapid reduction and subsequent plateau and showed a mild increase in 75% of combinations.

4.3. Interpretation of The Time-Kill Assays

These assays show the ability of V+, P+ and M+ to reduce a high microbial concentration over a 4-to-8-h period, with very little difference given each microorganism despite the 5-fold difference in the triclosan load and suture structure. The plateau and frequent mild increase are unexpected.

The key question is whether the predominant killing was by the triclosan concentration in the culture medium or by the contact killing at the suture surface or near it, where the triclosan concentration is high. Indeed, triclosan is a hydrophobic solid whose dissolution follows the Noyes & Whitney principle, which explains the triclosan gradient between the solid surface and the bulk of the solvent in a static tube. This gradient disappears when the tube is shaken. Therefore, its dissolution rate is slow around the suture, whose volume is 0.3 to 0.4% of the culture volume. Therefore, unless the tube is shaken, the triclosan diffusion layer forms a gradient with a maximum close to 10 µg/mL at the suture surface, decreasing with the distance from the source [7,26,36–39].

The culture conditions in static TSB (water + 3% organic nutrients and minerals) were closer to pure static water release kinetic determinations than to the ethanol/water 3.3% w/w solution with 24 rounds-per-minute constant rotation [26]. The triclosan release in 10 mL of pure static water is with V+ 2.3, 3.3 and 6.8 µg at 4, 8 and 24 h, respectively, with P+, 5.3, 6.7 and 6.1 µg and with M+, 7.6, 9.6 and 7.4 µg [26]. After sonication, the triclosan concentrations were homogeneous in the bulk of the cultures, i.e., with V+ 0.23, 0.33 and 0.68 µg/mL at 4, 8 and 24 h, respectively, with P+, 0.53, 0.67 and 0.61 µg/mL and with M+, 0.76, 0.96 and 0.74 µg/mL.

The triclosan minimal bactericidal concentrations (MBC) of *S. aureus* (0.03 to 2 µg/mL), *E. coli* (0.03 to 16 µg/mL) and *C. albicans* (0.12 to 16 µg/mL) are presented [40,41]. Therefore, the triclosan concentrations were within the MBC ranges, as the first sonication. This should have caused microbial killing through T24 in all TS ×microbial combinations. The plateau and a mild increase in 75% of combinations after T8 suggest that the triclosan concentration in the medium was too low to prevent microbial multiplication.

The available data do not provide proof of the mechanism. However, a potential explanation is that the dispersed triclosan after the T4 and T8 sonications were captured in the lipids of the killed bacteria. Therefore, the plateau between T8 and T24 is likely to be an experimental artefact when the tubes are still, and the two terms of the pharmacodynamic functions offset each other.

The obtained in vitro data do not show why the mild microbial concentration increased during the T8–T24 plateau in 75% of the assays. One potential explanation is the gradual decrease in the release rate while exponential microbial growth continued.

4.4. Comparison with Other Preclinical Studies

These in vitro pharmacodynamic models and underlying release kinetics are compatible with the absence of *E. coli* and *S. aureus* surface growth on agar plates with TS segments explanted from rodents after 48 h [26,28].

One study attempted to measure the duration and level of TS antimicrobial activity with a TS segment transferred consecutively from one static surface agar culture plate to another for up to 30 days [42]. The conclusions were that TSs display antimicrobial activity from about one week to one month depending on the TS × microorganism.

The methods of these in vitro experiments must be considered when interpreting the results. (1) The two-dimensional diffusion around the TS segments on the agar surface cultures represents, at most, the amount that would be contained in the three-dimension diffusion layer of the suspension culture. (2) The water/air surface diffusion meets less resistance than it does in full immersion. Therefore, ZOI overestimates, by several-fold, the antimicrobial volume of TS. (3) The TS triclosan release in static cultures is 15 to 60 times slower than that observed in large animal subcutaneous or intramuscular explants [26]. Therefore, the in vitro antimicrobial activity duration is also overestimated.

4.5. Translational Interpretation to Live Operated Human Tissues

The translational application of this pharmacodynamic study to operated human tissues is limited.

(1) The TS release rate in the time-kill assays is 15 to 60 times slower than it is in operated tissues. (2) Surgical sites are much larger than 10 mL tubes, so the suture volume is much less than 0.3 to 0.4% of the surgical site. (3) The permanent in vivo motion maintains a thin diffusion layer around the sutures, so the contact or near-contact volume with drifting microorganisms is probably negligible compared to the surgical site volume at risk. (4) When natural defences are functional, scattered bacteria are rapidly killed, and few encounter the TS. When natural defences are weak, scattered microbial multiplication requires antibiotics and/or reintervention

Therefore, preventing microbial colonisation while the triclosan release rate is efficacious, i.e., a few hours after implantation, is the only result of this pharmacodynamic study that can translate to a surgical site. However, that can relieve natural defences by reducing the risk of microbial colonisation of the suture. The three TS types share similar in vitro antimicrobial activity. There is no indication they would have significantly different in vitro antimicrobial activities in operated tissues. Measurements of microbial dose or concentration reduction cannot translate to operated tissues because they do not consider the complexity of surgical sites and natural defences.

5. Conclusions

This in vitro study shows that triclosan sutures kill susceptible microorganisms that come in direct contact or near contact with their surface. The in vitro antimicrobial profiles of the braided polyglactin-910, monofilament polydioxanone and monofilament poliglecaprone 25 sutures present no significant difference, and no difference in the operated tissues is predicted.

The in vitro pharmacodynamics suggest a significant reduction in the microbial dose close to the sutures as of implantation. Triclosan can minimise the suture colonisation risk early on, relieve natural defences and reduce the risk of surgical site infection.

Supplementary Materials: The following supporting information can be downloaded at: https://www.mdpi.com/article/10.3390/antibiotics11091195/s1. Figure S1: MRSA ATCC 33592 with V+; Figure S2: MRSA ATCC 33592 with P+; Figure S3: MRSA ATCC 33592 with M+; Figure S4: MRSA clinical with V+; Figure S5: MRSA clinical with P+; Figure S6: MRSA clinical with M+; Figure S7: C. albicans ATCC 10231 with V+; Figure S8: *C. albicans* ATCC 10231 with P+; Figure S9: *C. albicans* ATCC 10231 with M+; Figure S10: *C. albicans* clinical with V+; Figure S11: *C. albicans* clinical with P+; Figure S12: *C. albicans* clinical with M+; Table S1: Pharmacodynamics raw data.

Author Contributions: Conceptualisation, F.C.D., A.-M.R., N.M. and F.M.; methodology, F.M., A.Z. and F.C.D.; microbiological data assays, F.M. and A.Z.; quality assessment, F.M.; validation, A.-M.R. and F.C.D.; data analysis, F.C.D.; interpretation, F.C.D., F.M. and A.-M.R.; writing—original draft preparation, F.C.D.; writing—review and editing, F.M., A.Z., A.-M.R. and N.M.; supervision, N.M. and A.-M.R. All authors have read and agreed to the published version of the manuscript.

Funding: The project specification, organisation and analysis were funded by Université de Bordeaux, INSERM U1219 Bordeaux Population Health. The microbiology assays were performed at Université de Bordeaux, Aquitaine Microbiologie, UMR 5234 CNRS, and they were supported by an unrestricted grant from Ethicon, a division of Johnson and Johnson Medical Limited (investigator-initiated project ID.: IIS15-216/2016-11-09).

Institutional Review Board Statement: Not applicable.

Informed Consent Statement: Not applicable.

Data Availability Statement: All source data is in the Supplementary Materials File (raw data table).

Acknowledgments: Ethicon, a division of Johnson and Johnson Medical Limited, shared their experience with in vitro assays with those products without interfering with the protocol design, conduct and analysis.

Conflicts of Interest: The authors declare no conflict of interest.

References

1. Hochberg, J.; Meyer, K.M.; Marion, M.D. Suture Choice and Other Methods of Skin Closure. *Surg. Clin. N. Am.* **2009**, *89*, 627–641. [CrossRef] [PubMed]
2. Osunde, O.; Adebola, R.; Saheeb, B. A comparative study of the effect of suture-less and multiple suture techniques on inflammatory complications following third molar surgery. *Int. J. Oral Maxillofac. Surg.* **2012**, *41*, 1275–1279. [CrossRef]
3. Lam, K.Y.; Reardon, M.J.; Yakubov, S.J.; Modine, T.; Fremes, S.; Tonino, P.A.; Tan, M.E.; Gleason, T.G.; Harrison, J.K.; Hughes, G.C.; et al. Surgical Sutureless and Sutured Aortic Valve Replacement in Low-risk Patients. *Ann. Thorac. Surg.* **2021**, *113*, 616–622. [CrossRef]
4. Chisci, G.; Parrini, S.; Capuano, A. The use of suture-less technique following third molar surgery. *Int. J. Oral Maxillofac. Surg.* **2013**, *42*, 150–151. [CrossRef] [PubMed]
5. Chisci, G.; Capuano, A.; Parrini, S. Alveolar Osteitis and Third Molar Pathologies. *J. Oral Maxillofac. Surg.* **2018**, *76*, 235–236. [CrossRef]
6. Dhillon, G.S.; Kaur, S.; Pulicharla, R.; Brar, S.K.; Cledón, M.; Verma, M.; Surampalli, R.Y. Triclosan: Current Status, Occurrence, Environmental Risks and Bioaccumulation Potential. *Int. J. Environ. Res. Public Health* **2015**, *12*, 5657–5684. [CrossRef] [PubMed]
7. Scientific Committee on Consumer Safety (SCCS). Opinion on Triclosan 2010. European Commission—Directorate General for Health and Consumers. Available online: https://op.europa.eu/en/publication-detail/-/publication/3b684b59-27f7-4d0e-853c-4a9efe0fb4cb/language-en (accessed on 3 December 2021).
8. Villalaín, J.; Mateo, C.R.; Aranda, F.J.; Shapiro, S.; Micol, V. Membranotropic Effects of the Antibacterial Agent Triclosan. *Arch. Biochem. Biophys.* **2001**, *390*, 128–136. [CrossRef]
9. Petersen, R.C. Triclosan Computational Conformational Chemistry Analysis for Antimicrobial Properties in Polymers. *J. Nat. Appl. Sci.* **2015**, *1*, e54. Available online: https://www.ncbi.nlm.nih.gov/pubmed/25879080 (accessed on 15 October 2021).
10. Bernabeu, A.; Shapiro, S.; Bernabeu-Sanz, A. Location and orientation of Triclosan in phospholipid model membranes. *Eur. Biophys. J.* **2004**, *33*, 448–453. [CrossRef]
11. Hayami, M.; Okabe, A.; Kariyama, R.; Abe, M.; Kanemasa, Y. Lipid Composition of *Staphylococcus aureus* and Its Derived L-forms. *Microbiol. Immunol.* **1979**, *23*, 435–442. [CrossRef] [PubMed]

12. Russell, A.D. Similarities and differences in the responses of microorganisms to biocides. *J. Antimicrob. Chemother.* **2003**, *52*, 750–763. [CrossRef] [PubMed]

13. Heath, R.J.; Su, N.; Murphy, C.K.; Rock, C.O. The Enoyl-[acyl-carrier-protein] Reductases FabI and FabL from *Bacillus subtilis*. *J. Biol. Chem.* **2000**, *275*, 40128–40133. [CrossRef]

14. Schweizer, H.P. Triclosan: A widely used biocide and its link to antibiotics. *FEMS Microbiol. Lett.* **2001**, *202*, 1–7. [CrossRef] [PubMed]

15. Chuanchuen, R.; Karkhoff-Schweizer, R.R.; Schweizer, H.P. High-level triclosan resistance in *Pseudomonas aeruginosa* is solely a result of efflux. *Am. J. Infect. Control* **2003**, *31*, 124–127. [CrossRef]

16. Zhu, L.; Lin, J.; Ma, J.; Cronan, J.E.; Wang, H. Triclosan Resistance of *Pseudomonas aeruginosa* PAO1 Is Due to FabV, a Triclosan-Resistant Enoyl-Acyl Carrier Protein Reductase. *Antimicrob. Agents Chemother.* **2010**, *54*, 689–698. [CrossRef]

17. BASF Names West Coast Distributor for Care Chemicals Business. Available online: https://www.basf.com/us/en/media/news-releases/2013/10/p-13-441.html (accessed on 8 September 2021).

18. Heath, R.J.; Li, J.; Roland, G.E.; Rock, C.O. Inhibition of the *Staphylococcus aureus* NADPH-dependent Enoyl-Acyl Carrier Protein Reductase by Triclosan and Hexachlorophene. *J. Biol. Chem.* **2000**, *275*, 4654–4659. [CrossRef] [PubMed]

19. Heath, R.J.; Rubin, J.R.; Holland, D.R.; Zhang, E.; Snow, M.E.; Rock, C.O. Mechanism of Triclosan Inhibition of Bacterial Fatty Acid Synthesis. *J. Biol. Chem.* **1999**, *274*, 11110–11114. [CrossRef] [PubMed]

20. Levy, C.W.; Roujeinikova, A.; Sedelnikova, S.; Baker, P.J.; Stuitje, A.R.; Slabas, A.R.; Rice, D.W.; Rafferty, J.B. Molecular basis of triclosan activity. *Nature* **1999**, *398*, 383–384. [CrossRef] [PubMed]

21. Escalada, M.G.; Harwood, J.L.; Maillard, J.-Y.; Ochs, D. Triclosan inhibition of fatty acid synthesis and its effect on growth of Escherichia coli and *Pseudomonas aeruginosa*. *J. Antimicrob. Chemother.* **2005**, *55*, 879–882. [CrossRef] [PubMed]

22. Giuliana, G.; Pizzo, G.; Milici, M.E.; Musotto, G.C.; Giangreco, R. In Vitro Antifungal Properties of Mouthrinses Containing Antimicrobial Agents. *J. Periodontol.* **1997**, *68*, 729–733. [CrossRef]

23. *VICRYL Plus—389595.R04*; Ethicon Inc.: Raritan, NJ, USA, 2002.

24. *PDS Plus—389688.R02*; Ethicon Inc.: Raritan, NJ, USA, 2006.

25. *MONOCRYL Plus—389680.R02*; Ethicon Inc.: Raritan, NJ, USA, 2005.

26. Daoud, F.C.; Goncalves, R.; Moore, N. How Long Do Implanted Triclosan Sutures Inhibit *Staphylococcus aureus* in Surgical Conditions? A Pharmacological Model. *Pharmaceutics* **2022**, *14*, 539. [CrossRef]

27. Rothenburger, S.; Spangler, D.; Bhende, S.; Burkley, D. In Vitro Antimicrobial Evaluation of Coated VICRYL* Plus Antibacterial Suture (Coated Polyglactin 910 with Triclosan) using Zone of Inhibition Assays. *Surg. Infect.* **2002**, *3* (Suppl. 1), s79–s87. [CrossRef]

28. Ming, X.; Rothenburger, S.; Nichols, M.M. In Vivo and In Vitro Antibacterial Efficacy of PDS Plus (Polidioxanone with Triclosan) Suture. *Surg. Infect.* **2008**, *9*, 451–457. [CrossRef]

29. Ming, X.; Rothenburger, S.; Yang, D. In Vitro Antibacterial Efficacy of MONOCRYL Plus Antibacterial Suture (Poliglecaprone 25 with Triclosan). *Surg. Infect.* **2007**, *8*, 201–208. [CrossRef]

30. Storch, M.L.; Rothenburger, S.J.; Jacinto, G. Experimental Efficacy Study of Coated VICRYL plus Antibacterial Suture in Guinea Pigs Challenged with *Staphylococcus aureus*. *Surg. Infect.* **2004**, *5*, 281–288. [CrossRef] [PubMed]

31. Ahmed, I.; Boulton, A.J.; Rizvi, S.; Carlos, W.; Dickenson, E.; A Smith, N.; Reed, M. The use of triclosan-coated sutures to prevent surgical site infections: A systematic review and meta-analysis of the literature. *BMJ Open* **2019**, *9*, e029727. [CrossRef]

32. WHO. *World Health Organization Global Guidelines for the Prevention of Surgical Site Infection*, 2nd ed.; Licence: CC BY-NC-SA 3.0 IGO. 2018; WHO: Geneva, Switzerland, 2018. Available online: https://apps.who.int\T1\guilsinglrighthandle\T1\guilsinglright978 9241550475-eng.pdf (accessed on 15 October 2021).

33. Horan, T.C.; Gaynes, R.P.; Martone, W.J.; Jarvis, W.R.; Emori, T.G. CDC definitions of nosocomial surgical site infections, 1992: A modification of CDC definitions of surgical wound infections. *Am. J. Infect. Control.* **1992**, *20*, 271–274. [CrossRef]

34. Clinical and Laboratory Standards Institute. M26-A Methods for Determining Bactericidal Activity of Antimicrobial Agents; Approved Guideline. September 1999. Available online: https://clsi.org/standards/products/microbiology/documents/m26/ (accessed on 15 October 2021).

35. Halaby, C.N. Panel Models in Sociological Research: Theory into Practice. *Annu. Rev. Sociol.* **2004**, *30*, 507–544. [CrossRef]

36. Smith, B.T. *Solubility and Dissolution. Remington Education: Physical Pharmacy*; Pharmaceutical Press: London, UK, 2016; pp. 31–50. Available online: https://www.pharmpress.com/product/9780857112521/remington-education-physical-pharmacy-ebook (accessed on 1 August 2020).

37. Mircioiu, C.; Voicu, V.; Anuta, V.; Tudose, A.; Celia, C.; Paolino, D.; Fresta, M.; Sandulovici, R.; Mircioiu, I. Mathematical Modeling of Release Kinetics from Supramolecular Drug Delivery Systems. *Pharmaceutics* **2019**, *11*, 140. [CrossRef] [PubMed]

38. *The Merck Index—An Encyclopedia of Chemicals, Drugs, and Biologicals*, 15th ed.; O'Neil, M.J.; Maryadele, J. (Eds.) The Royal Society of Chemistry: Cambridge, UK, 2013; 1789p.

39. Yalkowsky, S.H.; Parijat, H.Y.; Triclosan, J. *Handbook of Aqueous Solubility Data*; CRC Press: Boca Raton, FL, USA, 2010; p. 844. Available online: https://pubchem.ncbi.nlm.nih.gov/compound/Triclosan#section=Melting-Point (accessed on 1 August 2020).

40. Ciusa, M.L.; Furi, L.; Knight, D.; Decorosi, F.; Fondi, M.; Raggi, C.; Coelho, J.R.; Aragones, L.; Moce, L.; Visa, P.; et al. A novel resistance mechanism to triclosan that suggests horizontal gene transfer and demonstrates a potential selective pressure for reduced biocide susceptibility in clinical strains of *Staphylococcus aureus*. *Int. J. Antimicrob. Agents* **2012**, *40*, 210–220. [CrossRef] [PubMed]

41. Morrissey, I.; Oggioni, M.R.; Knight, D.R.; Curiao, T.; Coque, T.; Kalkanci, A.; Martinez, J.L. The BIOHYPO Consortium Evaluation of Epidemiological Cut-Off Values Indicates that Biocide Resistant Subpopulations Are Uncommon in Natural Isolates of Clinically-Relevant Microorganisms. *PLoS ONE* **2014**, *9*, e86669. [CrossRef] [PubMed]
42. McCagherty, J.; Yool, D.A.; Paterson, G.K.; Mitchell, S.R.; Woods, S.; Marques, A.I.; Hall, J.L.; Mosley, J.R.; Nuttall, T.J. Investigation of the in vitro antimicrobial activity of triclosan-coated suture material on bacteria commonly isolated from wounds in dogs. *Am. J. Veter Res.* **2020**, *81*, 84–90. [CrossRef] [PubMed]

Article

Analysis of a Library of *Escherichia coli* Transporter Knockout Strains to Identify Transport Pathways of Antibiotics

Lachlan Jake Munro [1] and Douglas B. Kell [1,2,*]

[1] Novo Nordisk Foundation Center for Biosustainability, Technical University of Denmark, 2800 Lyngby, Denmark
[2] Institute of Systems, Molecular and Integrative Biology, University of Liverpool, Liverpool L69 3BX, UK
*   Correspondence: doukel@biosustain.dtu.dk or douglas.kell@liverpool.ac.uk

Abstract: Antibiotic resistance is a major global healthcare issue. Antibiotic compounds cross the bacterial cell membrane via membrane transporters, and a major mechanism of antibiotic resistance is through modification of the membrane transporters to increase the efflux or reduce the influx of antibiotics. Targeting these transporters is a potential avenue to combat antibiotic resistance. In this study, we used an automated screening pipeline to evaluate the growth of a library of 447 *Escherichia coli* transporter knockout strains exposed to sub-inhibitory concentrations of 18 diverse antimicrobials. We found numerous knockout strains that showed more resistant or sensitive phenotypes to specific antimicrobials, suggestive of transport pathways. We highlight several specific drug-transporter interactions that we identified and provide the full dataset, which will be a useful resource in further research on antimicrobial transport pathways. Overall, we determined that transporters are involved in modulating the efficacy of almost all the antimicrobial compounds tested and can, thus, play a major role in the development of antimicrobial resistance.

Keywords: transporters; antibiotics; *Escherichia coli*

check for **updates**

Citation: Munro, L.J.; Kell, D.B. Analysis of a Library of *Escherichia coli* Transporter Knockout Strains to Identify Transport Pathways of Antibiotics. *Antibiotics* **2022**, *11*, 1129. https://doi.org/10.3390/antibiotics11081129

Academic Editor: Dóra Kovács

Received: 12 July 2022
Accepted: 15 August 2022
Published: 19 August 2022

Publisher's Note: MDPI stays neutral with regard to jurisdictional claims in published maps and institutional affiliations.

1. Introduction

Antibiotic resistance is a major global healthcare burden, with over 1 million deaths attributable to antibiotic resistance in 2019, and with the World Health Organization listing antimicrobial resistance as one of the top 10 global public health threats facing humanity [1,2]. Despite a pressing need for novel antibiotics that can overcome resistance, few novel drugs have been developed and approved in the last 20 years, with a general downward trend in new approvals since 1983 [3], albeit the last decade has seen some increase in new approvals [4]. While systems have been proposed to overcome the economic obstacles to incentivize antibiotic development [5], other solutions are needed.

In order to exert their effects, most antibiotics must first cross the bacterial cell membrane. There is now strong evidence to suggest that in order to enter cells, nearly all compounds must pass through membrane transporters and that "passive bilayer diffusion is negligible" [6–8]. The importance of transporters is further highlighted by the well-established fact that a major mechanism of antibiotic resistance is through adaptations in transport systems. These adaptations reduce the intracellular concentration of antibiotics through either the increased efflux or the decreased influx of antibiotics through membrane transporters [9–11]. Antibiotic adjuncts that inhibit the efflux of the active drug have been proposed as a means to counteract resistance [12,13].

While modelling suggests that mutations that cause antibiotic resistance will predominantly be in exporters rather than importers [14], there are also examples where adaptations that reduce the import of antibiotics confer resistance. This has been observed in the case of fosfomycin [15], aminoglycosides [16] and chloramphenicol [17].

Of the ~4401 genes in the *Escherichia coli* K-12 chromosome, an estimated 598 encode established or predicted membrane transporters [18]. Despite extensive research, around a quarter of these transporters are orphans (also known as y-genes), in that they have no assigned substrate. Recent studies have illustrated that accumulation of cationic fluorophores involves multiple influx and efflux transporters, with a library of transporter knockouts spanning a 30–70-fold difference in fluorescence level when exposed to the fluorophores SYBR Green or diS-C3(5) [19]. We hypothesized that accumulation and excretion of antimicrobial agents may similarly involve numerous membrane transporters. Previously, we have used a high-throughput screening method to identify a melatonin exporter [20]. In this, growth of a library of knockout transporters was measured in the presence of melatonin, and inhibited growth was used as a proxy to identify when an exporter had been knocked out (conversely, a resistant strain potentially has an importer knocked out).

In the present study, which forms part of an ongoing project to deorphanize all orphan *E. coli* transporters, we investigated the previously developed library of 447 *E. coli* transporter knockouts [19,20] for growth, in the presence of 18 structurally diverse antimicrobial agents. We included novel and recently approved antibiotics, such as cefiderocol and flumequine, as well as drugs not traditionally used against Gram-negative bacteria. We also developed an automated workflow for the liquid-handing steps and data processing. The full dataset encompasses approximately 8000 growth curves (in duplicate), and the entire raw dataset and code used for processing are provided in the Supplementary Materials. We identified several transporter knockout strains that showed resistance or sensitivity to many of the antimicrobials tested, including previously unannotated transporters or y-genes. Finally, we illustrate the potential utility of large-scale analysis of the dataset for predicting substrates of orphan transporters.

2. Results

2.1. Antimicrobial Selection

We selected a range of compounds that were available and readily soluble in our LB media. We endeavored to include compounds from several major antibiotic classes as well as compounds with activity that have previously not been tested extensively against Gram-negative bacteria (ornidazole and paraquat). Cefiderocol was very recently approved, so was included due to novelty.

2.2. Determination of Antibiotic Concentrations for High-Throughput Screening

The workflow for the screening of each compound (Supplementary Figure S1) initially required identifying a sub-inhibitory concentration of the antibiotic in the wild-type strain BW1556 (WT, parent strain for the Keio collection). Minimum inhibitory concentrations (MICs) were determined by measurement of *E. coli* growth in a two-fold serial dilution of the relevant antibiotic in LB. We defined MIC values as those with OD levels at 48 h less than 10% of the antibiotic-free media condition. For screening of the transporter library, we selected concentrations that showed some inhibition of growth in the WT strain without causing full inhibition. Generally, this was a concentration half that of the MIC; however, when effects on growth (reduction in maximum OD, increased lag time, decreased growth rate, or some combination thereof) could be clearly seen at concentrations significantly lower than that, then these concentrations were used. The 18 antimicrobial compounds included in this study, as well as the MIC and screening concentrations, are listed in Table 1, and chemical structures of the antibiotics used are shown in Supplementary Figure S2.

2.3. Baseline Characteristics of Transporter Library

Generally, the growth of nearly all transporter knockout strains in LB without supplementation was similar. The empirical area under the curve (AUC) (a general measure of growth) had a mean of 30.0 and an interquartile range (IQR) of 1.90, while the maximum growth rate had a mean of 1.19 and an IQR of 0.14 h^{-1}. The WT growth rate was slightly higher than the mean growth rate, possibly due to the burden of expression of the constitutively active kanamycin cassette present in all knockout strains [21,22]. Histograms illustrating the distribution of the AUC and growth rate are shown in Figure 1A,B. We also observed good correlation between replicates with regards to growth rate and AUC (Figure 1C,D). It should be noted that the automated inoculation step occasionally did not dispense a droplet into the well. In cases where this was clearly the case (i.e., where we observed growth in one but not another replicate), the replicate missing growth was filtered from the results.

Table 1. Antibiotics and antimicrobials used in this study, with minimum inhibitory concentration (MIC) in the wild-type (WT) strain, and the concentrations used for high-throughput screening experiments are shown.

Antibiotic	WT MIC (mg/L)	Screening Concentration (mg/L)
Azithromycin	30	7.5
Cefiderocol	1	0.25
Cefoperazone	0.5	0.25
Chloramphenicol	2.5	1.25
Ceftriaxone	2.5	1
D-Cycloserine	12.5	6.25
Flumequine	>1.5 *	1.5
Gentamycin	4	2
Levofloxacin	0.25	0.125
Meropenem	0.0625	0.03125
Ofloxaxin	0.125	0.0625
Ornidazole	800	400
Paraquat	>268 *	67
Phosphomycin	50	25
Rifampicin	25	12.5
Streptomycin	25	6.25
Trimethoprim	2.3	0.25
Zidovudine	>100 *	100

* For these compounds, growth was detected at the highest concentration tested, despite clear inhibition at lower concentrations.

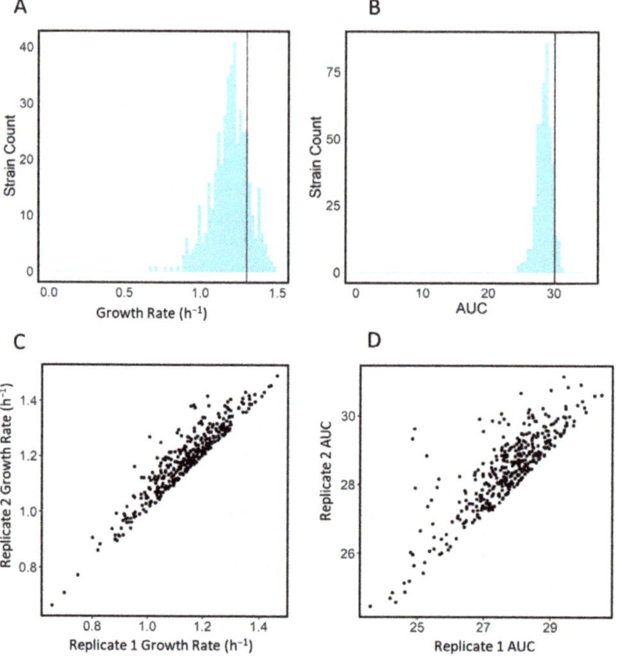

Figure 1. Histograms illustrating distributions of (**A**) growth rate and (**B**) area under the curve (AUC) for transporter library growth in LB in the absence of antibiotics. Scatter plots showing correlation between replicates for (**C**) growth rates and (**D**) AUC for transporter library growth in LB alone.

Compared to growth in LB, we saw a much larger degree of variation in growth in the presence of the antimicrobial compounds, consistent with multiple transporter knockouts influencing the level of intracellular antibiotic accumulation. Figure 2A shows rank-order curves for the mean AUC for growth of the transporter library in LB, while Figure 2B–D shows the same information for growth in the indicated antibiotics. In the presence of antibiotics, the rank-order curves show a much wider spread of values. Figure 3 shows a box and whisker plot illustrating the distribution of AUC values for all compounds tested.

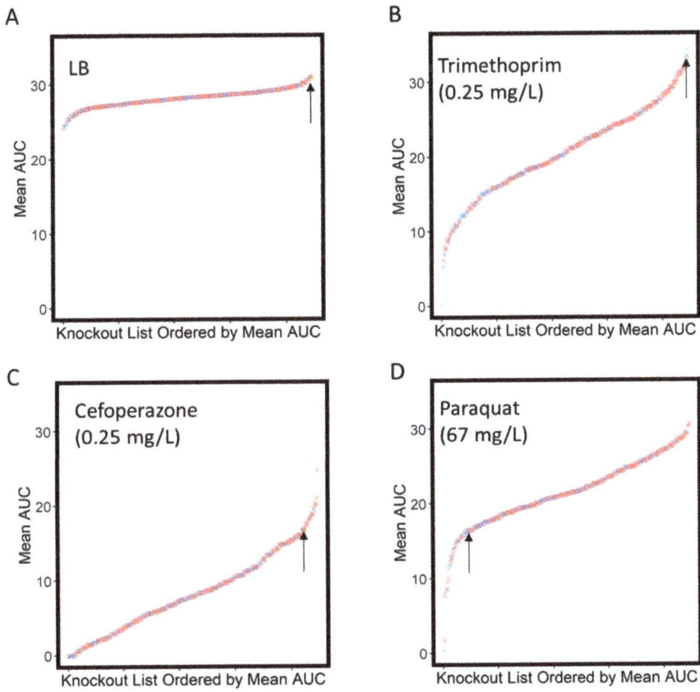

Figure 2. Transporter knockout library ordered by the mean area under the curve (AUC). Data shown for growth in LB (**A**) and subinhibitory concentrations of antibiotics/antimicrobials indicated (**B–D**). Y-genes are shown in blue, and annotated transporters are shown in red, while the arrow indicates the position of the WT strain.

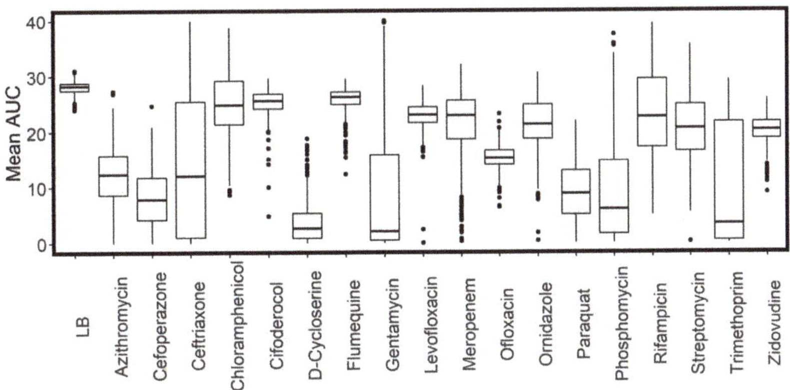

Figure 3. Box and whisker plot showing distribution of mean area under the curve (AUC) for growth of the transporter library in LB and antimicrobial compounds.

2.4. Resistant and Sensitive Transporter Results

We observed that many transporter knockouts had effects on the sensitivity to the compounds tested. Due to the large volume of data generated, detailed descriptions of all relevant results are beyond the scope of this paper. As we intend to perform targeted follow-up validation experiments, given that this is an initial exploratory study, we are hesitant to define specific criteria for sensitivity or resistance. Indeed, one advantage of our study over previous high-throughput growth assays is the generation of full growth curves rather than the reduction in growth of a single metric, as we elaborate on in the discussion.

Selected specific results are discussed below. Generally, when searching for novel results, we looked at strains ranked highest or lowest in normalized growth rate or AUC and then manually inspected the growth curves. We selected those highlighted in this paper based on interactions with y-genes or where we felt there were plausible and interesting mechanisms for discussion.

Our study replicated some previously established drug–transporter interactions. First, we observed that knockout of acrB, a promiscuous drug-efflux protein well-known to play a role in export of many xenobiotics [23], caused increased sensitivity to 10 out of the 18 antimicrobial compounds tested. Interestingly we saw the increased inhibition caused by knockout of acrB take different forms: growth rate decreases including increased lag time, decreased max OD and AUC and complete inhibition (Figure 4). Fosphomycin has been shown to gain access to the cell via the hexose-6-phosphate:phosphate antiporter uhpT and the glutamate/aspartate: H+ symporter gltP [15,24,25]. In our phosphomycin experiment, knockouts of each of these transporters were among the most resistant of the transporter knockout strains tested, consistent with these knockouts causing reduction in influx (Supplementary Materials). MacAB-Tolc has been reported to efflux macrolide antibiotics, and expression of this complex in a strain hypersensitive to antibiotics conferred resistance to macrolides including azithromycin [26]. In our study, the macB knockout strain (Δ*macB*) did not result in azithromycin sensitivity, and we also could not see evidence of increased sensitivity to azithromycin in a previously published dataset [27].

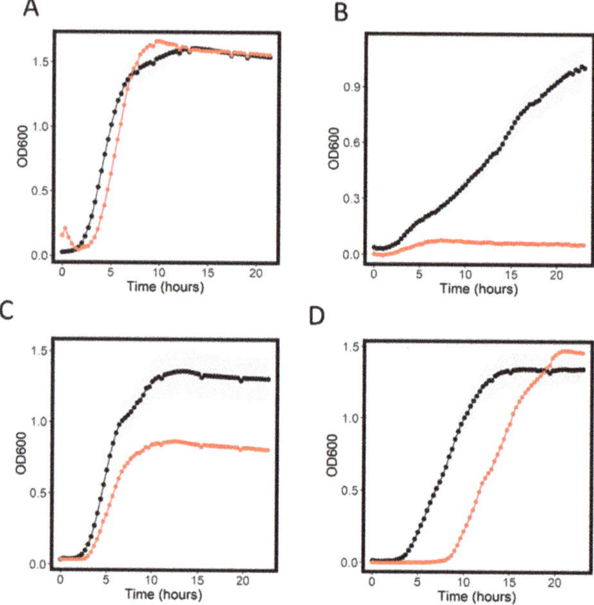

Figure 4. Growth curves for WT (black) and Δ*acrB* (red) in LB (**A**), azithromycin, (**B**) D-cycloserine (**C**) and chloramphenicol (**D**). Shaded areas represent the standard deviation for the WT growth curve.

Ornidazole is a nitroimidazole generally used to treat protozoan infections, and, in this study, we found it to have an MIC in *E. coli* BW1556 of 800 mg/L (Table 1). The transporter nimT (previously yeaN) has been shown to function as an exporter of 2-nitroimidazole, a compound structurally similar to ornidazole [28]. We saw only a modest increase in ornidazole sensitivity in Δ*nimT* compared to WT (Figure 5A). We did, however, observe a clear increased sensitivity to ornidazole in Δ*argO* (Figure 5B), a strain with a probable arginine exporter knocked out [29]. Some strains also showed an ornidazole-resistant phenotype. These included Δ*narU*, a nitrite/nitrate exchanger knockout (Figure 5C). Possibly, this transporter is interacting in some way with the nitro-group on ornidazole, and the knockout of this reduces influx into the cell. We also saw a resistant phenotype in Δ*ygaH* (Figure 5D), an uncharacterized orphan transporter. The other most resistant and most sensitive strains (by normalized AUC) are shown in Table 2.

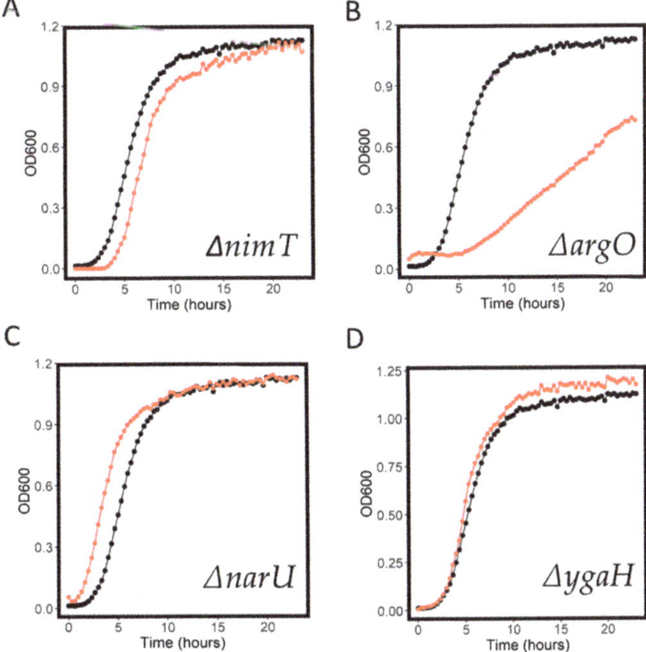

Figure 5. Growth curves for WT (black) and knockout strains (red) in 400 mg/L ornidazole. Δ*nimT* (**A**), Δ*argO* (**B**), Δ*narU* (**C**) and Δ*ygaH* (**D**) are shown. Shaded areas represent the standard deviation for the WT growth curve.

Azithromycin is a macrolide antibiotic, widely used in the treatment of clinical infections. We saw several transporter knockout strains that had dramatically decreased growth in the presence of 7.5 mg/L azithromycin compared to WT. These included the choline transporter knockout Δ*betT* and a tyrosine symporter knockout Δ*tyrP* (Figure 6A,B). Near-complete inhibition was also seen in the multidrug exporter knockout strain Δ*mdtN*. We also observed resistant phenotypes, specifically in the orphan knockout strains Δ*yhdW* and Δ*ydfJ* (Figure 6C,D). Growth parameters for the most resistant and sensitive strains are shown in Table 3.

Table 2. Growth parameters for the 15 most sensitive and most resistant strains in Ornidazole by normalized AUC. Normalized values show the value in ornidazole/value in LB for the given strain.

Strain	Max OD	Rate	AUC	Normalized Max OD	Normalized Rate	Normalized AUC
argO	0.74	0.35	6.36	0.29	0.45	0.23
ydjE	0.58	0.71	6.55	0.71	0.37	0.26
tolQ	0.60	1.05	7.98	1.16	0.38	0.29
glvB	0.75	0.57	9.78	0.54	0.44	0.34
sapB	0.64	0.73	9.05	0.74	0.41	0.35
yebQ	0.83	0.59	8.96	0.89	0.53	0.36
ddpF	0.76	0.80	11.43	0.64	0.44	0.37
uup	0.72	0.85	10.07	0.81	0.45	0.37
cysW	0.64	1.62	10.22	1.40	0.41	0.38
ydcS	0.65	0.54	9.25	0.70	0.39	0.38
ydeE	0.78	0.89	11.00	0.74	0.48	0.39
yicJ	0.81	0.84	11.78	0.64	0.49	0.39
guaB	0.69	1.08	11.22	1.18	0.42	0.40
gntU	0.87	1.04	12.13	0.83	0.52	0.40
ynfA	0.77	0.93	9.97	1.31	0.52	0.41
ybhF	1.15	0.95	19.30	0.96	0.66	0.65
arnE	1.14	1.03	17.97	1.14	0.70	0.66
sapC	1.11	0.74	17.12	0.77	0.67	0.66
cycA	1.17	1.07	19.65	0.83	0.69	0.66
narU	1.14	0.86	19.88	0.67	0.67	0.66
lacY	1.19	0.87	20.61	0.71	0.68	0.67
kup	1.16	0.90	18.67	0.68	0.70	0.67
mdtH	1.24	0.80	18.59	0.89	0.75	0.67
pitA	1.25	1.01	18.19	0.78	0.77	0.67
yadG	1.09	0.84	17.65	0.66	0.66	0.67
clcB	1.14	0.86	18.96	0.67	0.69	0.68
trkH	1.20	0.76	18.90	0.56	0.74	0.68
exbD	1.22	0.92	20.52	0.79	0.70	0.70
wt	1.04	1.10	21.76	0.77	0.58	0.70
dhaM	1.34	0.79	23.18	0.63	0.80	0.78

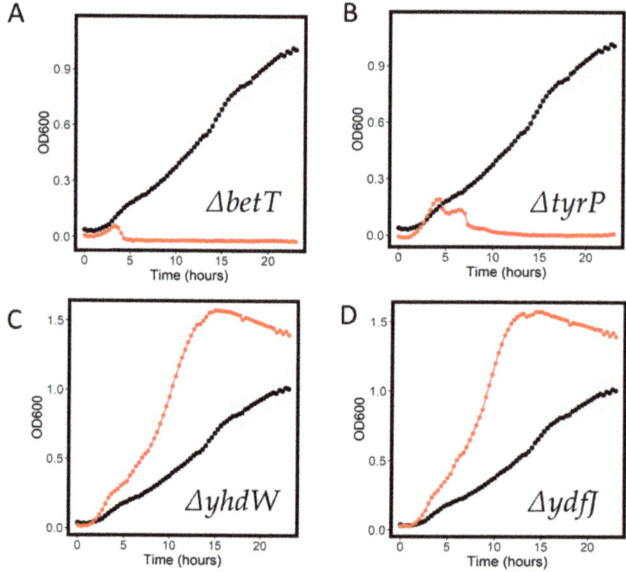

Figure 6. Growth curves for WT (black) and knockout strains (red) in 400 mg/L azithromycin. Δ*betT* (**A**), Δ*tyrP*(**B**), Δ*yhdW* (**C**) and Δ*ydfJ* (**D**) are shown. Shaded areas represent the standard deviation for the WT growth curve.

Table 3. Growth parameters for the 15 most sensitive and most resistant strains in ornidazole. Normalized values represent the value in ornidazole/value in LB for the given strain and values for the maximum OD, rate and area under the curve are shown.

Strain	Max OD	Rate	AUC	Normalized Max tOD	Normalized Rate	Normalized AUC
betT	0.06	0.02	0.30	0.02	0.04	0.01
glvB	0.11	0.03	0.32	0.03	0.07	0.01
glnP	0.11	0.02	0.33	0.02	0.06	0.01
citT	0.06	0.04	0.37	0.04	0.04	0.01
tyrP	0.24	0.01	0.56	0.01	0.15	0.02
corA	0.14	0.03	0.62	0.02	0.08	0.02
tolR	0.11	1.69	0.67	1.46	0.07	0.02
mdtN	0.67	0.50	0.70	0.40	0.40	0.02
ydjN	0.34	0.44	0.76	0.38	0.20	0.03
yibH	0.20	0.01	0.77	0.00	0.12	0.03
torT	0.18	0.04	0.80	0.03	0.11	0.03
acrB	0.09	0.66	1.35	0.54	0.05	0.05
kdgT	2.16	0.92	1.48	0.75	1.33	0.05
tolQ	0.74	1.04	1.76	1.16	0.47	0.06
ydcS	0.31	0.28	1.61	0.36	0.18	0.07
ddpD	1.51	0.39	19.71	0.30	0.92	0.71
ydiN	1.53	0.44	20.19	0.36	0.96	0.71
ydjX	1.58	0.43	20.37	0.35	0.96	0.72
yddB	1.59	0.40	19.66	0.28	1.00	0.72
atpB	1.57	0.44	21.09	0.35	0.96	0.73
yccA	1.54	0.43	19.30	0.37	0.99	0.73
ygaZ	1.50	0.58	18.50	0.56	1.09	0.73
cmtA	1.54	0.41	20.84	0.36	0.95	0.74
kgtP	1.52	0.47	21.29	0.38	0.88	0.74
pitA	1.51	0.46	20.38	0.36	0.93	0.75
ydfJ	1.58	0.55	23.16	0.41	0.93	0.76
yhdW	1.58	0.48	22.07	0.34	0.95	0.76
yraQ	2.07	0.81	24.46	0.75	1.23	0.87
yiaM	2.51	0.51	26.88	0.42	1.50	0.90
argO	1.92	0.94	27.30	0.78	1.17	0.97

We also included compounds not traditionally used as antibiotics, which have antimicrobial activity. Paraquat (also known as methyl viologen) is a widely used herbicide, which exerts toxic effects, after conversion to a superoxide radical, once inside the cell [30]. We found that concentrations up to 257 mg/L did not cause full inhibition of *E. coli* growth; however, there was evidence of growth inhibition at concentrations of 32 mg/L (Figure 7A). We screened the growth of the transporter knockout library at 67 mg/L and observed several strains that had increased sensitivity including the y-gene knockout Δ*ydcZ* and in two metal cation exporter knockouts Δ*cusA* (Figure 7B) and Δ*fieF*, suggestive of the ability of these transporters to export paraquat, which notably is also a cation. We also observed strains with resistance; interestingly, these included Δ*aroP*, an aromatic amino acid permease knockout. Given the aromatic structure of paraquat, it is very plausible that one of the means of its entry into the cell is via this aromatic amino acid permease (Figure 7C). The most sensitive and resistant strains in paraquat are shown in Table 4.

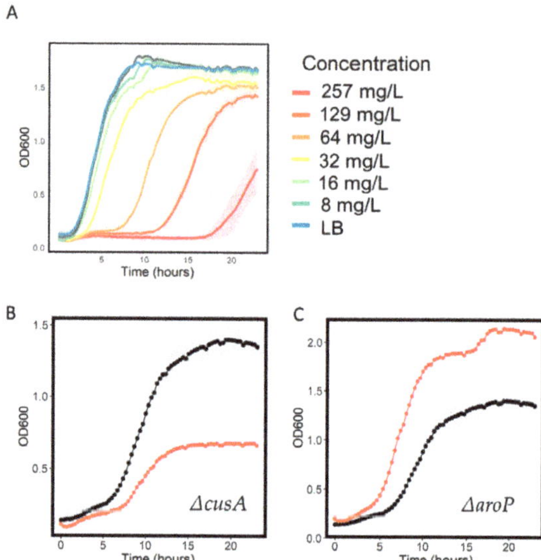

Figure 7. Growth features of paraquat. (**A**) Concentration inhibition data for growth of WT *E. coli* in the indicated concentration of paraquat. Shaded areas represent the standard deviation, n = 3. Growth curves for WT (black) and knockout strains (red) in 32 mg/L paraquat. Δ*cusB* (**B**) and Δ*aroP* (**C**) are shown.

Table 4. Growth parameters for the 15 most sensitive and most resistant strains in paraquat. Normalized values represent the value in paraquat/value in LB for the given strain and values for the maximum OD, rate and area under the curve are shown.

Strain	Max OD	Rate	AUC	Normalized Max OD	Normalized Rate	Normalized AUC
ydcZ	0.11	1.55	0.39	1.35	0.06	0.01
potH	0.74	0.57	1.79	0.50	0.44	0.06
fieF	1.46	1.20	7.66	0.92	0.85	0.26
yihO	1.35	1.07	8.11	0.88	0.79	0.29
kdgT	2.22	0.49	8.51	0.40	1.36	0.29
cusA	0.69	0.49	8.66	0.50	0.41	0.29
yfdV	0.78	0.60	8.42	0.64	0.47	0.30
uhpC	2.15	0.55	9.50	0.48	1.27	0.32
uacT	2.40	0.50	10.09	0.40	1.42	0.33
atoS	1.47	0.26	11.42	0.25	0.84	0.38
yihN	1.11	0.64	11.91	0.55	0.65	0.41
ygjI	0.97	0.53	11.79	0.41	0.58	0.42
ygaH	1.41	1.01	13.21	0.85	0.82	0.42
mdtN	0.96	0.46	12.28	0.37	0.57	0.43
yidK	1.03	1.04	12.84	0.75	0.60	0.43
aroP	2.15	0.60	28.71	0.51	1.30	1.03
gsiC	2.21	0.68	28.48	0.64	1.37	1.03
yaaU	2.15	0.55	28.40	0.47	1.33	1.03
copA	2.15	0.57	28.30	0.50	1.30	1.03
araJ	2.16	0.64	28.94	0.52	1.32	1.03
dinF	2.13	0.60	29.45	0.47	1.27	1.04
codB	2.14	0.58	29.43	0.47	1.27	1.04
sbp	2.01	0.67	27.00	0.53	1.27	1.04
amtB	2.29	0.60	29.30	0.53	1.41	1.05
kdpA	2.34	0.60	29.41	0.50	1.42	1.05
yfeO	2.01	0.64	28.17	0.49	1.22	1.06
uidB	2.25	0.55	30.25	0.44	1.36	1.06
caiT	2.34	0.57	30.46	0.45	1.41	1.06
acrB	2.15	0.52	29.25	0.43	1.28	1.07
focA	2.26	0.61	30.58	0.52	1.37	1.09

2.5. Large-Scale Data Analysis

The generation of a dataset such as this one also allows the large-scale investigation of relationships between transporters. The heat map of correlations between AUC values for strains across growth conditions is shown in Figure 8A. We also extracted the p values for the correlation and found a peak near zero (Figure 8B), indicating likely true positives in the significant results. We found 35 correlations that met the conditions for significance at $p < 0.05$ after full Bonferroni correction [31] (with a corrected p value of 2.5×10^{-7}). These are shown in Table 2. One possible application of this type of large-scale data analysis is in predicting substrates of unannotated transporters by their relationship to transporters with annotations (a "guilt by association" methodology) (Table 5). For instance, the growth parameters of $\Delta sapB$, a putrescine exporter knockout, have a high correlation with the knockout of orphan transporter knockout $\Delta ydjE$ (Figure 8C), suggesting transport of compounds similar to putrescine or at least some overlap in substrate selectivity and function.

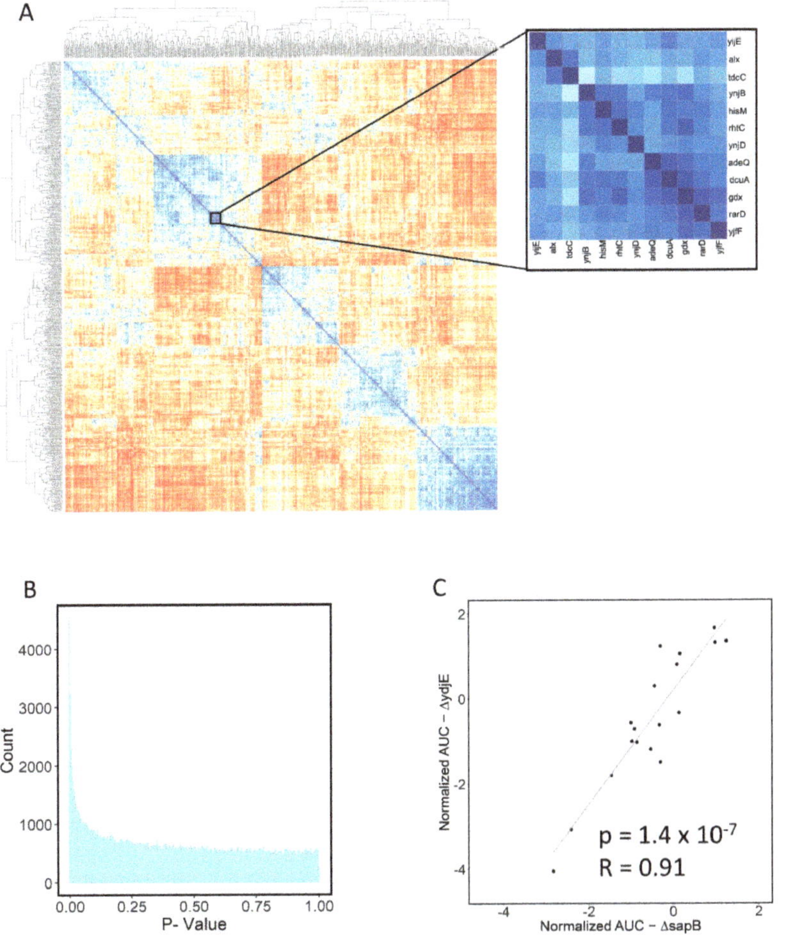

Figure 8. (**A**). Heatmap showing correlations between area under the curve values for transporters in all antibiotics. Zoom inset highlights a cluster of highly correlated transporter knockout strains. (**B**) Histogram of p-values for all correlations showing the peak around 0, indicating likely true positive results. (**C**) Scatter plot indicating high degree of correlation between $\Delta sapB$ and $\Delta ydjE$.

Table 5. Transporters that are highly correlated.

Transporter 1	Transporter 2	Correlation	p Value
yihP	yhdX	0.91	2.46×10^{-7}
ydjK	mdtB	0.91	2.36×10^{-7}
uup	gsiC	0.91	2.36×10^{-7}
yicL	yhdX	0.91	2.26×10^{-7}
btuC	araJ	0.91	2.22×10^{-7}
mdtG	copA	0.91	2.04×10^{-7}
rarD	adeQ	0.91	2.02×10^{-7}
mscK	ccmC	0.91	1.98×10^{-7}
cynX	btuC	0.91	1.72×10^{-7}
ybbA	cynX	0.91	1.65×10^{-7}
mdfA	chaA	0.91	1.49×10^{-7}
yjfF	gdx	0.91	1.24×10^{-7}
ynjB	gdx	0.91	1.18×10^{-7}
nhaB	ddpB	0.91	1.16×10^{-7}
ydjE	sapB	0.91	1.14×10^{-7}
yphD	gudP	0.91	1.07×10^{-7}
ybaT	btuC	0.92	9.3×10^{-8}
yjfF	adeQ	0.92	8.62×10^{-8}
ypjA	fryA	0.92	5.86×10^{-8}
rarD	gdx	0.92	5.63×10^{-8}
malX	amtB	0.92	5×10^{-8}
araJ	acrB	0.92	4.29×10^{-8}
gsiC	fetA	0.93	3.67×10^{-8}
yaaU	sapA	0.93	3.54×10^{-8}
uidB	mdtG	0.93	3.53×10^{-8}
yphD	ascF	0.93	2.77×10^{-8}
yegT	frwC	0.93	2.61×10^{-8}
ytfT	kup	0.93	1.47×10^{-8}
copA	ccmC	0.93	1.46×10^{-8}
yddB	ddpD	0.94	7.5×10^{-9}
ybbA	mscK	0.95	2.46×10^{-9}
mscK	fetA	0.95	8.93×10^{-10}
cynX	araJ	0.96	3.92×10^{-10}
gdx	adeQ	0.97	5.67×10^{-11}
yihP	yicL	0.97	7.51×10^{-12}

3. Discussion

Identifying substrates and describing the structure activity relationship of bacterial transporters is a challenging task. Given the importance of transport systems in the function of most antibiotics, understanding the influx and efflux pathways of antimicrobial agents is highly relevant to mitigating problem of drug resistant bacteria. In this study, we have sought to gain insights into the potential influx and efflux pathways of a set of compounds with antimicrobial activity in *E. coli*. We have used an automated method to generate a dataset of growth curves for 447 transporter knockout strains against subinhibitory concentrations of 18 structurally diverse compounds with antibacterial activity. Our data showed results consistent with some previously reported pathways for antibiotic influx and efflux. We also report on numerous novel specific observations of transporter knockouts that reduce or enhance growth in the presence of antimicrobials as well as discuss potentially novel observations from analysis of the full dataset. All data are provided as a resource in the Supplementary Materials.

There were numerous novel transporter–compound interactions identified in this study, which suggest the need for follow-up experiments. We have previously developed a workflow for confirmatory work on the results that involves: (1) PCR to confirm the strain is correctly labelled, and (2) MIC determination to validate the observed result [20]. Work to automate these processes is ongoing and will be detailed in future publications. Once a specific transporter–compound interaction has been confirmed, the task of finding other, potentially higher affinity substrates for the transporter is feasible, as often the identification of an initial substrate is the most difficult step. Tools that can identify compounds "closest" to the antimicrobial in chemical space [32] will allow detailed investigation of the structure activity relationship of transporters.

A limitation to the approach described in this study for transporter pathway identification is that growth is used as a proxy for transport with no direct measurement. While the simplest mechanism by which a transporter knockout changes growth in the presence of an antimicrobial is through the direct alteration of transport, there are other possible mechanisms. Gene knockouts are well-known to cause pleiotropic effects, with knockouts of a single gene often causing altered expression in many other genes that may affect transporter sensitivity [33]. Further validation of predicted effects can be achieved by use of the 'ASKA' overexpression collection [34], where, if the opposite effect is observed (e.g., resistance in an overexpression strain and sensitivity in a knockout strain), this could be seen as providing further evidence that the compound is indeed a substrate of the given transporter [19,35,36]. Given the large degree of redundancy in transporters (i.e., a given substrates may be transported by several transporters), more robust follow-up results could be achieved through investigation of double knockouts, as has recently been demonstrated in yeast [37].

There are studies that have investigated the growth of the full Keio collection of *E. coli* knockouts under different conditions (including numerous antibiotics) on solid media [27,38,39]. While the use of solid media and the application of advanced robotics allows a much higher throughput (with all of the above studies run in a 1536-well format), this does come with limitations. This high-density solid-media growth introduces a "neighbor effect", where crowding of the bacterial colonies causes growth inhibition that needs to be corrected for [38]. Additionally, these studies report one metric or at most two metrics for growth, which potentially obscures the relevant information only available from the full growth curve. It should also be noted that while the authors of the above papers all commendably shared their data, in two out of the above three publications, the provided links for accessing data are no longer active [27,38], fitting with an observed trend of data accessibility decreasing in the years following publication [40].

There is a large degree of homology found between transporters of different pathogenic bacterial families. Given this, as well as the recent advances in prediction of protein structure from sequence [41,42], we envision that the dataset generated in this paper will be of use for understanding and predicting the interaction between antibiotics and membrane transporters in other clinically relevant bacterial species.

4. Materials and Methods

Antibiotics were sourced from Sigma, apart from rifampicin, ceftriaxone and azithromycin, which were purchased from Tokyo Chemical Industries (Zwijndrecht, Belgium).

Plates were prepared by first inoculating deep-well plates with 1 mL Lucia Broth (LB) from glycerol stocks of the transporter library. The inoculated LB was grown overnight under agitation at 37 °C. Following overnight growth, the cultures were mixed with 1 mL 50% glycerol. CR1496c polystyrene plates (Enzyscreen, NL, Heemstede, The Netherlands) were prepared for growth assays, by dispensing a 3 µL droplet of the culture and glycerol mixture into the bottom of the well. These loaded plates were then stored at −20 °C for up to 3 weeks or −80 °C for longer storage.

The growth assays were initiated by adding 297 µL LB, containing the appropriate dose of antibiotics to the pre-inoculated plates, and sealing with CR1396b Sandwich covers (Enzyscreen, NL, Heemstede). Growth was assessed using the Growth Profiler 960 (Enzyscreen, NL, Heemstede), which uses camera-based measurements to estimate growth rates simultaneously in up to 10 96-well plates. As our library encompassed 5 plates, this allowed us to run the full library in duplicate. The Growth Profiler 960 was set to 37 °C with 225 rpm shaking (recommended settings for *E. coli*), with pictures taken every 20 min.

Inoculation and media loading were performed using an Opentrons OT2 robot fitted with a 20 µL multichannel pipette and a 300 µL multichannel pipette. Scripts used in operation can be found at github.com/ljm176/TransporterScreening (11 August 2022). Growth plates were sterilized between uses by washing and UV in accordance with the instructions of the manufacturer.

G-values were obtained from the plate images using the manufacturer's software. *G*-values were converted to *OD600* values using the formula:

$$OD_{600} = a * \left(G_{value} - G_{Blank}\right)^b$$

With the predetermined values $a = 0.0158$ and $b = 0.9854$, which were found by measurement of a standard curve in accordance with the instructions of the manufacturer.

Data analysis and generation of figures was performed in R (version 4.1.2). Growth rates were determined by using the R package Growthcurver [43]. Growthcurver fits growth curves to the equation:

$$N_t = \frac{K}{1 + \left(\frac{K - N0}{N0}\right)e^{-rt}}$$

where N_t is total cell population, K is the carrying capacity, initial cell population is $N0$ and r is the intrinsic growth rate. For strains that showed growth in only a single replicate, the replicate without growth was filtered from analysis, as this was determined to be the result of missed inoculation during automated loading.

Figures were generated using ggplot2 for R. The R script used in data analysis and figure generation is available at github.com/ljm176/TransporterScreening (11 August 2022).

Supplementary Materials: The following supporting information can be downloaded at: https://www.mdpi.com/article/10.3390/antibiotics11081129/s1. Figure S1: Workflow for high throughput screening of transporter knockout library against antibiotics. Figure S2: Structures of antibiotics used in this study; Table S1: Growth Parameters.

Author Contributions: Conceptualization L.J.M. and D.B.K.; study design and experimental work L.J.M.; writing, review and editing L.J.M. and D.B.K.; funding acquisition, D.B.K. All authors have read and agreed to the published version of the manuscript.

Funding: This research was funded by the University of Liverpool and Novo Nordisk Foundation (grant NNF20CC0035580).

Institutional Review Board Statement: Not applicable.

Informed Consent Statement: Not applicable.

Data Availability Statement: Datasets are available in the Supplementary Materials for this paper. The code used in analysis is available at github.com/ljm176/TransporterScreening.

Acknowledgments: We thank the University of Liverpool and the Novo Nordisk Foundation (grant NNF20CC0035580) for the financial support.

Conflicts of Interest: The authors declare no conflict of interest. The funders had no role in the design of the study; in the collection, analyses, or interpretation of data; in the writing of the manuscript; or in the decision to publish the results.

References

1. Murray, C.J.; Ikuta, K.S.; Sharara, F.; Swetschinski, L.; Aguilar, G.R.; Gray, A.; Han, C.; Bisignano, C.; Rao, P.; Wool, E.; et al. Global burden of bacterial antimicrobial resistance in 2019: A systematic analysis. *Lancet* **2022**, *399*, 629–655. [CrossRef]
2. WHO. Ten Threats to Global Health in 2019. Available online: https://www.who.int/news-room/spotlight/ten-threats-to-global-health-in-2019 (accessed on 8 July 2022).

3. Spellberg, B.; Powers, J.H.; Brass, E.P.; Miller, L.G.; Edwards, J.J.E. Trends in Antimicrobial Drug Development: Implications for the Future. *Clin. Infect. Dis.* **2004**, *38*, 1279–1286. [CrossRef] [PubMed]

4. Kalra, B.S.; Batta, A.; Khirasaria, R. Trends in FDA drug approvals over last 2 decades: An observational study. *J. Fam. Med. Prim. Care* **2020**, *9*, 105–114. [CrossRef]

5. Glover, R.E.; Manton, J.; Willcocks, S.; Stabler, R.A. Subscription model for antibiotic development. *BMJ* **2019**, *366*, l5364. [CrossRef]

6. Dobson, P.D.; Kell, D. Carrier-mediated cellular uptake of pharmaceutical drugs: An exception or the rule? *Nat. Rev. Drug Discov.* **2008**, *7*, 205–220. [CrossRef]

7. Kell, D.B.; Oliver, S.G. How drugs get into cells: Tested and testable predictions to help discriminate between transporter-mediated uptake and lipoidal bilayer diffusion. *Front. Pharmacol.* **2014**, *5*, 231. [CrossRef]

8. Kell, D.B.; Dobson, P.D.; Bilsland, E.; Oliver, S.G. The promiscuous binding of pharmaceutical drugs and their transporter-mediated uptake into cells: What we (need to) know and how we can do so. *Drug Discov. Today* **2013**, *18*, 218–239. [CrossRef]

9. Li, X.-Z.; Plésiat, P.; Nikaido, H. The Challenge of Efflux-Mediated Antibiotic Resistance in Gram-Negative Bacteria. *Clin. Microbiol. Rev.* **2015**, *28*, 337–418. [CrossRef]

10. Poole, K. Efflux pumps as antimicrobial resistance mechanisms. *Ann. Med.* **2007**, *39*, 162–176. [CrossRef]

11. Willers, C.; Wentzel, J.; du Plessis, L.; Gouws, C.; Hamman, J.H. Efflux as a mechanism of antimicrobial drug resistance in clinical relevant microorganisms: The role of efflux inhibitors. *Expert Opin. Ther. Targets* **2017**, *21*, 23–36. [CrossRef]

12. Nakashima, R.; Sakurai, K.; Yamasaki, S.; Hayashi, K.; Nagata, C.; Hoshino, K.; Onodera, Y.; Nishino, K.; Yamaguchi, A. Structural basis for the inhibition of bacterial multidrug exporters. *Nature* **2013**, *500*, 102–106. [CrossRef] [PubMed]

13. Nishino, K.; Yamasaki, S.; Nakashima, R.; Zwama, M.; Hayashi-Nishino, M. Function and Inhibitory Mechanisms of Multidrug Efflux Pumps. *Front. Microbiol.* **2021**, *12*, 737288. [CrossRef] [PubMed]

14. Mendes, P.; Girardi, E.; Superti-Furga, G.; Kell, D.B. Why Most Transporter Mutations That Cause Antibiotic Resistance Are to Efflux Pumps Rather than to Import Transporters. *bioRxiv* **2020**. [CrossRef]

15. Castañeda-García, A.; Blázquez, J.; Rodríguez-Rojas, A. Molecular Mechanisms and Clinical Impact of Acquired and Intrinsic Fosfomycin Resistance. *Antibiotics* **2013**, *2*, 217–236. [CrossRef] [PubMed]

16. Taber, H.W.; Mueller, J.P.; Miller, P.F.; Arrow, A.S. Bacterial uptake of aminoglycoside antibiotics. *Microbiol. Rev.* **1987**, *51*, 439–457. [CrossRef] [PubMed]

17. Prabhala, B.K.; Aduri, N.G.; Sharma, N.; Shaheen, A.; Sharma, A.; Iqbal, M.; Hansen, P.R.; Brasen, C.; Gajhede, M.; Rahman, M.; et al. The prototypical proton-coupled oligopeptide transporter YdgR from *Escherichia coli* facilitates chloramphenicol uptake into bacterial cells. *J. Biol. Chem.* **2018**, *293*, 1007–1017. [CrossRef]

18. Elbourne, L.D.H.; Tetu, S.G.; Hassan, K.A.; Paulsen, I.T. TransportDB 2.0: A database for exploring membrane transporters in sequenced genomes from all domains of life. *Nucleic Acids Res.* **2017**, *45*, D320–D324. [CrossRef]

19. Jindal, S.; Yang, L.; Day, P.J.; Kell, D.B. Involvement of multiple influx and efflux transporters in the accumulation of cationic fluorescent dyes by *Escherichia coli*. *BMC Microbiol.* **2019**, *19*, 195. [CrossRef]

20. Yang, L.; Malla, S.; Özdemir, E.; Kim, S.H.; Lennen, R.; Christensen, H.B.; Christensen, U.; Munro, L.J.; Herrgård, M.J.; Kell, D.B.; et al. Identification and Engineering of Transporters for Efficient Melatonin Production in *Escherichia coli*. *Front. Microbiol.* **2022**, *13*, 880847. [CrossRef]

21. Yamamoto, N.; Nakahigashi, K.; Nakamichi, T.; Yoshino, M.; Takai, Y.; Touda, Y.; Furubayashi, A.; Kinjyo, S.; Dose, H.; Hasegawa, M.; et al. Update on the Keio collection of *Escherichia coli* single-gene deletion mutants. *Mol. Syst. Biol.* **2009**, *5*, 335. [CrossRef]

22. Baba, T.; Ara, T.; Hasegawa, M.; Takai, Y.; Okumura, Y.; Baba, M.; Datsenko, K.A.; Tomita, M.; Wanner, B.L.; Mori, H. Construction of *Escherichia coli* K-12 in-frame, single-gene knockout mutants: The Keio collection. *Mol. Syst. Biol.* **2006**, *2*, 2006.0008. [CrossRef] [PubMed]

23. Kobylka, J.; Kuth, M.S.; Müller, R.T.; Geertsma, E.R.; Pos, K.M. AcrB: A mean, keen, drug efflux machine. *Ann. N. Y. Acad. Sci.* **2020**, *1459*, 38–68. [CrossRef] [PubMed]

24. Kahan, F.M.; Kahan, J.S.; Cassidy, P.J.; Kropp, H. The mechanism of action of fosfomycin (phosphonomycin). *Ann. N. Y. Acad. Sci.* **1974**, *235*, 364–386. [CrossRef] [PubMed]

25. Silver, L.L. Fosfomycin: Mechanism and Resistance. Cold Spring Harb. *Perspect. Med.* **2017**, *7*, a025262. [CrossRef]

26. Kobayashi, N.; Nishino, K.; Yamaguchi, A. Novel Macrolide-Specific ABC-Type Efflux Transporter in *Escherichia coli*. *J. Bacteriol.* **2001**, *183*, 5639–5644. [CrossRef]

27. Nichols, R.J.; Sen, S.; Choo, Y.J.; Beltrao, P.; Zietek, M.; Chaba, R.; Lee, S.; Kazmierczak, K.M.; Lee, K.J.; Wong, A.; et al. Phenotypic Landscape of a Bacterial Cell. *Cell* **2011**, *144*, 143–156. [CrossRef]

28. Ogasawara, H.; Ohe, S.; Ishihama, A. Role of transcription factor NimR (YeaM) in sensitivity control of *Escherichia coli* to 2-nitroimidazole. *FEMS Microbiol. Lett.* **2015**, *362*, 1–8. [CrossRef]

29. Nandineni, M.R.; Gowrishankar, J. Evidence for an Arginine Exporter Encoded by yggA (argO) That Is Regulated by the LysR-Type Transcriptional Regulator ArgP in *Escherichia coli*. *J. Bacteriol.* **2004**, *186*, 3539–3546. [CrossRef]

30. Carr, R.J.; Bilton, R.F.; Atkinson, T. Toxicity of paraquat to microorganisms. *Appl. Environ. Microbiol.* **1986**, *52*, 1112–1116. [CrossRef]

31. Broadhurst, D.I.; Kell, D.B. Statistical strategies for avoiding false discoveries in metabolomics and related experiments. *Metabolomics* **2006**, *2*, 171–196. [CrossRef]

32. O'Hagan, S.; Kell, D.B. Structural Similarities between Some Common Fluorophores Used in Biology, Marketed Drugs, Endogenous Metabolites, and Natural Products. *Mar. Drugs* **2020**, *18*, 582. [CrossRef] [PubMed]

33. Featherstone, D.E.; Broadie, K. Wrestling with pleiotropy: Genomic and topological analysis of the yeast gene expression network. *BioEssays* **2002**, *24*, 267–274. [CrossRef] [PubMed]

34. Kitagawa, M.; Ara, T.; Arifuzzaman, M.; Ioka-Nakamichi, T.; Inamoto, E.; Toyonaga, H.; Mori, H. Complete set of ORF clones of *Escherichia coli* ASKA library (A Complete Set of *E. coli* K-12 ORF Archive): Unique Resources for Biological Research. *DNA Res.* **2005**, *12*, 291–299. [CrossRef]

35. Salcedo-Sora, J.E.; Robison, A.T.R.; Zaengle-Barone, J.; Franz, K.J.; Kell, D.B. Membrane Transporters Involved in the Antimicrobial Activities of Pyrithione in *Escherichia coli*. *Molecules* **2021**, *26*, 5826. [CrossRef] [PubMed]

36. Salcedo-Sora, J.E.; Jindal, S.; O'Hagan, S.; Kell, D.B. A palette of fluorophores that are differentially accumulated by wild-type and mutant strains of *Escherichia coli*: Surrogate ligands for profiling bacterial membrane transporters. *Microbiology* **2021**, *167*, 001016. [CrossRef] [PubMed]

37. Almeida, L.D.; Silva, A.S.F.; Mota, D.C.; Vasconcelos, A.A.; Camargo, A.P.; Pires, G.S.; Furlan, M.; da Cunha Freire, H.M.R.; Klippel, A.H.; Silva, S.F.; et al. Yeast Double Transporter Gene Deletion Library for Identification of Xenobiotic Carriers in Low or High Throughput. *mBio* **2021**, *12*, e0322121. [CrossRef]

38. French, S.; Mangat, C.; Bharat, A.; Côté, J.-P.; Mori, H.; Brown, E.D. A robust platform for chemical genomics in bacterial systems. *Mol. Biol. Cell* **2016**, *27*, 1015–1025. [CrossRef]

39. Takeuchi, R.; Tamura, T.; Nakayashiki, T.; Tanaka, Y.; Muto, A.; Wanner, B.L.; Mori, H. Colony-live—A high-throughput method for measuring microbial colony growth kinetics—Reveals diverse growth effects of gene knockouts in *Escherichia coli*. *BMC Microbiol.* **2014**, *14*, 171. [CrossRef]

40. Vines, T.H.; Albert, A.; Andrew, R.; Débarre, F.; Bock, D.G.; Franklin, M.T.; Gilbert, K.; Moore, J.-S.; Renaut, S.; Rennison, D. The Availability of Research Data Declines Rapidly with Article Age. *Curr. Biol.* **2014**, *24*, 94–97. [CrossRef]

41. Varadi, M.; Anyango, S.; Deshpande, M.; Nair, S.; Natassia, C.; Yordanova, G.; Yuan, D.; Stroe, O.; Wood, G.; Laydon, A.; et al. AlphaFold Protein Structure Database: Massively expanding the structural coverage of protein-sequence space with high-accuracy models. *Nucleic Acids Res.* **2022**, *50*, D439–D444. [CrossRef]

42. Jumper, J.; Evans, R.; Pritzel, A.; Green, T.; Figurnov, M.; Ronneberger, O.; Tunyasuvunakool, K.; Bates, R.; Žídek, A.; Potapenko, A.; et al. Highly accurate protein structure prediction with AlphaFold. *Nature* **2021**, *596*, 583–589. [CrossRef] [PubMed]

43. Sprouffske, K.; Wagner, A. Growthcurver: An R package for obtaining interpretable metrics from microbial growth curves. *BMC Bioinform.* **2016**, *17*, 172. [CrossRef] [PubMed]

Article

Effects of Levofloxacin, Aztreonam, and Colistin on Enzyme Synthesis by *P. aeruginosa* Isolated from Cystic Fibrosis Patients

Arianna Pani *, Valeria Lucini, Silvana Dugnani, Alice Schianchi and Francesco Scaglione

Oncology and Hemato-Oncology, University of Milan, 20122 Milan, Italy
* Correspondence: arianna.pani@unimi.it

Abstract: (1) Background: Cystic fibrosis (CF) is characterized by chronic pulmonary inflammation and persistent bacterial infections. *P. aeruginosa* is among the main opportunistic pathogens causing infections in CF. *P. aeruginosa* is able to form a biofilm, decreasing antibiotic permeability. LOX, a lipoxygenase enzyme, is a virulence factor produced by *P. aeruginosa* and promotes its persistence in lung tissues. The aim of this study is to evaluate if antibiotics currently used for aerosol therapy in CF are able to interfere with the production of lipoxygenase from open isolates of *P. Aeruginosa* from patients with CF. (2) Methods: Clinical isolates of *P. aeruginosa* from patients with CF were grown in Luria broth (LB). Minimum inhibitory concentration (MIC) was performed and interpreted for all isolated strains according to the European Committee on Antimicrobial Susceptibility Testing (EUCAST) guidelines. We selected four antibiotics with different mechanisms of action: aztreonam, colistin, amikacin, and levofloxacin. We used human pulmonary epithelial NCI-H929 cells to evaluate LOX activity and its metabolites according to antibiotic action at increasing concentrations. (3) Results: there is a correlation between LOX secretion by clinical isolates of *P. aeruginosa* and biofilm production. Levofloxacin exhibits highly significant inhibitory activity compared to the control. Amikacin also exhibits significant inhibitory activity against LOX production. Aztreonam and colistin do not show inhibitory activity. These results are also confirmed for LOX metabolites. (4) Conclusions: among the evaluated antibiotics, levofloxacin and amikacin have an activity on LOX secretion.

Keywords: cystic fibrosis; *P. aeruginosa*; antibiotics

Citation: Pani, A.; Lucini, V.; Dugnani, S.; Schianchi, A.; Scaglione, F. Effects of Levofloxacin, Aztreonam, and Colistin on Enzyme Synthesis by *P. aeruginosa* Isolated from Cystic Fibrosis Patients. *Antibiotics* **2022**, *11*, 1114. https://doi.org/10.3390/antibiotics11081114

Academic Editors: Dóra Kovács and Jeffrey Lipman

Received: 8 June 2022
Accepted: 9 August 2022
Published: 17 August 2022

Publisher's Note: MDPI stays neutral with regard to jurisdictional claims in published maps and institutional affiliations.

1. Introduction

Bacterial infection and linked chronic pulmonary inflammation are a pathological trait associated with the genetic disease cystic fibrosis (CF) [1]. These bacterial infections are characterized by robust inflammatory responses, with an elevation in the levels of proinflammatory cytokines and neutrophil accumulation in the CF airway. Unfortunately, responses triggered by the inflammation are not effective in clearing pathogenic microbes in the CF lung [2], creating instead a hyperinflammatory status which can lead to the damage of host tissues and respiratory failure, leading to transplantation or death [3].

Most adult patients with CF have a chronic infection of the airways caused by the opportunistic bacterial pathogen *P. aeruginosa*, which is frequently associated with morbidity and mortality. *P. aeruginosa* grows in the context of hyperinflammation in CF lungs and is able to form mechanically robust biofilms which are resistant to clinically achievable levels of antibiotics [4].

Antibiotic classes currently approved in many countries for use by inhalation include the aminoglycosides (tobramycin and amikacin), monobactams (aztreonam), polymyxins (colistimethate) and fluoroquinolones (levofloxacin). Although all of the antibiotics used are beneficial, none of them are capable of completely eradicating *P. aeruginosa* from bronchial secretions in CF. However, relatively recent studies have reported that levofloxacin, administered by aerosol in CF, produces positive effects that go beyond its simple antibacterial effect [5–7].

P. aeruginosa also persists in the airways by interfering with host defense via secreted bacterial virulence factors and small molecules. Recently, it has been demonstrated that several *P. aeruginosa* clinical isolates express *LoxA*, a gene which encodes for the lipoxygenase enzyme (LOX) that oxidizes polyunsaturated fatty acids [8]. In the lungs, LOX is able to process a wide range of host polyunsaturated fatty acids, with a consequent production of bioactive lipid mediators (including lipoxin A4). LOX is also able to inhibit major chemokine expression, such as macrophage inflammatory proteins (MIPs) and keratinocytes-derived chemokines (KCs), and to recruit leukocytes. Importantly, LOX is able to promote *P. aeruginosa* survival in lung tissues, suggesting a LOX-dependent interference between the host lipid pathways and *P. aeruginosa* lung pathogenesis.

In this study, we wanted to verify whether the antibiotics currently used via aerosols in CF could interfere with the production of LOX from isolates of *P. aeruginosa* from patients with CF.

2. Results

2.1. Antibiotic Intrinsic Activity (MICs)

In order to evaluate the activity of antibiotics approved for aerosol use in cystic fibrosis on the production of LOX by *P. aeruginosa*, we chose four antibiotics representing different mechanisms of action: levofloxacin, amikacin, aztreonam, and colistin. First, we evaluated the antibiotic intrinsic activity against 12 clinically isolated strains of *P. aeruginosa*.

Table 1 shows the minimum inhibitory concentrations (MICs) of the levofloxacin, amikacin, aztreonam, and colistin studied against the clinically collected strains of *P. aeruginosa*. Strains 1, 2, 3, 9, 10, 11, and 12 were resistant to levofloxacin (MIC > 2). Only strain 12 was resistant to amikacin (MIC > 16) and strains 9 and 10 to aztreonam (MIC > 16). Strains 1, 2, and 11 were resistant to colistin (MIC > 4). Low MICs ($\leq$0.25 mg/L) were detected for levofloxacin, amikacin, aztreonam, and colistin in strains 4 and 8.

Table 1. MICs of levofloxacin, amikacin, aztreonam and colistin against *P. aeruginosa* strains isolated *from patients with cystic fibrosis.*

	Levofloxacin	Amikacin	Aztreonam	Colistin
Strain 1	8	8	<0.25	>64
Strain 2	4	1	<0.25	8
Strain 3	4	2	<0.25	0.5
Strain 4	<0.25	<0.25	<0.25	<0.25
Strain 5	<0.25	1	4	0.5
Strain 6	<0.25	1	4	0.5
Strain 7	0.25	1	4	1
Strain 8	0.25	<0.25	<0.25	<0.25
Strain 9	4	1	16	<0.25
Strain 10	8	4	32	<0.25
Strain 11	8	8	8	16
Strain 12	16	32	8	2

2.2. Biofilm Formation

Secondly, we evaluated the ability of the 12 clinically isolated strains to produce biofilm.

Figure 1 shows the biofilm formation by the 12 selected strains. As expected, there are different abilities to form biofilms. Strains 2, 4, 6, and 10 demonstrated the highest ability to produce biofilms at 12, 24, and 48 h. Strain 2 showed a CV absorbance of 0.620 (0.042) at 12 h, 1.040 (0.085) at 24 h, and 1.600 (0.141) at 48 h. Strain 4 showed a CV absorbance of 0.440 (0.127) at 12 h, 0.980 (0.170) at 24 h, and 1.600 (0.141) at 48 h. Strain 6 showed a CV absorbance of 0.400 (0.014) at 12 h, 0.995 (0.148) at 24 h, and 1.700 (0.141) at 48 h. Strain 10 showed a CV absorbance of 0.445 (0.134) at 12 h, 1.150 (0.636) at 24 h, and 2.020 (0.113) at 48 h.

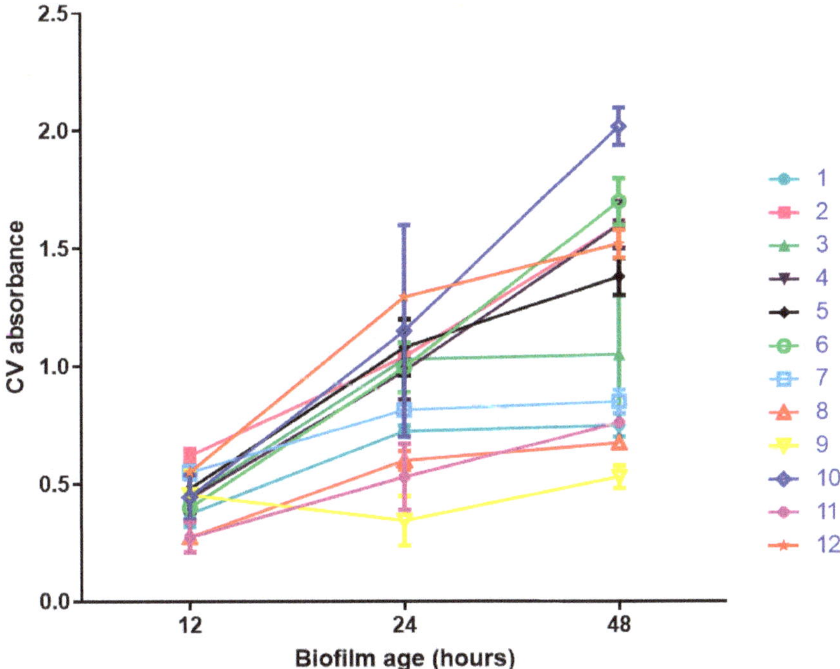

Figure 1. Ability to form biofilm for the 12 clinically isolated *P. aeruginosa* strains. Biofilm was evaluated at 12, 24, and 48 h. CV absorbance measures biofilm formation. Each colored line (numbers from 1 to 12, see Table 1) refers to a different *P. aeruginosa* strain.

2.3. LOX Activity

Since the expression of the loxA gene differs from one *P. aeruginosa* strain to another, especially in biofilm growth conditions [9], we measured the LOX activity of the 12 clinically isolated strains. The aim of our evaluations of biofilm and LOX activity was to select the best strain on which to continue our evaluation on the impact of antibiotics on LOX production.

Figure 2 shows the LOX activity of the 12 selected strains. From the comparison of Figures 1 and 2 it emerges that the strains with the greatest ability to make biofilms are those that have the greatest LOX activity. By performing a one-way ANOVA with Bonferroni's correction, the LOX activity differs statistically significantly between the 12 strains ($p < 0.001$). In fact, strains 2, 4, 6, and 10 demonstrated the highest LOX activity at 12, 24, and 48 h. Mean absorbance of strain 2 was 1.040 (0.085) at 12 h, 1.200 (0.141) at 24 h, and 1.450 (0.071) at 48 h. Mean absorbance of strain 4 was 1.350 (0.212) at 12 h, 1.500 (0.141) at 24 h, and 1.550 (0.212) at 48 h. Mean absorbance of strain 6 was 1.350 (0.212) at 12 h, 1.450 (0.212) at 24 h, and 1.500 (0.141) at 48 h. Mean absorbance of strain 10 was 1.550 (0.071) at 12 h, 1.400 (0.283) at 24 h, and 2.050 (0.212) at 48 h.

Considering these results, the number 10 strain of *P. aeruginosa* was chosen to perform the subsequent experiments.

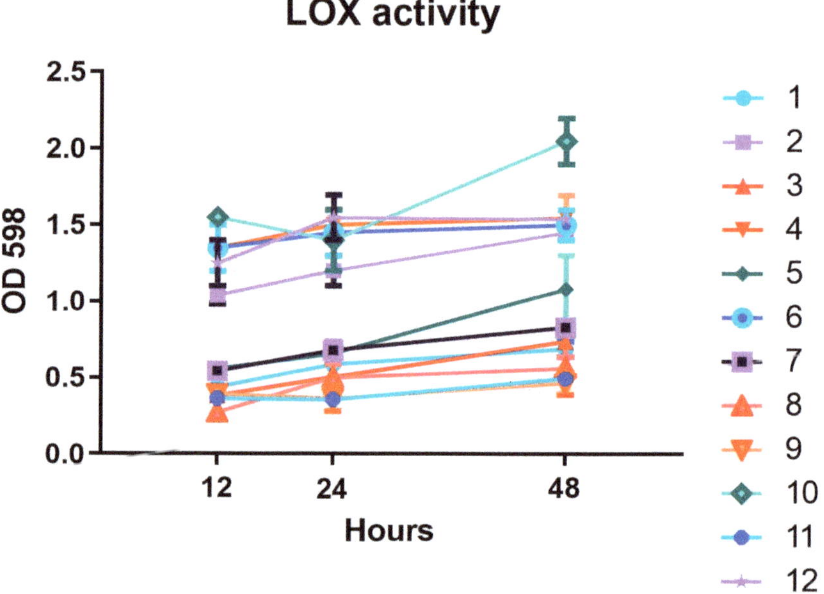

Figure 2. Lipoxygenase (LOX) activity of the 12 selected *P. aeruginosa* strains. LOX activity was evaluated at 12, 24, and 48 h. Each colored line (numbers from 1 to 12, see Table 1) refers to a different *P. aeruginosa* strain.

2.4. Effects of Antibiotics on LOX Activity

We then evaluated the effect of the levofloxacin, aztreonam, amikacin, and colistin on the LOX activity of strain 10 of *P. aeruginosa*. Results are shown in Figure 3.

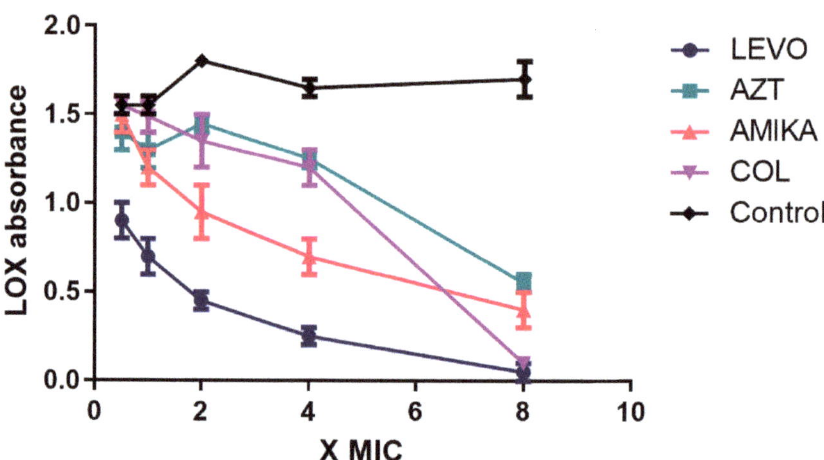

Figure 3. Effects of levofloxacin, aztreonam, amikacin, and colistin on the LOX activity of a clinically isolated strain of *P. aeruginosa* with high LOX activity. Antibiotics were tested at increasing concentrations (0.5- to 8-fold the MICs). (LEVO = levofloxacin, AZT = aztreonam, AMIKA = amikacin, COL = colistin, Control = untreated strain).

Levofloxacin exhibits highly significant inhibitory activity compared to the control (mean absorbance at four times the MIC—0.250 (0.071) vs. 1.650 (0.071)). To a lesser extent, amikacin also exhibits significant inhibitory activity (mean absorbance at four times the MIC—0.700 (0.141) vs. 1.650 (0.071)), while aztreonam and colistin do not show inhibitory activity (mean absorbance at four times the MIC, respectively 1.250 (0.071) and 1.200 (0.141)). It should be noted that at eight times the MIC all antibiotics show inhibitory activity. This is most likely related to bactericidal activity on the strain rather than to the activity inhibiting LOX synthesis.

We compared the effects of the four different antibiotics on LOX activity. Table 2 reports the results of the analysis. Differences between levofloxacin and aztreonam, colistin and aztreonam, and colistin were statistically significant (Table 2).

Table 2. One-way ANOVA effects of different antibiotics on LOX activity.

Bonferroni's Multiple Comparison Test	Mean Difference	t	$p < 0.05$	95% CI
Levofloxacin vs. aztreonam	−0.7200	4.237	Yes	−1.273 to −0.1673
Levofloxacin vs. amikacin	−0.4800	2.824	No	−1.033 to 0.07266
Levofloxacin vs. colistin	−0.6680	3.931	Yes	−1.221 to −0.1153
Levofloxacin vs. control	−1.180	6.943	Yes	−1.733 to −0.6273
Aztreonam vs. amikacin	0.2400	1.412	No	−0.3127 to 0.7927
Aztreonam vs. colistin	0.05200	0.3060	No	−0.5007 to 0.6047
Aztreonam vs. control	−0.4600	2.707	No	−1.013 to 0.09266
Amikacin vs. colistin	−0.1880	1.106	No	−0.7407 to 0.3647
Amikacin vs. control	−0.7000	4.119	Yes	−1.253 to −0.1473
Colistin vs. control	−0.5120	3.013	No	−1.065 to 0.04066

2.5. Production of 15-LOX-Dependent Metabolites in Lung Epithelial Cells Infected by P. aeruginosa

To evaluate LOX effects on the host response, we measured LOX metabolites in human lung epithelial NCI-H292 cells challenged by *P. aeruginosa* strain 10. We measured 15-Hydroxyeicosatetraenoic acid (15-HETE), 17-hydroxydocosahexaenoic acid (17-HDoHE) and lipoxin A4 production (LXA4). Figure 4 shows the results relating to the effect of the various antibiotics on the production of 15-HETE. Levofloxacin exhibits significant inhibitory activity compared to the control (mean difference −112.1, 95% CI 27.99–196.2, $p < 0.005$), while amikacin aztreonam, and colistin show no significant inhibitory activity. In this case it should also be noted that at eight times the MIC all antibiotics show inhibitory activity. This is most likely related to bactericidal activity on the strain rather than to the activity inhibiting LOX synthesis.

Figure 5 shows the results relating to the effect of the various antibiotics on the production of 17-HDoHE. Only levofloxacin exhibits significant inhibitory activity compared to the control (mean difference −112 95% CI 55.4–514.8, $p < 0.005$), while amikacin, aztreonam, and colistin show no significant inhibitory activity. In this case it should also be noted that at eight times the MIC all antibiotics show inhibitory activity.

Figure 6 shows the results relating to the effect of various antibiotics on LXA44 production. Only Levofloxacin exhibits significant inhibitory activity compared to the control (mean difference −794.9, 95%CI 225–1364, $p < 0.005$). While amikacin aztreonam and colistin show no significant inhibitory activity. In this case it should also be noted that at eight times the MIC all antibiotics show inhibitory activity.

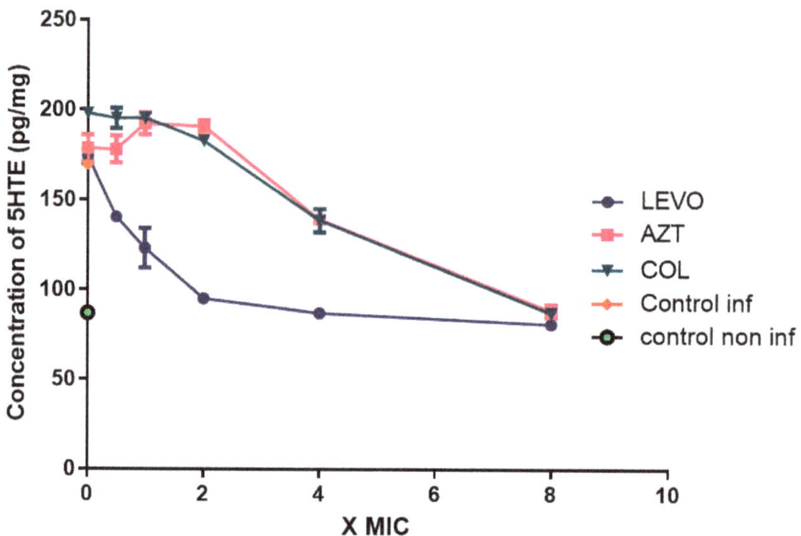

Figure 4. Concentration of 15-HETE (pg/mg of protein) in extracts of human lung epithelial NCI-H292 cells infected or noninfected with *P. aeruginosa*, 24 h post-infection. (LEVO = levofloxacin, AZT = aztreonam, COL = colistin, Control inf = control in infected cells, Control non inf = control in noninfected cells).

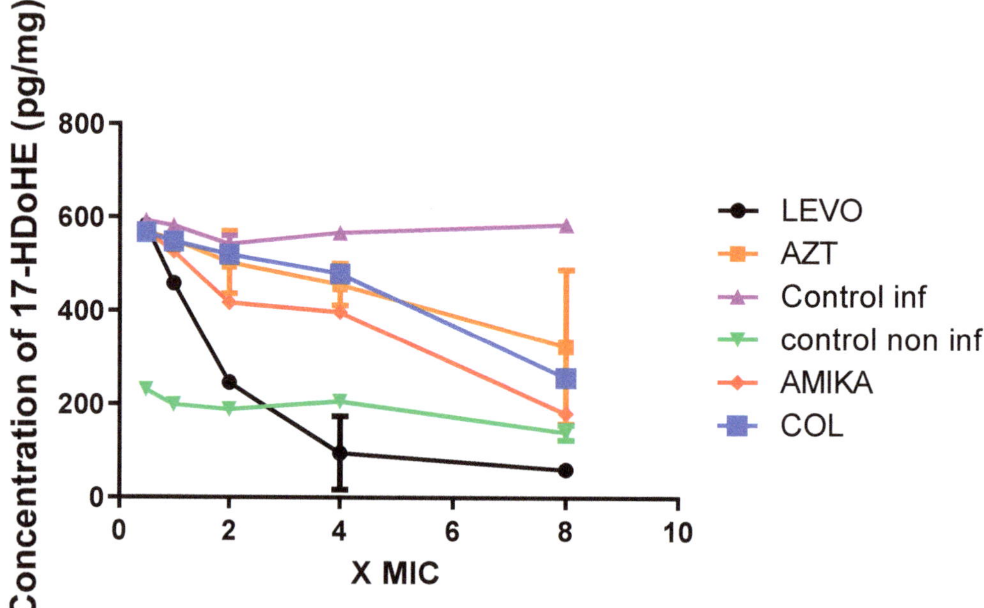

Figure 5. Concentration of 17-HDoHE (pg/mg of protein) in extracts of human lung epithelial NCI-H292 cells non-infected and treated with antibiotics, 24 h. (LEVO = levofloxacin, AZT = aztreonam, AMIKA = amikacin, COL = colistin, Control inf = control in infected cells, Control non-inf = control in noninfected cells).

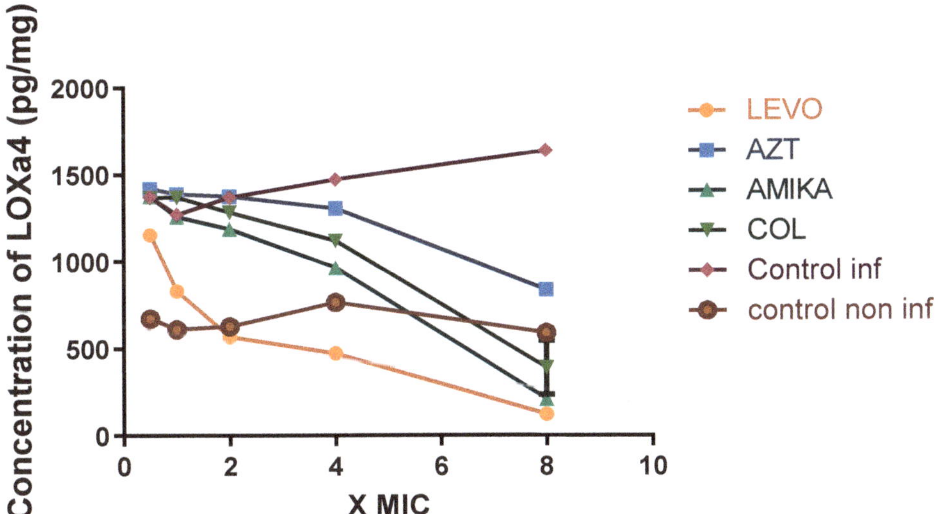

Figure 6. Concentration of LOXa4 (pg/mg of protein) in extracts of human lung epithelial NCI-H292 cells infected, noninfected, and treated with antibiotics after 24 h. (LEVO = levofloxacin, AZT = aztreonam, AMIKA = amikacin, COL = colistin, Control inf = control in infected cells, Control non inf = control in noninfected cells).

3. Discussion

Pseudomonas aeruginosa is a highly versatile bacterium. One basis for its versatility is the arsenal of enzymes that helps this pathogen to adapt to its environment. Our study shows that lipoxygenase (LOX) is secreted by clinical isolates of *P. aeruginosa*, producing biofilm, and this enzyme may contribute to the lung pathogenesis triggered by this opportunistic pathogen. In a series of elegant experiments, Morello et al. [8] have shown that LOX activity decreases release of chemokines such as KC (CXCL-1) and macrophage inflammatory proteins (MIP-1α/CCL-3, MIP-1β/CCL-4, and MIP-2/CXCL-2), producing a lower recruitment of immune cells in the airspaces. More importantly, they also observed that LOX activity promotes the spread of bacteria in lung tissues.

The role of 15-LOX has been implicated in various inflammation-related diseases. Increasing evidence highlights the controversial nature of 12/15-LOX in inflammation, as its metabolites have been shown to have both pro- and anti-inflammatory properties [10].

The pro-inflammatory role of 15-LOX and its metabolite 15(S)-HETE was demonstrated by various studies [11,12]. It is important to note that expression of pro-inflammatory cytokines IL-6, IL-12, CXCL9, and CXCL10, which is LPS-induced, is reduced by inhibition of 12/15-LOX in macrophages [13]. Furthermore, 15-LOX is able to regulate the expression of pro-inflammatory eoxins in epithelial airway cells and eosinophils. Eoxins can cause endothelial cell dysfunction and enhance vascular permeability [14]. Additionally, the increased expression of some of these inflammatory molecules is dependent on nuclear factor κB activation [15,16]. Moreover, recently it has been reported that 15-LOX may interact with secretory phospholipase A2 (sPLA2), contributing to sterile inflammation in chronic conditions with differences from classical inflammation on the cytokine level [17].

On the other hand, demonstrations of the anti-inflammatory properties of 12/15-LOX and its metabolites have been reported. The exact mechanisms by which 12/15-LOX explicates its anti-inflammatory effects are not fully understood, but they may be due to its pro-resolving mediators such as lipoxins, resolvins, and protectins, which are able to induce a potent and direct anti-inflammatory response in various cell types. In summary, it is becoming clear that 12/15-LOX and its metabolites have both pro- and anti-inflammatory effects, and some of these differential effects of the same metabolites could be due to their different concentrations [10].

In this study, we analyzed the activity of three antibiotics currently used in cystic fibrosis as antipseudomonal inhalation therapy. Among the antibiotics used via aerosol in CF, only levofloxacin exhibited significant inhibitory activity of LOX activity compared to the control, while aztreonam and colistin show no significant inhibitory activity. Levofloxacin acting on the synthesis of enzymes by *P. aeruginosa* determines an indirect anti-inflammatory effect, which is very useful in CF. These data may explain the positive effects exceeding those attributable to the antibacterial activity alone, which have been obtained with levofloxacin during clinical trials in CF [6,18].

4. Materials and Methods

4.1. Bacterial Strains

Nonduplicate clinical isolates of *P. aeruginosa* from a permanent collection for the storage of bacterial strains of unrecognizable patients with CF were grown in Luria broth (LB) (Sigma-Aldrich, Milan, Italy). Minimum inhibitory concentrations (MICs) were determined using the reference broth microdilution methodology according to the Clinical and Laboratory Standards Institute (CLSI) guidelines [19] for levofloxacin, colistin, aztreonam, and amikacin for all isolated strains. EUCAST v.12 clinical breakpoints were used for interpretation of susceptibility data, where available. After 18 h of incubation at 37 °C, the MICs of the aforementioned antibiotics were defined by no visible growth on plates. Since it has been shown that LOX is mainly produced in Pseudomonas strains that produce biofilms, we have preliminarily evaluated the ability of selected strains to produce biofilm [20].

4.2. Biofilm Quantification

One microliter of a late-log-phase culture was added to 99 µL LB in a 96-well microtiter plate, incubated for 10 h at 37 °C. Biofilm production was quantified by measuring the absorbance of crystal violet (Sigma-Aldrich-, Italy). Biofilms were fixed by heat at 60 °C for about 1 h for the crystal violet assay. Subsequently, wells with 150 µL of crystal violet solution (2.3% prepared in 20% ethanol) were incubated for 15 min at room temperature. Excesses of crystal violet were removed by washing and then 200 µL of 33% glacial acetic acid was used to dissolve dye fixed to the biofilm and it was incubated for 1 h at room temperature. CV absorbance was measured at 570 nm using a microplate spectrophotometer (BioTek™ Epoch 2 Microplate Spectrophotometer; BioTek Instruments, Winooski, VT, USA). Biofilm formation was evaluated at 12, 24, and 48 h.

4.3. Antibiotics

To evaluate the effect of antibiotics, we decided to select three antibiotics with different mechanisms of action. Aztreonam inhibits bacterial cell wall formation, colistin interferes with membrane phospholipids, and levofloxacin inhibits protein synthesis. Aztreonam, colistin, and levofloxacin were purchased from the market (colistin sulfate, aztreonam and levofloxacin for microbiological assay, European Pharmacopoeia (EP) Reference Standard, Sigma Aldrich S.r.l., St. Louis, MO, USA). Antibiotic activity was evaluated against LOX activity and lipid production in a medium containing antibiotics at increasing concentrations (0.5- to 8-fold the MICs in broth).

4.4. Lipoxygenase Assay

P. aeruginosa strains were grown as single colonies on LB agar overnight. Then, a single colony was inoculated in 10 mL of LB broth and grown to stationary phase under static conditions at 28 °C. Subsequently the culture broth was centrifuged at $8000 \times g$ for 15 min and the supernatant was collected in a syringe, sterilized, and stored at −80 °C. 10 mL of sample (concentrated supernatants) was mixed with 100 mL of solution A (0.5 mM purified lipoxygenase assay, 10 mM (dimethyl-amino)-benzoic acid (DMAB) (Sigma-Aldrich-Italy) prepared in 100 mM phosphate buffer (pH 6) and incubated for 20 min before the addition of 100 mL of supernatant from each well was transferred into a new plate, and their absorbance was measured at 598 nm.

4.5. Culture and Infection of Pulmonary Epithelial Cells

Human pulmonary epithelial NCI-H929 cells were obtained from the American Type Culture Collection (ATCC, Manassas, VA, USA). Cells were grown in RPMI 1640 Glutamax (Gibco, Life Technologies, Rodano, Italy) medium supplemented with 10% heat-inactivated Fetal Bovine Serum (FBS) (Sigma-Aldrich-Italy) in a humidified incubator with 5% CO_2 at 37 C. For infection experiments, cells were cultivated in 24-well plates until confluence (5.105 cells per well). Exponential growth phase bacteria (LB, 37C, 180 rpm, OD 600 nm) were washed twice in ice-cold PBS before addition to freshly dispensed cell culture medium to obtain MOI = 0.1. After 20 h, supernatants were then centrifuged at $8000 \times g$ for 10 min, collected in a syringe, and sterilized with a 0.22 mm polymer filter before immediate snap-freezing and storage in liquid nitrogen until lipid mediator extraction.

4.6. Lipid Extraction and Liquid Chromatography/Tandem Mass Spectrometry (LC-MS/MS)

We then evaluated the 15-LOX-dependent metabolites, 15-hydroxy-octadecadienoic acid (15-HETE) and 17-HDoHE, and lipoxin A4 (LXA4) in cell lysates and supernatants. Solid phase extraction was performed with HRX-50 mg 96-well plates. Simultaneous separation of the lipids of interest was performed as well as LC-MS/MS analysis on an ultra-high-performance liquid chromatography system (Q Exactive™ Plus Hybrid Quadrupole-Orbitrap™ Mass Spectrometer Thermo Scientific™, Waltham, MA, USA).

4.7. Statistical Methods

Continuous variables are expressed as mean and standard deviations (sd). To compare the effects on LOX, 15-HETE, 17-HDoHE, and LOXa4 activity according to the different antibiotics at increasing concentrations, we used a two-way analysis of variance (ANOVA) followed by post hoc comparison using Bonferroni *t*-test. *p*-values of <0.05 were considered significant.

Statistical analyses were performed using GraphPad InStat 8 (GraphPad Software Inc., La Jolla, CA, USA).

5. Conclusions

Among clinically isolated strains of *P. aeruginosa* from patients with CF, we selected the strain with highest activity of biofilm production and LOX activity.

Testing levofloxacin, aztreonam, colistin, and amikacin at increasing concentrations, only levofloxacin showed a significant activity on the secretion of LOX. Levofloxacin also showed a significant activity on 15-HETE, 17-HDoHE, and LXA4 production. In conclusion, levofloxacin is the only antibiotic, among those studied according to the mechanism of action and the approval for aerosol use, which can impact the secretion of LOX in a strain with a high ability of biofilm production.

Author Contributions: Conceptualization, A.P. and F.S.; methodology, F.S., V.L. and S.D.; formal analysis, F.S.; investigation, V.L.; data curation, S.D.; writing—original draft preparation, A.P., F.S. and A.S.; writing—review and editing, A.P. and A.S.; visualization, A.S.; supervision, F.S.; project administration, F.S.; funding acquisition, F.S. All authors have read and agreed to the published version of the manuscript.

Funding: An unrestricted grant has been provided by Chiesi Farmaceutici S.p.A to partially fund research activities. Grant number is not applicable.

Institutional Review Board Statement: Not applicable.

Informed Consent Statement: Not applicable.

Data Availability Statement: Data available on request from the authors upon reasonable request.

Conflicts of Interest: All authors declare no conflict of interest.

References

1. Dp, N.; Jf, C. Inflammation and its genesis in cystic fibrosis. *Pediatr. Pulmonol.* **2015**, *50* (Suppl. 40), S39–S56. [CrossRef]
2. Gifford, A.M.; Chalmers, J.D. The role of neutrophils in cystic fibrosis. *Curr. Opin. Hematol.* **2014**, *21*, 16–22. [CrossRef] [PubMed]
3. Cantin, A.M.; Hartl, D.; Konstan, M.W.; Chmiel, J.F. Inflammation in cystic fibrosis lung disease: Pathogenesis and therapy. *J. Cyst. Fibros. Off. J. Eur. Cyst. Fibros. Soc.* **2015**, *14*, 419–430. [CrossRef] [PubMed]
4. Cohen, T.S.; Prince, A. Cystic fibrosis: A mucosal immunodeficiency syndrome. *Nat. Med.* **2012**, *18*, 509–519. [CrossRef] [PubMed]
5. Flume, P.A.; VanDevanter, D.R.; Morgan, E.E.; Dudley, M.N.; Loutit, J.S.; Bell, S.C.; Kerem, E.; Fischer, R.; Smyth, A.R.; Aaron, S.D.; et al. A phase 3, multi-center, multinational, randomized, double-blind, placebo-controlled study to evaluate the efficacy and safety of levofloxacin inhalation solution (APT-1026) in stable cystic fibrosis patients. *J. Cyst. Fibros. Off. J. Eur. Cyst. Fibros. Soc.* **2016**, *15*, 495–502. [CrossRef]
6. Elborn, J.S.; Flume, P.A.; Cohen, F.; Loutit, J.; VanDevanter, D.R. Safety and efficacy of prolonged levofloxacin inhalation solution (APT-1026) treatment for cystic fibrosis and chronic *Pseudomonas aeruginosa* airway infection. *J. Cyst. Fibros. Off. J. Eur. Cyst. Fibros. Soc.* **2016**, *15*, 634–640. [CrossRef] [PubMed]
7. Geller, D.E.; Flume, P.A.; Staab, D.; Fischer, R.; Loutit, J.S.; Conrad, D.J. Mpex 204 Study Group Levofloxacin inhalation solution (MP-376) in patients with cystic fibrosis with *Pseudomonas aeruginosa*. *Am. J. Respir. Crit. Care Med.* **2011**, *183*, 1510–1516. [CrossRef] [PubMed]
8. Morello, E.; Pérez-Berezo, T.; Boisseau, C.; Baranek, T.; Guillon, A.; Bréa, D.; Lanotte, P.; Carpena, X.; Pietrancosta, N.; Hervé, V.; et al. *Pseudomonas aeruginosa* Lipoxygenase LoxA Contributes to Lung Infection by Altering the Host Immune Lipid Signaling. *Front. Microbiol.* **2019**, *10*, 1826. [CrossRef] [PubMed]
9. Starkey, M.; Hickman, J.H.; Ma, L.; Zhang, N.; De Long, S.; Hinz, A.; Palacios, S.; Manoil, C.; Kirisits, M.J.; Starner, T.D.; et al. *Pseudomonas aeruginosa* rugose small-colony variants have adaptations that likely promote persistence in the cystic fibrosis lung. *J. Bacteriol.* **2009**, *191*, 3492–3503. [CrossRef] [PubMed]
10. Singh, N.K.; Rao, G.N. Emerging role of 12/15-Lipoxygenase (ALOX15) in human pathologies. *Prog. Lipid Res.* **2019**, *73*, 28–45. [CrossRef] [PubMed]
11. Cole, B.K.; Lieb, D.C.; Dobrian, A.D.; Nadler, J.L. 12- and 15-lipoxygenases in adipose tissue inflammation. *Prostaglandins Other Lipid Mediat.* **2013**, *104–105*, 84–92. [CrossRef] [PubMed]
12. Kühn, H.; O'Donnell, V.B. Inflammation and immune regulation by 12/15-lipoxygenases. *Prog. Lipid Res.* **2006**, *45*, 334–356. [CrossRef] [PubMed]
13. Namgaladze, D.; Snodgrass, R.G.; Angioni, C.; Grossmann, N.; Dehne, N.; Geisslinger, G.; Brüne, B. AMP-activated protein kinase suppresses arachidonate 15-lipoxygenase expression in interleukin 4-polarized human macrophages. *J. Biol. Chem.* **2015**, *290*, 24484–24494. [CrossRef] [PubMed]
14. Feltenmark, S.; Gautam, N.; Brunnström, Å.; Griffiths, W.; Backman, L.; Edenius, C.; Lindbom, L.; Björkholm, M.; Claesson, H.-E. Eoxins are proinflammatory arachidonic acid metabolites produced via the 15-lipoxygenase-1 pathway in human eosinophils and mast cells. *Proc. Natl. Acad. Sci. USA* **2008**, *105*, 680–685. [CrossRef] [PubMed]
15. Natarajan, R.; Reddy, M.A.; Malik, K.U.; Fatima, S.; Khan, B.V. Signaling mechanisms of nuclear factor-kappab-mediated activation of inflammatory genes by 13-hydroperoxyoctadecadienoic acid in cultured vascular smooth muscle cells. *Arterioscler. Thromb. Vasc. Biol.* **2001**, *21*, 1408–1413. [CrossRef] [PubMed]
16. Anderson, S.D.; Kippelen, P. Airway injury as a mechanism for exercise-induced bronchoconstriction in elite athletes. *J. Allergy Clin. Immunol.* **2008**, *122*, 225–235. [CrossRef] [PubMed]
17. Ha, V.T.; Lainšček, D.; Gesslbauer, B.; Jarc-Jovičić, E.; Hyötyläinen, T.; Ilc, N.; Lakota, K.; Tomšič, M.; van de Loo, F.A.J.; Bochkov, V.; et al. Synergy between 15-lipoxygenase and secreted PLA2 promotes inflammation by formation of TLR4 agonists from extracellular vesicles. *Proc. Natl. Acad. Sci. USA* **2020**, *117*, 25679–25689. [CrossRef] [PubMed]
18. Machado, F.S.; Johndrow, J.E.; Esper, L.; Dias, A.; Bafica, A.; Serhan, C.N.; Aliberti, J. Anti-inflammatory actions of lipoxin A4 and aspirin-triggered lipoxin are SOCS-2 dependent. *Nat. Med.* **2006**, *12*, 330–334. [CrossRef] [PubMed]
19. M07: Dilution AST for Aerobically Grown Bacteria—CLSI. Available online: https://clsi.org/standards/products/microbiology/documents/m07/ (accessed on 5 August 2022).
20. Deschamps, J.D.; Ogunsola, A.F.; Jameson, J.B.; Yasgar, A.; Flitter, B.A.; Freedman, C.J.; Melvin, J.A.; Nguyen, J.V.M.H.; Maloney, D.J.; Jadhav, A.; et al. Biochemical and Cellular Characterization and Inhibitor Discovery of *Pseudomonas aeruginosa* 15-Lipoxygenase. *Biochemistry* **2016**, *55*, 3329–3340. [CrossRef] [PubMed]

Article

Survey of the Knowledge, Attitudes and Practice towards Antibiotic Use among Prospective Antibiotic Prescribers in Serbia

Olga Horvat [1], Ana Tomas Petrović [1,*], Milica Paut Kusturica [1], Dragica Bukumirić [2], Bojana Jovančević [1] and Zorana Kovačević [3]

1   Faculty of Medicine, Department of Pharmacology Toxicology and Clinical Pharmacology, University of Novi Sad, 21000 Novi Sad, Serbia
2   Institute of Public Health "Dr Milan Jovanovic Batut", 11000 Belgrade, Serbia
3   Faculty of Agriculture, Department of Veterinary Medicine, University of Novi Sad, 21000 Novi Sad, Serbia
*   Correspondence: ana.tomas@mf.uns.ac.rs

Abstract: The complex issue of antibacterial resistance (ABR) requires actions taken with the One Health approach, involving both human and veterinarian medicine. It can spread from animals to humans through the food chain or through direct contact. Health profession students, as the future antibiotic providers, can greatly impact antibiotic-related issues in the future. The study was conducted to evaluate knowledge, attitudes and practice of future antibiotic prescribers in relation to judicious use of antibiotics. This cross-sectional, questionnaire-based study was performed on 400 students of health professions who were allowed to prescribe antibiotics of the University of Novi Sad, Serbia. Students of medicine and students of dentistry showed a significantly higher knowledge score compared to students of veterinary medicine ($p = 0.001$). Multivariate regression identified predictors of adequate antibiotic knowledge: being a female student (B = 0.571; $p = 0.020$), higher grade average (B = 1.204; $p = 0.001$), students of medicine (B = 0.802; $p = 0.006$) and dentistry (B = 0.769; $p = 0.026$), and students who used a complete package of antibiotics during the last infection (B = 0.974; $p = 0.001$) or for the period recommended by the doctor (B = 1.964; $p = 0.001$). Out of the total sample, self-medication was reported among 42.8% of students. The identified predictors of self-medication were: more frequent (B = 0.587; $p = 0.001$) and irregular (B = 0.719; $p = 0.007$) antibiotic use, taking antibiotics until symptoms disappeared (B = 2.142; $p = 0.001$) or until the bottle was finished (B = 1.010; $p = 0.001$) during the last infection. It seems prudent to reevaluate the educational curricula regarding antibiotic use and ABR of prospective prescribers in Serbia.

Keywords: antibiotics; self-medication; students; habits

Citation: Horvat, O.; Petrović, A.T.; Paut Kusturica, M.; Bukumirić, D.; Jovančević, B.; Kovačević, Z. Survey of the Knowledge, Attitudes and Practice towards Antibiotic Use among Prospective Antibiotic Prescribers in Serbia. *Antibiotics* **2022**, *11*, 1084. https://doi.org/10.3390/antibiotics11081084

Academic Editor: Dóra Kovács

Received: 15 July 2022
Accepted: 8 August 2022
Published: 10 August 2022

1. Introduction

Although antibacterial resistance is one of the most serious global public health threats in this century, antibacterial resistance (ABR) represents at the moment the major problem, both for the high rates of resistance observed in bacteria that cause common infections and for the complexity of the consequences of ABR [1].

The One Health concept has appeared to go hand in hand with the issue of antibiotic resistance as the most comprehensive and global solution, since there are many ways in which ABR can be transferred between humans and animals, via close contact, through the food chain or indirectly via the environment [2,3]

Better managing the problem of ABR includes taking steps to preserve the continued effectiveness of existing antibacterials such as trying to eliminate their inappropriate use, especially to reduce the number of unnecessary prescriptions, since new antibiotics coming to the market have not kept pace with the increasing need for improvements in antibiotic treatment [4]. What is particularly important is that antibiotics used in human medicine are to some extent the same as those used in veterinary medicine, and evidence is emerging that antibiotics critical for human health are being used among veterinary professionals [5,6]. Besides the widely implemented antimicrobial stewardship programs for improving the antimicrobial prescription in human and veterinary practice, the World Health Organization advocates to implement strategies that allow the next generation of doctors to be better prepared to appropriately use antibiotics and combat bacterial resistance [4,7,8].

Whereas several studies have tried to measure the knowledge, attitude and practice (KAP) of prospective doctors towards antibiotics, the existing literature is mainly focused on antibiotic prescribing practices rather than on KAP about antibiotic consumption. In addition, many of them show a lack of KAP on the importance of judicious use of antibiotics and optimal antibiotic prescription practices [9–13].

Serbia belongs to a group of European countries with the highest rates of resistance, as well as a high antibiotic consumption rate [14,15]. Based on a 2020 report by the Central Asian and European Surveillance of Antimicrobial Resistance (CAESAR) network, which includes 18 non-European Union countries and areas, high percentages of resistance in *P. aeruginosa*, *Acinetobacter* spp. and *E. faeciumi* in Serbia are concerning, in addition to the moderately high resistance for third-generation cephalosporins, aminoglycosides and fluoroquinolones in *E. coli* and for penicillin and macrolides in *S. pneumoniae* [14]. Furthermore, Serbia with the consumption of 31.57 DDD per 1000 inhabitants per day is together with Greece (33.85) and Turkey (38.18) among the three countries with the highest overall antibiotic consumption in Europe according to the WHO Europe Antimicrobial Medicines Consumption Network report for the period 2016–2018 [15]. In 2019, following the resolution of the United Nations by all countries, the Republic of Serbia formulated its National Antimicrobial Resistance Control Program for 2019–2023, where among other strategies, special attention is given on increasing awareness among those who prescribe antimicrobials [16]. In Serbia, the medical education of prospective medical doctors, dentists and veterinarians offers specialized curricula including pharmacology and microbiology, but not courses of antimicrobial use and resistance. Therefore, the aim of this study was to evaluate the knowledge, attitude and practice of students of medicine, dentistry and veterinary medicine who will be the only prospective prescribers of antibiotics in Serbia.

2. Results

2.1. Sociodemographic and Academic Characteristics

Almost two-thirds (62.5%) of the total surveyed students were female (Table 1). The mean age of the subjects was 23.80 ± 1.53 years. The observed students of the fifth and sixth year of medicine were on average older compared to the students of the fourth and fifth years of dentistry and veterinary medicine. Students of medicine (89.5%) and students of dentistry (92%) had a higher average grade (8.00–10.00) compared to students of veterinary medicine (61%) ($p <0.005$). There was no statistically significant difference between these groups of students in relation to place of residence ($p = 0.007$), number of visits to general practitioners in the past 12 months ($p = 0.107$) and in relation to whether they have a family member who is a healthcare worker ($p = 0.162$).

Table 1. Socio-demographic and academic characteristics of students of medicine (M), dentistry (D) and veterinary medicine (V) of the University of Novi Sad.

		M		D		V		χ^2	p	Total	
		n	%	n	%	n	%			n	%
Gender	Male	75	37.5	26	26	49	49	11.285	0.004	150	37.5
	Female	125	62.5	74	74	51	51			250	62.5
Average grade	6.00–7.99	21	10.5	8	8	39	39	46.031	0.001	68	17
	8.00–10.00	179	89.5	92	92	61	61			332	83
Place of living	with parents	64	32	43	43	38	38	17.568	0.007	145	36.3
	in university dormitories	33	16.5	7	7	14	14			54	13.5
	in rented apartments	67	33.5	32	32	43	43			142	35.5
	in own apartment	36	18	18	18	5	5			59	14.8
Number of visits to GP in the last 12 months	None	109	54.5	46	46	39	39.8	4.467	0.107	194	48.7
	1 4	82	41	52	52	57	58.2			191	48
	5–10	7	3.5	2	2	2	2			11	2.8
	>10	2	1	0	0	0	0			2	0.5
Having a family member who is healthcare worker	Yes	70	35	38	38	26	26	3.636	0.162	134	33.5
	No	130	65	62	62	74	74			266	66.5

There were no statistically significant differences among the examined groups of students in relation to the frequency of taking antibiotics (p = 0.318). The majority of surveyed students (46%) claimed to take antibiotics once every 2–3 years (Figure 1).

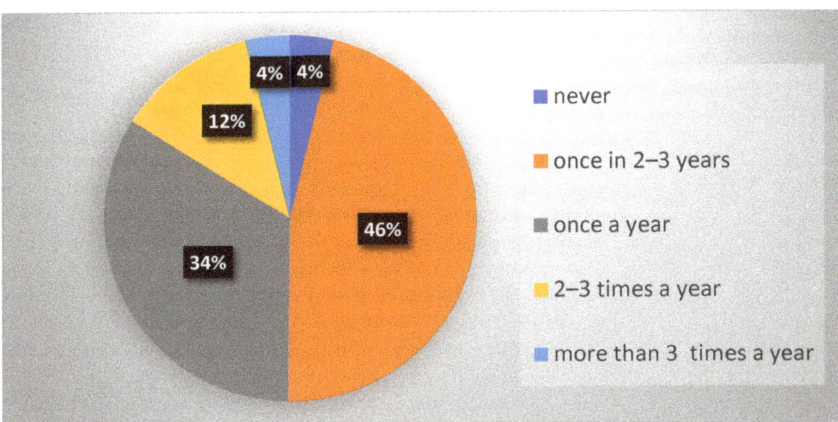

Figure 1. Frequency of taking antibiotics in all surveyed students.

2.2. Students' Knowledge Regarding the Antibiotics Use

Respondents showed good knowledge regarding the use of antibiotics (Table 2). The percentage of correct answers was higher than 80% in 9 of the 12 proposed claims. The lowest percentage of correct answers was recorded for the statement "Antibiotics are taken until the whole package is consumed". Of the 400 students surveyed, only 64% gave the correct answer. Of the 400 respondents, 250 (62.5%) showed adequate knowledge while 150 (37.5%) had inadequate knowledge. The median knowledge of all surveyed students was 11 (interquartile range, 5–11). Students of medicine and students of dentistry had a significantly higher knowledge score compared to students of veterinary medicine (p < 0.001). The median knowledge of students of medicine was 11 (interquarter range, 10–12), the median knowledge of students of dentistry was 11 (interquarter range,

10–12), while the median knowledge of student students of veterinary medicine was 10 (interquarter range, 9–11).

Table 2. Knowledge of students of medicine (M), dentistry (D) and veterinary medicine (V) use antibiotic use.

	M		D		V		χ^2	p
	n	%	*n*	%	*n*	%		
Antibiotics are used to decrease fever	174	87	91	91	75	75	11.234	0.004
Antibiotics are used to decrease pain	190	95	95	95	88	88	5.839	0.054
Antibiotics are used to overcome malaise and fatigue	197	98.5	97	97	97	97	1.023	0.501
Antibiotics are used for common cold	171	85.5	91	91	63	63	30.474	<0.001
Antibiotic treatment begins with an antibiotic found at home in order not to waste time	180	90	95	95	94	94.9	3.551	0.169
Antibiotic treatment is started after a visit to doctor and with a doctor's prescription	194	97	97	97	97	97	0.125	1
Antibiotic treatment is started when it is advised by a pharmacist	171	85.5	79	79	67	67	13.88	<0.001
Antibiotic is used until the symptoms disappear	167	83.5	88	88	75	75	6.13	0.047
Antibiotic is used until the bottle finishes	136	68	65	65	55	55	4.98	0.084
Antibiotic is used as long as the doctor prescribes	194	97	97	97	96	96	0.239	0.933
Taking the medicine twice a day means after waking up and before going to bed	136	81.5	80	80	64	64	12.231	0.002
Frequent and improper use of antibiotics is harmful and dangerous	194	97	98	98	95	95	1.511	0.501

2.3. Practice and Attitudes of Respondents toward Antibiotics Use

A lower share of students of medicine and students of dentistry compared to students of veterinary medicine irregularly used antibiotics prescribed by doctors ($p < 0.032$) (Table 3). To the question "What do you do when you think the antibiotics you are taking are ineffective?", 72% of students of medicine and 65% of students of dentistry stated that they continue to take antibiotics recommended by a doctor. The share of students of veterinary medicine who stated the same, as well as those who reported that they stop taking antibiotics and go to the doctor, was the same (43%) ($p < 0.001$). The share of students of medicine (74.5%) who used antibiotics as recommended by their doctors during their last infection was higher compared to students of dentistry (61%) and students of veterinary medicine (63%) ($p < 0.047$).

Among sociodemographic and academic characteristics, and responses related to student practice and attitudes, six predictors of adequate knowledge were statistically significant in univariate models ($p < 0.05$). The multivariate logistic regression model that predicts adequate knowledge showed that female students have 1.8 times higher chance of adequate knowledge (OR = 1.8; B = 0.571; $p = 0.020$) than male students (Table 4). Students of medicine (OR = 2.2; B = 0.802; $p = 0.006$) and students of dentistry (OR = 2.2; B = 0.769; $p = 0.026$) have a 2.2 times higher chance of adequate knowledge compared to students V as a reference category. Students with an average grade of 8 and higher have an over 3 times higher chance of adequate knowledge (OR = 3.3; B = 1.204; $p < 0.001$) compared to students with a lower average grade. Students who took antibiotics during the last infection for a period the doctor recommended (OR = 7.1; B = 1.964; $p < 0.001$) and those who took antibiotics until the whole package was finished have a better chance of adequate knowledge (OR = 2.6; B = 0.974; $p < 0.001$) compared to those students who used antibiotics until their symptoms relieved as a reference category (7.1 times and 2.6 times the chance, respectively).

Table 3. Attitudes and practice of students of medicine (M), dentistry (D) and veterinary medicine (V) about antibiotic use.

	M		D		V		χ^2	p
	n	%	*n*	%	*n*	%		
Have you ever used antibiotics in order not to get ill?								
No	200	100	100	100	100	100		1
Have you ever started antibiotics on your own when you got ill?								
Yes	75	37.5	37	37	32	32	0.933	0.627
Have you ever used antibiotics prescribed by the doctor irregularly?								
Yes	42	21	21	21	34	34	6.9	0.032
What do you do when you think that antibiotic you are taking is not effective?								
I stop taking it and go to the doctor	36	18	22	22	43	43		
I stop taking it and go to another doctor	8	4	1	1	6	6	32.444	<0.001
I use it for the recommended period	144	72	65	65	43	43		
Other	12	6	12	12	8	8		
How did you use antibiotics during your last infection?								
Until the bottle is finished	38	19	27	27	22	22		
Until the symptoms disappeared	13	6.5	12	12	15	15	9.634	0.047
As advised by the doctor	149	74.5	61	61	63	63		
How did you get antibiotics during your last infection?								
I used the antibiotic previously used or as advised by my friends or relatives	5	2.5	8	8	5	5		
I used the antibiotic previously prescribed by my doctor	17	8.5	4	4	9	9		
I visited my doctor and used the prescribed antibiotic	144	72	73	73	69	69	8.354	0.392
I asked the pharmacist and used the antibiotic recommended	8	4	4	4	2	2		
I do not remember	26	13	11	11	15	15		

Table 4. Multivariate logistic regression with adequate knowledge of antibiotics as a dependent variable.

Independent Variables	B	p	OR	95% Confidence Intervals	
				Lower	Upper
Male gender	Reference category				
Female gender	0.571	0.02	1.8	1.095	0.571
V	Reference category				
M	0.082	0.006	2.23	1.25	3.96
D	0.769	0.026	2.16	1.09	4.24
How often do you take antibiotics?	−0.214	0.132	0.087	0.611	1.07
Average grade	1.204	<0.001	3.332	1.77	6.28
Have you ever used antibiotics prescribed by the doctor irregularly?	−0.376	0.174	0.687	0.39	1.18
How did you use antibiotics during your last infection?					
Until the symptoms disappeared	Reference category				
As advised by the doctor	1.964	<0.001	7.13	3.01	16.83
Until the bottle is finished	0.974	0.036	2.649	1.06	6.58

M—students of medicine; D—students of dentistry; V—students of veterinary medicine. The multivariate logistic regression model included 3 predictors of self-medication, which were analyzed on 400 subjects (of whom 171 reported self-medication). The tendency for practicing self-medication significantly increased with more frequent use of antibiotics ($p < 0.001$) as shown in Table 5. Students who irregularly took antibiotics prescribed by their doctor had twice the chance of self-medication (OR = 2.1; B = 0.719; $p = 0.007$). Students who took antibiotics in the last infection until they finished the whole package (OR = 2.7;

B = 1.010; $p < 0.001$) or until symptoms disappeared (OR = 8.5; B = 2.142; $p < 0.001$) were more likely prone to self-medication in relation to those students who used antibiotics for the period the doctor recommended as a reference category (2.1 and 8.5 times higher chance, respectively).

Table 5. Multivariate logistic regression model with self-medication as a dependent variable.

Independent Variables	B	*p*	OR	95% Confidence Intervals	
				Lower	Upper
How often do you take antibiotics?	0.587	<0.001	1.798	1.35	2.38
Have you ever used antibiotics prescribed by the doctor irregularly?	0.719	0.007	2.052	1.21	3.47
As advised by the doctor	Reference category				
Until the bottle is finished	1.01	<0.001	2.746	1.62	4.65
Until the symptoms disappeared	2.142	<0.001	8.514	3.33	21.76

Reasons for Self-Medication with Antibiotics

The most common reason for self-medication was sore throat (28% for students of medicine, 22% for students of veterinary medicine and 21% for students of dentistry). In addition to this cause for practicing self-medication, students of veterinary medicine and students of dentistry were most likely to list common cold as a reason for self-medication, whereas students of medicine were most prone to self-medication in case of cystitis and cough (15.5% and 15.0%, respectively) (Figure 2).

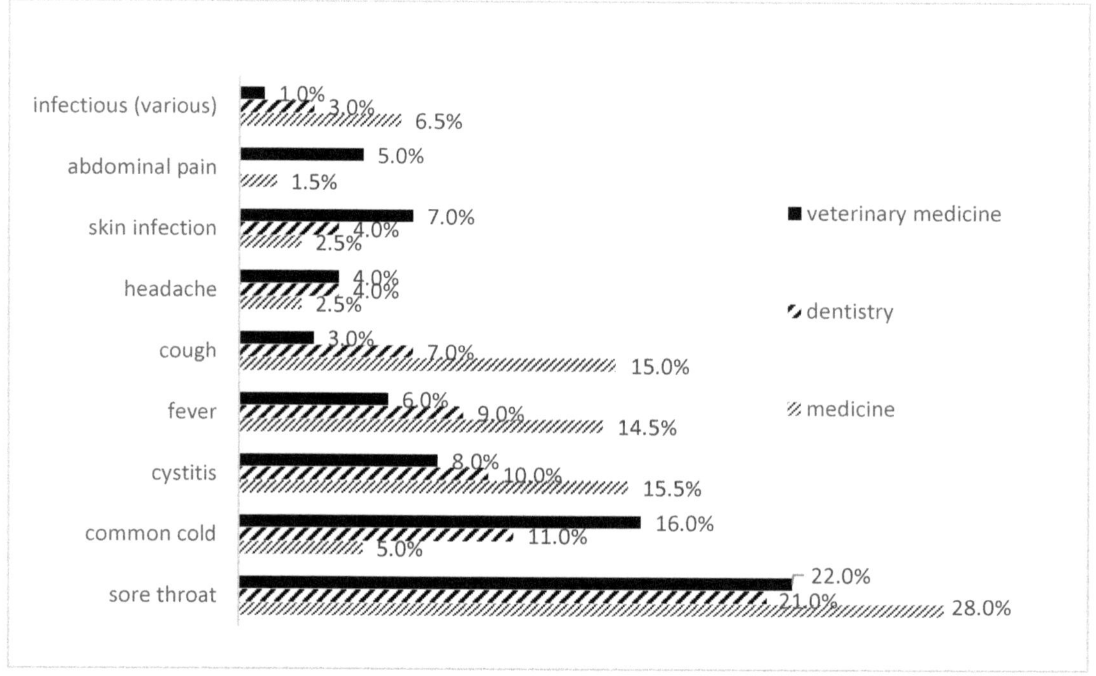

Figure 2. Indications listed as reasons for self-medication with antibiotics among medical, dental and veterinary medicine students in 2019 in Serbia.

3. Discussion

No detailed study assessing KAP of Serbian students of medicine, dentistry and veterinary medicine about antibiotics has to our knowledge been published so far. This is also a unique study that simultaneously surveys students on three different courses, gaining valuable insight into this topic among prospective prescribers.

Our findings showed how the healthcare profession students had a fairly good knowledge about antibiotics, although the percentage of students with adequate knowledge was only slightly higher than among the general population in Serbia (62.5% vs. 61.6%, respectively) [17]. More than 80% of the sample answered correctly to all the knowledge-related items administered, with two important exceptions. Indeed, almost 40% of the sample stated that antibiotics should be taken until the whole package is consumed and around 25% of students reported that taking antibacterial drugs twice a day means after waking up and before going to bed. Previous studies have found similar misconceptions regarding adherence to antibiotic regimens among the general population in Serbia, but we found a small reduction in these beliefs among students who were at the final stages of their course [17]. Given the crucial role that healthcare professionals play as daily communicators with the general public, we think it is important that students completing their healthcare course are aware of and able to communicate certain basic concepts, for example, that inappropriate use of antibiotics could harm both humans and animals and increase the prevalence of resistant strains. In addition, our results agree with other findings about knowledge of antibiotic use reported among students in China Ethiopia, Poland and Italy, as well as among adults in Jordan [18–22]. Regarding the Balkan region, single-center studies have found gaps in knowledge on antibiotic use among students of medicine and pharmacy in Croatia and among adults in Albania, Kosovo and the Republic of Macedonia [23–26]. However, some studies showed that knowledge of antibiotic use was moderate to poor among medical and veterinary students [13,27,28].

Although 81.3% of the total sample stated antibiotics cannot be used for the treatment of common cold, it is surprising that this percentage was significantly lower among students of veterinary medicine (63%) compared to students of medicine (85.1%) and dentistry (91%) ($p < 0.001$). Similar results have been observed among students of veterinary medicine in Bangladesh, where about 30% did not know that antibiotics were ineffective against viruses and only 56.79% believed antibiotics could not speed up the recovery of common cold. On the contrary, most students (92%) on selected healthcare courses in the United Kingdom agreed that most coughs, colds and sore throats get better on their own without the need for antibiotics, but 25% of students D still thought that antibiotics were effective against colds [29]. Similar findings have been reported among students of medicine by many studies, which could indicate their greater familiarity with the indications [21,30,31]. Considering the impact of veterinary students' future professional practices on ABR, it would be relevant to conduct similar surveys in this group, because a large proportion of these students were unaware that antibiotics do not cure viral infections such as the common cold. Evidence of improper antibiotic behavior patterns were also found among future prescribers of antibiotics in our survey. Whereas only 21% of students of medicine and students of dentistry reported irregular antibiotic use, among students of veterinary medicine 34% stated this practice, which is near to the percentage recorded among the general population in Serbia (40%) [17]. There are many conditions in which stopping antibiotics upon resolution of symptoms increases the risk of the patient experiencing a relapse with resistant pathogenic bacteria [32]. Interestingly, in our study, students of dentistry and veterinary medicine more frequently used antibiotics only until their symptoms resolved compared to students of medicine ($p = 0.047$). This practice, besides leading to relapse of infection and lack of treatment efficacy, also contributes to the emergence and spread of antibiotics resistance, a very current problem in Serbia [14,33].

It is additionally worth noting that students of dentistry (61%) and students of veterinary medicine (63%) showed decreased willingness to use antibiotics during their last infection as advised by the doctor compared to students of medicine (74.5%) ($p = 0.047$).

It may be a symptom of lack of trust in colleagues and lack of humility, even though it is increasingly apparent that One Health approaches are essential to tackling ABR [17,34]. As with prescribing doctors and veterinarians, who often lay the blame for ABR and poor behaviors on other groups, students of veterinary medicine in the study of Golding et al. (2022) also consider their own (future) prescribing to be less of a contributor to ABR than that of other prescribers [35]. Furthermore, the results of the present study are encouraging if compared to the Serbian general population. As much as 7.5% of the prospective healthcare professionals in our study used leftover antibiotics during their last infections, which is considerably lower than among the general population (17%) [17]. This practice was higher among medical students in Italy and Turkey and veterinary students in Portugal and Bangladesh [12,21,36,37]. An effective strategy to combat incorrect antibiotics use consists of continuing education to improve knowledge and awareness of both professionals and the general population [21,38,39].

Some of the most interesting contributions of the present study are the conclusions derived from the logistic regression analysis conducted to determine predictors of adequate knowledge among prospective prescribers of antibiotics.

Firstly, this study revealed a sex difference in KAP toward knowledge on antibiotic use. Female students showed almost a two times higher chance of adequate knowledge than male students. This is similar to a study performed in Italy among medical, dental and nursing students as well as among university students in the United Arab Emirates, whereas this difference was not recorded among students in Mumbai [21,40,41] The fact that gender plays a vital role in antibiotics is well-documented [42]. In the survey among the outpatients in England and Wales, females were prescribed higher antibiotics than males, and they suffered the most from antibiotic misuse as they showed lack of knowledge regarding the proper use of antibiotics in comparison to their male counterparts [43]. Secondly, students of medicine and dentistry have a more than two times higher chance of adequate knowledge compared to students of veterinary medicine.

This agrees with other findings about adequate knowledge on antibiotic use among prospective healthcare professionals reported in the literature [9,18,19,21]. On the other hand, studies among final-year veterinary students reported knowledge gaps on antibiotic use [12,31]. There is some evidence that, as would be expected, medical and veterinary curricula do have the desired effect of improving knowledge about responsible use of antimicrobials [44,45]. Although general practitioners and dentists are responsible for nearly 80% of all antibiotic prescriptions written, efforts should also be intensified for greater exposure to the prudent use of antibiotics during clinical teaching in the final year of veterinary school, which may lead to greater awareness and perceptions of ABR [46].

Our results showed that the most common reason for self-medication with antibiotics reported by all students' groups was sore throat. However, the most reported condition for students of veterinary medicine and students of dentistry was common cold, whereas students of medicine were most likely to practice self-medication with antibiotics in case of cystitis and cough. This is comparable with the results of some previous studies. For example, a recent study performed in Saudi Arabia among future prescribers showed that sore throat was the most common reported symptom, followed by fever and cough [47]. In addition, studies conducted in Nepal, India, Sri Lanka and China all showed that the common cold was the most common medical condition for which medical students self-medicated with antibiotics [18,48,49]. Slightly different results were observed in the Sudanese study, where the most common symptoms treated by medical students with non-prescription antibiotics were acute respiratory tract symptoms, cough and common cold [50]. Furthermore, the findings from the UK study showed that 25% of dentistry students were of the opinion that antibiotics were effective against colds [51].

Logistic regression models also showed interesting results in terms of identifying the most relevant factors influencing self-medication with antibiotics of interviewed students. Students who more frequently used antibiotics and who irregularly took them showed a higher probability for practicing self-medication. Similarly, among 1042 medical students from two universities in Saudi Arabia, 97.2% of them had used antibiotics in the previous year, whereas half of them reported self-medication [52]. Several scientists agree that it is essential to provide more information about antibiotics and the possible adverse effects of their improper use because such information helps decrease the frequency of use and encourages the correct use of these drugs. However, it has also been documented that knowledge does not always correlate with practice [53].

Finally, the present study also highlighted that students who used antibiotics during the last infection until the bottle was finished or those who reported cessation of antibiotic use after their symptoms resolved were less likely to have adequate knowledge and were more likely to self-medicate than those who reported antibiotic usage for the period advised by the doctor. These findings are in line with those obtained in other studies, which points to the need for more education on antibiotics use in degree courses, especially as students are future clinicians, teachers and researchers who will be at the forefront in educating the general public [9,35,37].

The strength of the present study is that this is the first survey exploring the knowledge, attitude and practice toward antibiotics among prospective prescribers in Serbia, and is thus a very valuable contribution to the current literature. Another strength of the present study that should be highlighted is a large sample size. However, despite the large sample size, interpretation of the findings should be done with caution due to the self-administered questionnaire and the possibility of the participants to over- or under-report socially desirable behaviors. The second limitation that should be mentioned is that the study was conducted at only one university in Serbia, and thus the findings may not be with certainty extrapolated to the national level. Notwithstanding these limitations, we have succeeded in identifying some key knowledge gaps among the medical students who participated in our study and provided useful findings for public health policymakers in Serbia.

4. Materials and Methods

4.1. Study Design

The study was conducted in January and February 2019 at the University of Novi Sad, one of the largest Serbian universities. The study population comprised a total of 400 final-year students in the 2018/2019 academic year of the integrated academic study programs of medicine (200), dentistry (100) and veterinary medicine (100). The selected groups of students are prospective healthcare professionals who will be permitted to prescribe antibiotics in the Republic of Serbia. The sample size, calculated based on the estimate that 81.3% of medical science students practice self-medication, was 234, with a relative accuracy of 10% and a confidence interval of 95% [16,17]. The sample size was intentionally exceeded, taking into account the possibility of excluding respondents from the study, withdrawal and the need for subgroup analysis. After signing the informed consent, the respondents filled out the questionnaire. The research was approved by the Ethical Committee of the Faculty of Medicine in Novi Sad (approval number 01-39/290/1).

4.2. Questionnaire Design

The questionnaire used in this study was based on the questionnaire of Buke et al. [38] (Annex 1). Necessary modifications have been made to provide correct answers to questions and claims. The content, comprehension, readability and appearance of the questionnaire were previously tested on 30 students at the University of Novi Sad who were not included in the final analyses.

The questionnaire was divided into three parts. The first part, which contained 8 questions, included demographic characteristics of respondents and some general questions (age, gender, year of study, field of study, average grade, place of residence, number of visits to a general practitioner in the past 12 months, is anyone in the family a healthcare worker). The second part referred to the respondents' knowledge regarding the use of antibiotics. This part contained 12 statements with the possibility of answering with true or false. The questions were about the conditions that require the use of antibiotics, how to start antibiotic treatment, the duration of antibiotic therapy, the meaning of the claim "take twice a day", and whether frequent and inadequate use of antibiotics can be harmful. The knowledge score was determined by giving one point for each correct answer and the maximum score was 12.

The third part, which consisted of 15 questions with the possibility of answering yes or no, as well as multiple-choice answers, referred to the attitudes and practice of students toward antibiotics. This part was focused on the prophylactic use of antibiotics, self-medication with antibiotics, adherence to the antibiotic therapy, on whose recommendation the students take antibiotics, and how they used them during the previous infection.

4.3. Data Analysis

Descriptive and comparative statistical analysis of the results was performed with IBMSPSS Statistics 22 (IBM Corporation, Armonk, NY, USA) software. In addition to descriptive statistical methods, measurements of central tendency (arithmetic mean, median), measures of variability (standard deviation) and frequency were also used. Two variables were created, antibiotic knowledge status (adequate–inadequate) and self-medication (yes–no). The categories of knowledge were determined according to the median recent knowledge of the total sample (11, interquartile range 5–12). Respondents were grouped into those with adequate (knowledge score ≥ 11) and those with inadequate knowledge score (knowledge score ≤ 11). Self-medication was analyzed based on questions from the third part of the questionnaire. The variable self-medication was formed by combining the answers to the questions "Have you ever used antibiotics to avoid getting sick" (those who answered "yes"), "Have you ever started taking antibiotics on your own" (those who answered "yes") and "The last time you used antibiotic, who recommended it to you?"(those who answered any of the following—"I used previously used antibiotic or advised by a friend; I used antibiotics previously prescribed by my doctor; I asked a pharmacist and used the antibiotics recommended").

For testing statistical hypotheses, the χ^2 test, Mann-Whitney U test, analysis of variance with post hoc test, the Kruskal-Wallis test and Fisher test of exact probability were used. All p values less than 0.05 were considered significant.

The association of the students' characteristics with adequate knowledge about antibiotics and self-medication was primarily evaluated using univariate logistic regression. Multivariate logistic regression also included predictors of all variables with $p < 0.05$ in univariate analysis. The results are presented as odds ratio (OR) with 95% confidence interval (CI). All p values less than 0.05 were considered significant.

Although students of medicine and students of dentistry demonstrated a higher level of knowledge about antibiotic use than students of veterinary medicine, as far as attitude and practices regarding ABR and usage are concerned, there is a significant need for improvement among all three groups of future prescribers. It is a reasonable starting point as a way to tackle the problem from both human and animal welfare standpoints. Therefore, our study provides an important insight regarding their knowledge, attitude and practice, which can be considered, in order to plan for an effective undergraduate curriculum regarding antibiotic resistance and usage. The solution to this problem requires a concerted One Health approach to mitigate future risks to humans, animals and the environment.

Author Contributions: Conceptualization, O.H.; methodology, O.H.; software, D.B.; validation, A.T.P., M.P.K. and B.J.; formal analysis, D.B.; investigation, A.T.P., M.P.K. and Z.K.; resources, O.H.; data curation, A.T.P. and Z.K.; writing—original draft preparation, O.H. and B.J.; writing—review and editing, M.P.K. and Z.K.; visualization, A.T.P.; supervision, O.H.; project administration, O.H.; funding acquisition, O.H. All authors have read and agreed to the published version of the manuscript.

Funding: This work was supported by the Provincial Secretariat for Higher Education and Scientific Research project 142-451-2331/2022-01.

Institutional Review Board Statement: The study was conducted according to the guidelines of the Declaration of Helsinki, and approved by the Ethical Committee of the Faculty of Medicine in Novi Sad (approval number 01-39/290/1).

Informed Consent Statement: Informed consent was obtained from all subjects involved in the study.

Data Availability Statement: All data are available upon a reasonable request from the authors.

Acknowledgments: The authors would like to thank all the respondents who participated in this survey.

Conflicts of Interest: The authors declare no conflict of interest.

References

1. McEwen, S.A.; Collignon, P.J. Antimicrobial Resistance: A One Health Perspective. *Microbiol. Spectr.* **2018**, *6*, 2. [CrossRef]
2. Lloyd, D.H.; Page, S.W.; Aarestrup, F.M.; Schwarz, S.; Shen, J.; Cavaco, L. Antimicrobial stewardship in veterinary medicine. *Microbiol. Spectr.* **2018**, *6*. [CrossRef]
3. Graham, D.W.; Bergeron, G.; Bourrassa, M.W.; Dickson, J.; Gomes, F.; Howe, A.; Kahn, L.H.; Morley, P.S.; Scott, H.M.; Simjee, S.; et al. Complexities in understanding antimicrobial resistance across domesticated animal, human, and environmental systems. *Ann. N. Y. Acad. Sci.* **2019**, *1441*, 17–30. [CrossRef]
4. The Evolving Threat of Antimicrobial Resistance-Options for Action. Available online: http://apps.who.int/iris/bitstream/1066 5/44812/1/9789241503181_eng.pdf (accessed on 20 March 2022).
5. Cantas, L.; Suer, K. The important bacterial zoonoses in "one health" concept. *Front. Public Health* **2014**, *2*, 144. [CrossRef] [PubMed]
6. van den Bogaard, A.E.; Stobberingh, E.E. Epidemiology of resistance to antibiotics. Links between animals and humans. *Int. J. Antimicrob. Agents* **2000**, *14*, 327–335. [CrossRef]
7. CDC. Core Elements of Hospital Antibiotic Stewardship Programs. Centers for Disease Control and Prevention. 2014. Available online: http://www.cdc.gov/getsmart/healthcare/pdfs/checklist.pdf (accessed on 22 September 2021).
8. Gozdzielewska, L.; King, C.; Flowers, P.; Mellor, D.; Dunlop, P.; Price, L. Scoping review of approaches for improving antimicrobial stewardship in livestock farmers and veterinarians. *Prev. Vet. Med.* **2020**, *180*, 105025. [CrossRef]
9. Suaifan, G.A. A cross-sectional study on knowledge, attitude and behavior related to antibiotic use and resistance among medical and non-medical university students in Jordan. *Afr. J. Pharm. Pharmacol.* **2012**, *6*. [CrossRef]
10. Jairoun, A.; Hassan, N.; Ali, A.; Jairoun, O.; Shahwan, M. Knowledge, attitude and practice of antibiotic use among university students: A Cross Sectional Study in UAE. *BMC Public Health* **2019**, *19*, 518. [CrossRef]
11. Sannathimmappa, M.B.; Nambiar, V.; Aravindakshan, R. A cross-sectional study to evaluate the knowledge and attitude of medical students concerning antibiotic usage and antimicrobial resistance. *Int. J. Acad. Med.* **2021**, *7*, 113–119. [CrossRef]
12. Chapot, L.; Sarker, M.S.; Begum, R.; Hossain, D.; Akter, R.; Hasan, M.M.; Bayzid, M.; Salauddin, M.; Parvej, M.S.; Uddin, A.M.; et al. Knowledge, attitudes and practices regarding antibiotic use and resistance among veterinary students in Bangladesh. *Antibiotics* **2021**, *10*, 332. [CrossRef] [PubMed]
13. Pena Betancourt, S.D.; Posadas Pena, S.D.; Parra-Forero, L.Y. The knowledge of antibiotics in veterinary students and repercution in human health. *Health* **2020**, *12*, 1632–1639. [CrossRef]
14. WHO. Central Asian and Eastern European Surveillance of Antimicrobial Resistance. Annual Report. 2020. Available online: http://www.euro.who.int/__data/assets/pdf_file/0007/386161/52238-WHO-CAESAR-AR2018_low_V11_web.pdf?ua=1 (accessed on 22 September 2021).
15. WHO. *Report on Surveillance of Antibiotic Consumption: 2016–2018. Early Implementation*; World Health Organization: Geneva, Switzerland, 2018; Available online: https://www.who.int/medicines/areas/rational_use/who-amr-amc-report20181109.pdf (accessed on 16 September 2021).
16. Republic of Serbia: National Antibiotic Resistance Control Programme for the Period 2019–2021. Available online: https://www.who.int/publications/m/item/republic-of-serbia-national-antibiotic-resistance-control-programme-for-the-period-2019-2021 (accessed on 22 September 2021).
17. Horvat, O.J.; Tomas, A.D.; Paut Kusturica, M.M.; Savkov, A.V.; Bukumiric, D.U.; Tomic, Z.S.; Sabo, A.J. Is the level of knowledge a predictor of rational antibiotic use in Serbia? *PLoS ONE* **2017**, *12*, e0180799. [CrossRef]
18. Huang, Y.; Gu, J.; Zhang, M.; Ren, Z.; Yang, W.; Chen, Y.; Fu, Y.; Chen, X.; Cals, J.W.; Zhang, F. Knowledge, attitude and practice of antibiotics: A questionnaire study among 2500 Chinese students. *BMC Med. Educ.* **2013**, *13*, 163. [CrossRef] [PubMed]
19. Ganesh, M.; Sridevi, S.; Paul, C. Antibiotic use among medical and Para medical students: Knowledge, attitude and its practice in a tertiary health care Centre in Chennai—A scientific insight. *Int. J. Sci. Res.* **2014**, *3*, 332–335.
20. Shehadeh, M.; Suaifan, G.; Darwish, R.M.; Wazaify, M.; Zaru, L.; Alja'fari, S. Knowledge, attitudes and behavior regarding antibiotics use and misuse among adults in the community of Jordan. A pilot study. *Saudi Pharm. J.* **2012**, *20*, 125–133. [CrossRef] [PubMed]

21. Scaioli, G.; Gualano, M.R.; Gili, R.; Masucci, S.; Bert, F.; Siliquini, R. Antibiotic use: A cross-sectional survey assessing the knowledge, attitudes and practices amongst students of a school of medicine in Italy. *PLoS ONE* **2015**, *10*, e0122476. [CrossRef]
22. Sobierajski, T.; Mazińska, B.; Chajęcka-Wierzchowska, W.; Smiałek, M.; Hryniewicz, W. Antimicrobial and Antibiotic Resistance from the Perspective of Polish Veterinary Students: An Inter-University Study. *Antibiotics* **2022**, *11*, 115. [CrossRef]
23. Humphreys, H.; Dillane, T.; O'Connell, B.; Luke, L.C. Survey of recent medical graduates' knowledge and understanding of the treatment and prevention of infection. *Ir. Med. J.* **2006**, *99*, 58–59.
24. Rusic, D.; Bozic, J.; Vilovic, M.; Bukic, J.; Zivkovic, P.M.; Leskur, D.; Seselja Perisin, A.; Tomic, S.; Modun, D. Attitudes and Knowledge Regarding Antimicrobial Use and Resistance Among Pharmacy and Medical Students at the University of Split, Croatia. *Microb. Drug Resist.* **2018**, *24*, 1521–1528. [CrossRef]
25. Kaae, S.; Malaj, A.; Hoxha, I. Antibiotic knowledge, attitudes and behaviours of Albanian health care professionals and patients —A qualitative interview study. *J. Pharm. Policy. Pract.* **2017**, *10*, 13. [CrossRef]
26. Zajmi, D.; Berisha, M.; Begolli, I.; Hoxha, R.; Mehmeti, R.; Mulliqi-Osmani, G.; Kurti, A.; Loku, A.; Raka, L. Public knowledge, attitudes and practices regarding antibiotic use in Kosovo. *Pharm. Pract.* **2017**, *15*, 827. [CrossRef] [PubMed]
27. Alili-Idrizi, E.; Dauti, M.; Malaj, L. Validation of the parental knowledge and attitude towards antibiotic usage and resistance among children in Tetovo, the Republic of Macedonia. *Pharm. Pract.* **2014**, *12*, 467. [CrossRef] [PubMed]
28. Azevedo, M.M.; Pinheiro, C.; Yaphe, J.; Baltazar, F. Portuguese students' knowledge of antibiotics: A cross-sectional study of secondary school and university students in Braga. *BMC Public Health* **2009**, *9*, 359. [CrossRef]
29. Dyar, O.J.; Howard, P.; Nathwani, D.; Pulcini, C. Knowledge, attitudes, and beliefs of French medical students about antibiotic prescribing and resistance. *Med. Mal. Infect.* **2013**, *43*, 423–430. [CrossRef]
30. Jairoun, A.; Hassan, N.; Ali, A.; Jairoun, O.; Shahwan, M. A cross sectional study on knowledge, attitude and practice of medical students toward antibiotic resistance and its prescription, Iran. *Adv. Environ. Biol.* **2014**, *8*, 675–681.
31. Jamshed, S.Q.; Elkalmi, R.; Rajiah, K.; Al-Shami, A.K.; Shamsudin, S.H.; Siddiqui, M.J.; Abdul Aziz, M.A.; Hanafi, M.B.; Mohammad Shariff, N.I.; Ramlan, N.H.; et al. Understanding of antibiotic use and resistance among final-year pharmacy and medical students: A pilot study. *J. Infect. Dev. Ctries* **2014**, *8*, 780–785. [CrossRef]
32. Adembri, C.; Novelli, A. Pharmacokinetic and pharmacodynamic parameters of antimicrobials: Potential for providing dosing regimens that are less vulnerable to resistance. *Clin. Pharmacokinet.* **2009**, *48*, 517–528. [CrossRef]
33. Popović, R.; Tomić, Z.; Tomas, A.; Anđelić, N.; Vicković, S.; Jovanović, G.; Bukumirić, D.; Horvat, O.; Sabo, A. Five-year surveillance and correlation of antibiotic consumption and resistance of Gram-negative bacteria at an intensive care unit in Serbia. *J. Chemother.* **2020**, *32*, 294–303. [CrossRef]
34. Sobierajski, T.; Mazińska, B.; Wanke-Rytt, M.; Hryniewicz, W. Knowledge-based attitudes of medical students in antibiotic therapy and antibiotic resistance. A cross-sectional study. *Int. J. Environ. Res. Public Health* **2021**, *18*, 3930. [CrossRef]
35. Golding, S.E.; Higgins, H.M.; Ogden, J. Assessing Knowledge, Beliefs, and Behaviors around Antibiotic Usage and Antibiotic Resistance among UK Veterinary Students: A Multi-Site, Cross-Sectional Survey. *Antibiotics* **2022**, *11*, 256. [CrossRef]
36. Marta-Costa, A.; Miranda, C.; Silva, V.; Silva, A.; Martins, Â.; Pereira, J.E.; Maltez, L.; Capita, R.; Alonso-Calleja, C.; Igrejas, G.; et al. Survey of the Knowledge and Use of Antibiotics among Medical and Veterinary Health Professionals and Students in Portugal. *Int. J. Environ. Res. Public Health* **2021**, *18*, 2753. [CrossRef] [PubMed]
37. Buke, C.; Hosgor-Limoncu, M.; Ermertcan, S.; Ciceklioglu, M.; Tuncel, M.; Kose, T.; Eren, S. Irrational use of antibiotics among university students. *J. Infect.* **2005**, *51*, 135–139. [CrossRef] [PubMed]
38. Alzahrani, A.A.H.; Alzahrani, M.S.A.; Aldannish, B.H.; Alghamdi, H.S.; Albanghali, M.A.; Almalki, S.S.R. Inappropriate dental antibiotic prescriptions: Potential driver of the antimicrobial resistance in Albaha region, Saudi Arabia. *Risk Manag. Healthc. Policy* **2020**, *13*, 175–182. [CrossRef] [PubMed]
39. Aljadeeah, S.; Wirtz, V.J.; Nagel, E. Outpatient Antibiotic Dispensing for the Population with Government Health Insurance in Syria in 2018–2019. *Antibiotics* **2020**, *9*, 570. [CrossRef]
40. Jairoun, A.; Hassan, N.; Ali, A.; Jairoun, O.; Shahwan, M.; Hassali, M. University students' knowledge, attitudes, and practice regarding antibiotic use and associated factors: A cross-sectional study in the United Arab Emirates. *Int. J. Gen. Med.* **2019**, *12*, 235–246. [CrossRef]
41. Limaye, D.; Naware, S.; Bare, P.; Dalvi, S.; Dhurve, K.; Sydymanov, A.; Limaye, V.; Pitani, R.S.; Kanso, Z.; Fortwengel, G. Knowledge, attitude and practices of antibiotic usage among students from Mumbai University. *Int. J. Res. Med. Sci.* **2018**, *6*, 1908–1912. [CrossRef]
42. Schröder, W.; Sommer, H.; Gladstone, B.P.; Foschi, F.; Hellman, J.; Evengard, B.; Tacconelli, E. Gender differences in antibiotic prescribing in the community: A systematic review and meta-analysis. *J. Antimicrob. Chemother.* **2016**, *71*, 1800–1806. [CrossRef]
43. Majeed, A.; Moser, K. Age-and sex-specific antibiotic prescribing patterns in general practice in England and Wales in 1996. *Br. J. Gen. Pract.* **1999**, *49*, 735–736.
44. Odetokun, I.A.; Akpabio, U.; Alhaji, N.B.; Biobaku, K.T.; Oloso, N.O.; Ghali-Mohammed, I.; Biobaku, A.J.; Adetunji, V.O.; Fasina, F.O. Knowledge of antimicrobial resistance among veterinary students and their personal antibiotic use practices: A national cross-sectional survey. *Antibiotics* **2019**, *8*, 243. [CrossRef]
45. Lv, B.; Zhou, Z.; Xu, G.; Yang, D.; Wu, L.; Shen, Q.; Jiang, M.; Wang, X.; Zhao, G.; Yang, S.; et al. Knowledge, attitudes and practices concerning self-medication with antibiotics among university students in western China. *Trop. Med. Int. Health* **2014**, *19*, 769–779. [CrossRef]

46. Hardefeldt, L.; Nielsen, T.; Crabb, H.; Gilkerson, J.; Squires, R.; Heller, J.; Sharp, C.; Cobbold, R.; Norris, J.; Browning, G. Veterinary students' knowledge and perceptions about antimicrobial stewardship and biosecurity—A national survey. *Antibiotics* **2018**, *7*, 34. [CrossRef] [PubMed]

47. Benameur, T.; Al-Bohassan, H.; Al-Aithan, A.; Al-Beladi, A.; Al-Ali, H.; Al-Omran, H.; Saidi, N. Knowledge, attitude, behaviour of the future healthcare professionals towards the self-medication practice with antibiotics. *J. Infect. Dev. Ctries* **2019**, *13*, 56–66. [CrossRef] [PubMed]

48. Pant, N.; Sagtani, R.A.; Pradhan, M.; Bhattarai, A.; Sagtani, A. Self-medication with antibiotics among dental students of Kathmandu-prevalence and practice. *Nepal Med. Coll. J.* **2015**, *17*, 47–53.

49. Banerjee, I.; Bhadury, T. Self-medication practice among undergraduate medical students in a tertiary care medical college, West Bengal. *J. Postgrad. Med.* **2012**, *58*, 127–131. [CrossRef] [PubMed]

50. Elmahi, O.; Musa, R.; Shareef, A.; Omer, M.; Elmahi, M.; Altamih, R.; Mohamed, R.; Alsadig, T. Perception and practice of self-medication with antibiotics among medical students in Sudanese universities: A cross-sectional study. *PLoS ONE* **2022**, *17*, e0263067. [CrossRef]

51. Dyar, O.J.; Hills, H.; Seitz, L.T.; Perry, A.; Ashiru-Oredope, D. Assessing the Knowledge, Attitudes and Behaviors of Human and Animal Health Students towards Antibiotic Use and Resistance: A Pilot Cross-Sectional Study in the UK. *Antibiotics* **2018**, *7*, 10. [CrossRef]

52. Harakeh, S.; Almatrafi, M.; Ungapen, H.; Hammad, R.; Olayan, F.; Hakim, R.; Ayoub, M.; Bakhsh, N.; Almasaudi, S.B.; Barbour, E.; et al. Perceptions of medical students towards antibiotic prescribing for upper respiratory tract infections in Saudi Arabia. *BMJ Open Respir. Res.* **2015**, *2*, e000078. [CrossRef]

53. Sarahroodi, S.; Arzi, A.; Sawalba, A.F.; Ashtarinezhad, A. Antibiotiotics self-medication among southern Iranian university students. *Int. J. Pharmacol.* **2010**, *6*, 48–52. [CrossRef]

antibiotics

Article

Identification of Therapeutic Targets in an Emerging Gastrointestinal Pathogen *Campylobacter ureolyticus* and Possible Intervention through Natural Products

Kanwal Khan [1,†], Zarrin Basharat [2,†], Khurshid Jalal [3,*,†], Mutaib M. Mashraqi [4], Ahmad Alzamami [5], Saleh Alshamrani [4] and Reaz Uddin [1]

1   PCMD, International Center for Chemical and Biological Sciences, University of Karachi, Karachi 75270, Pakistan; khankanwal011@gmail.com (K.K.); riaasuddin@yahoo.com (R.U.)
2   Jamil-ur-Rahman Center for Genome Research, Dr. Panjwani Center for Molecular Medicine and Drug Research, International Center for Chemical and Biological Sciences, University of Karachi, Karachi 75270, Pakistan; zarrin.iiui@gmail.com
3   HEJ Research Institute of Chemistry, International Center for Chemical and Biological Sciences, University of Karachi, Karachi 75270, Pakistan
4   Department of Clinical Laboratory Sciences, College of Applied Medical Sciences, Najran University, Najran 61441, Saudi Arabia; mmmashraqi@nu.edu.sa (M.M.M.); saalshamrani@nu.edu.sa (S.A.)
5   Clinical Laboratory Science Department, College of Applied Medical Science, Shaqra University, Al-Quwayiyah 11961, Saudi Arabia; aalzamami@su.edu.sa
*   Correspondence: khurshid@iccs.edu
†   These authors contributed equally to this work.

Citation: Khan, K.; Basharat, Z.; Jalal, K.; Mashraqi, M.M.; Alzamami, A.; Alshamrani, S.; Uddin, R. Identification of Therapeutic Targets in an Emerging Gastrointestinal Pathogen *Campylobacter ureolyticus* and Possible Intervention through Natural Products. *Antibiotics* 2022, 11, 680. https://doi.org/10.3390/antibiotics11050680

Academic Editors: Dóra Kovács and Simone Carradori

Received: 8 April 2022
Accepted: 16 May 2022
Published: 18 May 2022

Publisher's Note: MDPI stays neutral with regard to jurisdictional claims in published maps and institutional affiliations.

Abstract: *Campylobacter ureolyticus* is a Gram-negative, anaerobic, non-spore-forming bacteria that causes gastrointestinal infections. Being the most prevalent cause of bacterial enteritis globally, infection by this bacterium is linked with significant morbidity and mortality in children and immunocompromised patients. No information on pan-therapeutic drug targets for this species is available yet. In the current study, a pan-genome analysis was performed on 13 strains of *C. ureolyticus* to prioritize potent drug targets from the identified core genome. In total, 26 druggable proteins were identified using subtractive genomics. To the best of the authors' knowledge, this is the first report on the mining of drug targets in *C. ureolyticus*. UDP-3-O-acyl-N-acetylglucosamine deacetylase (LpxC) was selected as a promiscuous pharmacological target for virtual screening of two bacterial-derived natural product libraries, i.e., postbiotics ($n = 78$) and streptomycin ($n = 737$) compounds. LpxC inhibitors from the ZINC database ($n = 142$ compounds) were also studied with reference to LpxC of *C. ureolyticus*. The top three docked compounds from each library (including ZINC26844580, ZINC13474902, ZINC13474878, Notoginsenoside St-4, Asiaticoside F, Paraherquamide E, Phytoene, Lycopene, and Sparsomycin) were selected based on their binding energies and validated using molecular dynamics simulations. To help identify potential risks associated with the selected compounds, ADMET profiling was also performed and most of the compounds were considered safe. Our findings may serve as baseline information for laboratory studies leading to the discovery of drugs for use against *C. ureolyticus* infections.

Keywords: pan-genome; *Campylobacter ureolyticus*; UDP-3-O-acyl-N-acetylglucosamine deacetylase; LpxC; campylobacteriosis

1. Introduction

Campylobacter ureolyticus is a Gram-negative bacterium previously classified as *Bacteroides ureolyticus* and belongs to a class of pathogens that cause gastroenteritis [1]. *Campylobacter* outnumbers *Shigella*, *E. coli O157*, and *Salmonella* as the most common cause of bacterial enteritis [2]. Previously, the most common *Campylobacter* species linked with human illness were *C. jejuni* and *C. coli*, but breakthroughs in molecular diagnostics paired with the development of innovative culture methods have helped identify and isolate a range of under-reported and fastidious *Campylobacter* species, including *C. ureolyticus*. Farm animals are the primary reservoir for *Campylobacter* sp. infections and the primary cause of campylobacteriosis. Farm animals are also the leading source of bacterial food poisoning and *Campylobacter* gastrointestinal illnesses worldwide. Campylobacter foodborne illness is a concern and an expensive burden for the human population, accounting for 8.4% of all diarrhea cases worldwide. In most cases, *C. ureolyticus* has been recovered from human samples, with just one report of *C. ureolyticus* isolation from healthy horse endometria. Following that, a retrospective investigation of over 7000 patients with diarrhea found *C. ureolyticus* in 23.8% of *Campylobacter*-positive samples, marking the first discovery of *C. ureolyticus* in the faeces of gastroenteritis patients and highlighting the species' status as an emerging enteric pathogen [3–5].

C. ureolyticus has been accompanied by a variety of illnesses in the past, including bacterial vaginosis [6], gangrenous lesions [7], superficial ulcers [1,8], nongonococcal urethritis [1,9], male infertility [10] and more recently, ulcerative colitis [11]. Furthermore, *C. ureolyticus* has been associated with periodontal diseases, such as gingivitis and periodontitis, similar to numerous other new and atypical *Campylobacter* species [9]. Recent research has led to the discovery and isolation of *C. ureolyticus* as the sole pathogen in the feces of many diarrheal patients. It is now thought to be the second most common *Campylobacter* species found in patients suffering from diarrhea, surpassing the established pathogens *C. coli* and *C. jejuni*. Furthermore, an examination of infectivity data indicates the predominance of this pathogen in diarrheal patients between the ages of 5 and 70, suggesting that it impacts the pediatric population more than adults. Patient data for gastrointestinal illnesses suggest that *C. ureolyticus* is an emerging gastrointestinal pathogen [11].

Its pathogenesis is based on invasion, colonization, adhesion, and toxin release. Previously, the absence of genetic data hindered the understanding of in-depth pathogenic mechanisms and virulence, but the democratization of next-generation sequencing (NGS) has made it easier to explore the species at genome scale [12]. Pan-genomic studies help exploit the entire repertoire of genes from all strains within a species, which enables investigation of genomic diversity and similarity [13]. This type of information aids the understanding of species evolution and pathology and helps identify drug targets for pan-treatment of *C. ureolyticus* [7,14]. In light of this, the current study aimed to investigate multiple isolates of *C. ureolyticus* to better understand its pathogenesis and characterize the core and pan-genome subsets to prioritize novel therapeutic targets from shared core genes. To the best of our knowledge, this is the first pan-genome study on *C. ureolyticus*, as well as the first study to make use of core gene data to exploit therapeutic targets. Virtual screening was also carried out against a selected target, using natural product libraries to identify potent inhibitors that could prove useful for campylobacteriosis treatment.

2. Results

2.1. Pan-Genome and Resistome Analysis

The pan-genome of 13 *C. ureolyticus* strains was composed of ~5500 accessory, 1182 core, and 400 unique genes. The pan-genome curve showed a B_{pan} value of 0.17 (i.e., <1), representing the open nature of the pan-genome of this pathogen (Figure 1A). The comparative study showed that each strain shared ~300–500 genes in the accessory genome fraction, with the highest number retrieved for the strain NCTC10941 ($n = 535$ genes), while strain ACS-301-V-Sch3b harbored the maximum number of unique genes ($n = 114$ genes). Additionally, five strains (ACS-301-V-Sch3b, LFYP111, LMG 6451, RIGS 9880, and UMB0112) lacked certain genes that were present in other strains (Supplementary File S2). The COG functional analysis provided insights into the conserved proteins and their specific metabolic pathways. The core genome showed the highest number of genes pertaining to metabolism, while the accessory genome contained the largest fraction of genes related to cellular processes and signaling. Further analysis revealed that the core genome was enriched in amino acid transport/metabolism, translational, ribosomal structure/biogenesis, and energy production/conversion apparatus, while the accessory genome was enriched in the cell wall, membrane, envelope biogenesis and inorganic ion transport metabolism. The unique genome comprised the main fraction of genes related to information processing and storage and those of unknown function (Figure 1B).

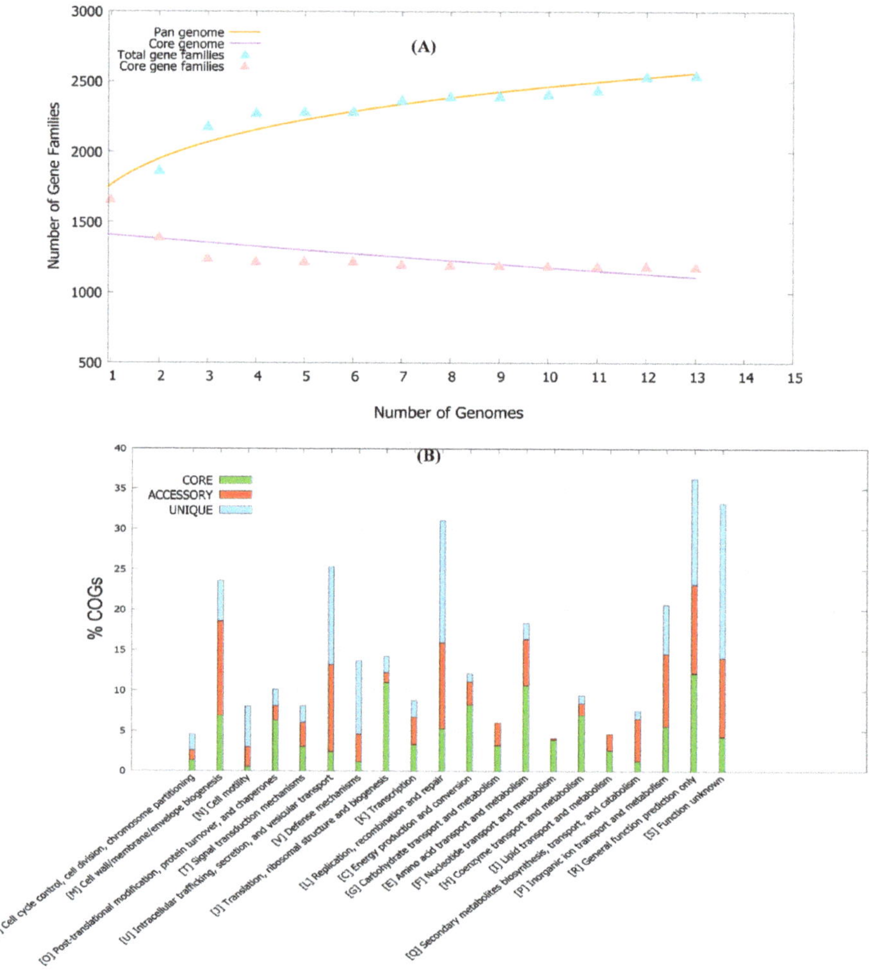

Figure 1. (**A**). A dot plot of the pan-genome vs. the core genome of 13 *C. ureolyticus* strains. (**B**) COG distribution of genome fractions for *C. ureolyticus* strains.

Additionally, the phylogenetic tree generated for the pan- and core genomes highlighted monophyly in all 13 strains (Figure 2). Strains with greater proportions of shared genes have been found in close lineages, hence the proportion of pan-genes between them reveals their evolutionary closeness. In *C. ureolyticus*, the average proportion of common genes was ~61.86 percent. Phylogenetically, all strains represented in the pan- and core genomes were found to be a group in almost the same clade, indicating similarities.

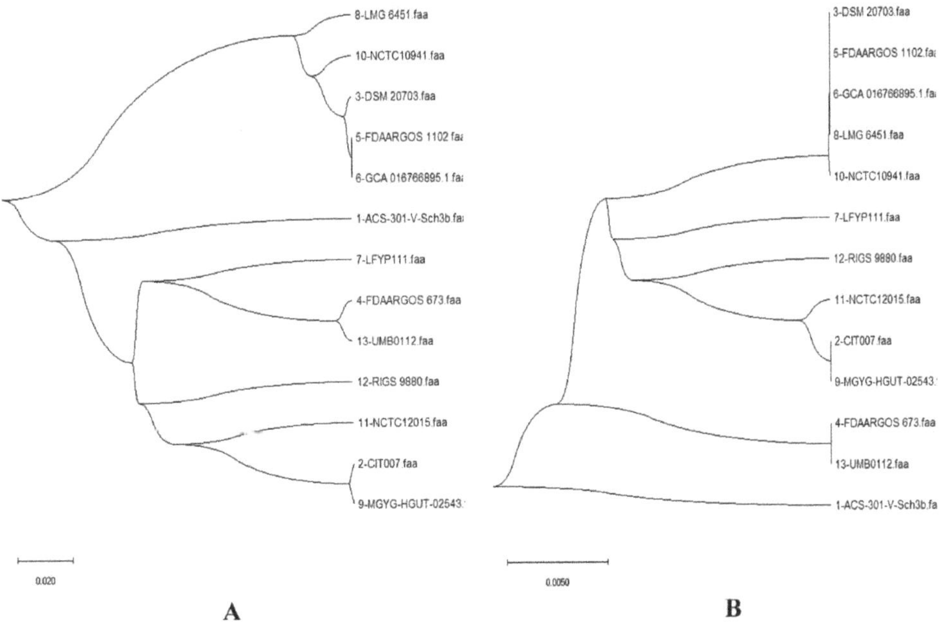

Figure 2. Curved phylogenomic trees based on: (**A**) the pan-genome of 13 *C. ureolyticus* strains (SBL is 0.71, while the distance in millions of years is 0.02); and (**B**) the core genome of the studied strains (SBL value is 0.08, while the distance in millions of years is 0.005).

The resistome analysis of these 13 strains resulted in the identification of certain antibiotic resistance genes (ARGs), found in the core and accessory genomes but absent in the unique genome fraction. The core genome had only the *gyrA* resistance gene, which encodes for fluoroquinolone via antibiotics that target the alteration mechanism [15]. Additionally, it was observed that the *gyrA* gene contained a single-nucleotide polymorphism at position T86I. This mutation has previously been well characterized in *Campylobacter* species [16–19] and has been tested as a biomarker to detect fluoroquinolone resistance in *C. jejuni* [20]. However, no such report for *C. ureolyticus* exists yet. In the accessory genome, two resistance genes, OXA-85 and TetM, were identified, which encode for OXA-beta-lactamases and tetracycline-resistant ribosomal protection protein. These proteins showed resistance to carbapenem, cephalosporin, penam class, and tetracycline antibiotics through antibiotic inactivation. OXA-85, encoded by FUS-1, is a narrow-spectrum beta-lactamase. It has been isolated previously from *Fusobacterium nucleatum* subsp. *polymorphum* [21] but not yet reported in any *Campylobacter* spp. This means that this gene could have been introduced in *C. ureolyticus* via horizontal gene transfer. TetM has been widely reported in many species before [22–27] and may be transposon- or chromosomal-mediated [25,27].

2.2. Differential Sequence Mining and Therapeutic Target Identification

The identified 1882 core genes of *C. ureolyticus* were further used for downstream drug target identification. To prioritize potent drug targets, paralogous, essential, and non-homologous sequence identifications are the crucial steps. CD-Hit redundancy analysis was used to eliminate paralogous sequences from the identified core fraction data, resulting in the discovery of 1178 paralogous sequences. The role of essential proteins helps in the survival of the pathogen. We used BLASTp with an E-value of 10^{-5} to locate the translated gene product against the DEG and CEG databases. This resulted in the shortlisting of 757 and 659 essential proteins, respectively. The intersection of this dataset revealed 647 common proteins. Furthermore, there were 109 *C. ureolyticus* proteins non-homologous to the gut bacterial proteome. These proteins were involved in critical cellular and metabolic processes. Selective targeting of these proteins could prevent cytotoxic reactions and harmful effects in the human host.

The ability of a small-molecule medication to regulate the activity of a therapeutic target is known as druggability. Proteins with a high frequency of sequence similarities (80% or more) were considered as druggable targets. Among 109 proteins, 35 were identified as druggable. These targets were then examined for their virulence as well. Among these, 26 proteins were found to be virulent (Table 1) and belonged to either metabolism, information or signaling, and cellular process families of proteins.

Table 1. Drug targets of *C. urolyticus* that are virulent in nature. The selected target is shown in bold.

Serial No.	Protein	Functional Category	Accession	No. of Amino Acids
1	Multidrug efflux RND transporter permease subunit	Signaling and cellular processes	WP_016646469.1	1053
2	Formate dehydrogenase subunit alpha	Carbohydrate metabolism	WP_081617940.1	974
3	Formate dehydrogenase subunit alpha	Energy metabolism	WP_081617935.1	935
4	Thiosulfate reductase PhsA		WP_024962542.1	761
5	Urease subunit alpha	Nucleotide metabolism	WP_050333258.1	571
6	TolC family protein	Cellular process	WP_018713635.1	481
7	Murein biosynthesis integral membrane Protein MurJ	Metabolism	WP_101637374.1	470
8	3-deoxy-7-phosphoheptulonate synthase class II		WP_018713201.1	447
9	Glutamyl-tRNA reductase		WP_101637465.1	436
10	Type II secretion system F family protein	Environmental information processing	WP_101636691.1	415
11	Lipid IV(A) 3-deoxy-D-manno-octulosonic acid transferase	Metabolism	WP_018712363.1	382
12	Efflux RND transporter periplasmic adaptor subunit	Genetic information processing	WP_101637141.1	371
13	Lipid-A-disaccharide synthase		WP_018713606.1	347
14	KpsF/GutQ family sugar-phosphate isomerase		WP_018713574.1	320
15	UDP-3-O-(3-hydroxymyristoyl)glucosamine N-acyltransferase	Metabolism	WP_101637361.1	317
16	UDP-3-O-acyl-N-acetylglucosamine deacetylase (LpxC)		WP_050333632.1	296
17	c-type cytochrome		WP_101637480.1	287
18	Pantoate–beta-alanine ligase		WP_016646546.1	273
19	3-deoxy-manno-octulosonate cytidylyltransferase		WP_018713833.1	240
20	Carbonic anhydrase		WP_018713612.1	227
21	MotA/TolQ/ExbB proton channel family protein	Signaling and cellular processes	WP_018713282.1	191
22	YceI family protein	-	WP_024962547.1	188
23	Biopolymer transporter ExbD	Signaling and cellular processes	WP_018713283.1	130
24	Aspartate 1-decarboxylase	Metabolism	WP_018713496.1	115
25	Urease subunit beta	Nucleotide metabolism	WP_018713462.1	104
26	Urease subunit gamma	Nucleotide metabolism	WP_018713461.1	100

UDP-3-O-Acyl-N-acetylglucosamine deacetylase (LpxC) was selected as a therapeutic target. It is a zinc-dependent metalloamidase and catalyzes the second and final step of lipid A production [28]. It eradicates the acetyl group from UDP-(3-O-(R-3-hydroxymyristoyl))-N-acetylglucosamine and results in the production of UDP-(3-O-(R-3-hydroxymyristoyl))-glucosamine and acetate. The ensuing enzymatic reaction converts UDP-(3-O-(R-3-hydroxymyristoyl))-glucosamine to lipid A, which is later integrated into lipopolysaccharides [29]. It is an appealing and validated target for the development of novel antibacterial medicines to treat Gram-negative infections because of its crucial role in lipid A production. Due to the absolute dependency of the microbes on this biosynthetic pathway, along with its nonexistence in humans [30], this potential drug target was chosen for further processing in *C. ureolyticus*.

2.3. Structure Prediction and Inhibitor Screening

The structure of *E. coli* LpxC (PDB ID: 4MDT) was utilized as a template due to its high sequence identity of 44 percent with the LpxC of *C. ureolyticus*. A 3D stereochemical model (Figure 3A) revealed 95% residues in the core favorable area (Figure 3B) and 2.1% outliers, with a Z-score of 0.78. Around 26% of the proteins consisted of alpha helices and 38% of beta-sheets, while 8% were disordered. Since the metal cofactor is important for LpxC catalytic activity, the zinc ion was also kept bound with the structure for docking.

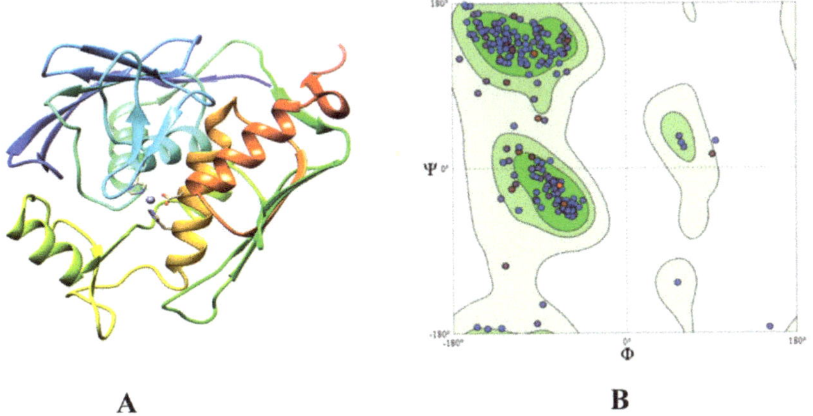

A   **B**

Figure 3. (**A**) Modeled structure of the LpxC protein. (**B**) Ramachandran plot using PROCHECK, showing 93% residues in the favored region.

Many effective LpxC inhibitors have been identified [31,32], with a range of chemical scaffolds and antibiotic profiles [31], but none of them has attained approval as an antibacterial moiety [31]. LpxC inhibitors have not been reported for *C. ureolyticus*. For this purpose, therefore, structure-based inhibitor screening was attempted using a molecular docking strategy. This is an excellent method for determining how drugs/compounds interact with a biological target. To comprehend the LpxC-drug binding mechanism and energy, analyses were conducted in MOE software. Compound libraries included LpxC inhibitors from the ZINC database, metabolites from *Streptomyces* spp., and postbiotics. *Streptomyces* spp. produce the largest number of antibiotics among all bacteria [33]. They decrease the fitness costs of secreted secondary metabolites, leading to enhanced yield and product diversity [34]. Postbiotics are the healthful metabolites of probiotics [35]. The use of such metabolites has been reported against *Vibrio campbelli* infection [36] and SARS-CoV-2 [37]. Docking revealed the binding interactions for the top selected compounds from each library (Table 2).

ZINC26844580 (S-value: −7.42), ZINC13474902 (S-value: −7.05), and ZINC13474878 (S-value: −6.90) were prioritized from the ZINC database LpxC inhibitors. Notoginsenoside St-4 (from *Lactobacillus caseii*, S-value: -8.59), Asiaticoside F (from *Lactobacillus caseii*, S-value: -8.43), and Paraherquamide E (from *Lactobacillus plantarum*, S-value: -8.02) were shortlisted from the postbiotics library, whereas ZINC08219868 (Phytoene, S-value: -7.20), ZINC08214943 (Lycopene, S-value: -7.03), and ZINC04742519 (Sparsomycin, S-value: -7.01) were selected from the streptomycin library. Figure 4 shows the 2D bonding interactions between the shortlisted compounds and the LpxC protein. Among the prioritized compounds, Notoginsenoside St-4, a dammarane-type saponin from the root of *Panax notoginseng* has been previously reported to hamper herpes simplex virus entry [38].

Table 2. Molecular docking analysis of shortlisted compounds from the studied libraries.

S. No.	Compounds	Ligand	Receptor	Interaction	Distance	E (kcal/mol)	S-Score
			LpxC ZINC database inhibitors				
		N12	O PHE187	H-donor	3.43	−0.8	
1	ZINC26844580	N18	O MET59	H-donor	2.95	−2.3	−7.42
		O19	NE2 HIS233	H-donor	2.92	−3.1	
2	ZINC13474902		Hydrophobic interactions				−7.05
3	ZINC13474878	O1	CA SER259	H-acceptor	3.44	−0.8	−6.90
			Postbiotics				
		O63	O LYS157	H-donor	2.89	−2.6	
		O64	O LYS157	H-donor	3.12	−1.0	
1	Notoginsenoside St-4	O35	NE2 HIS233	H-acceptor	2.93	−2.3	−8.59
		O53	NZ LYS140	H-acceptor	3.12	−0.5	
		O55	NZ LYS140	H-acceptor	3.00	−5.7	
		C26	6-ring PHE187	H-pi	4.32	−0.6	
		O79	OD1 ASP139	H-donor	2.89	−2.0	
2	Asiaticoside F	O94	O GLN58	H-donor	3.26	−0.5	−8.43
		C112	O PHE187	H-donor	3.30	−0.5	
		O94	N HIS260	H-acceptor	3.40	−0.7	
3	Paraherquamide E	O43	CA SER259	H-acceptor	3.43	−0.6	−8.02
		C18	6-ring PHE 187	H-pi	3.81	−0.7	
			Streptomycin compounds				
1	ZINC08219868 (Phytoene)		Hydrophobic interactions				−7.20
2	ZINC08214943 (Lycopene)		Hydrophobic interactions				−7.03
	ZINC04742519 (Spar-somycin)	O27	O MET59	H-donor	2.84	−0.7	
3		S37	CA GLY205	H-acceptor	4.06	−0.7	−7.01
		C24	5-ring HIS18	H-pi	4.82	−0.5	

Asiaticoside F has also been reported from the leaves of *Centella asiatica* and is known to inhibit tumor necrosis factor-α [39]. Paraherquamide E has been isolated from the fungus *Penicillium charlesh* and is a known anti-nematicidal agent [40]. These postbiotics have previously been isolated from plants or fungi, but bacterial production of this compound heralds an exciting use of biogenic metabolites of probiotic strains.

Phytoene is produced by *Streptomyces scabrisporus* NF3 [41], *Streptomyces griseus* [42], archaeon *Thermococcus kodakarensis* [43], algae *Dunaliella* sp., [44], yeast *Xanthophyllomyces dendrorhous* [45] and halophilic bacteria [46], among other microorganisms. Lycopene has also been produced by microorganisms [47]. Phytoene and lycopene have radical scavenging activity [48]. Sparsomycin has also been previously isolated from *Streptomyces* spp. and possesses anti-tumor activity [49]. Inhibition of LpxC by these compounds is of interest, making for welcome additions to the class of existing inhibitors of this enzyme.

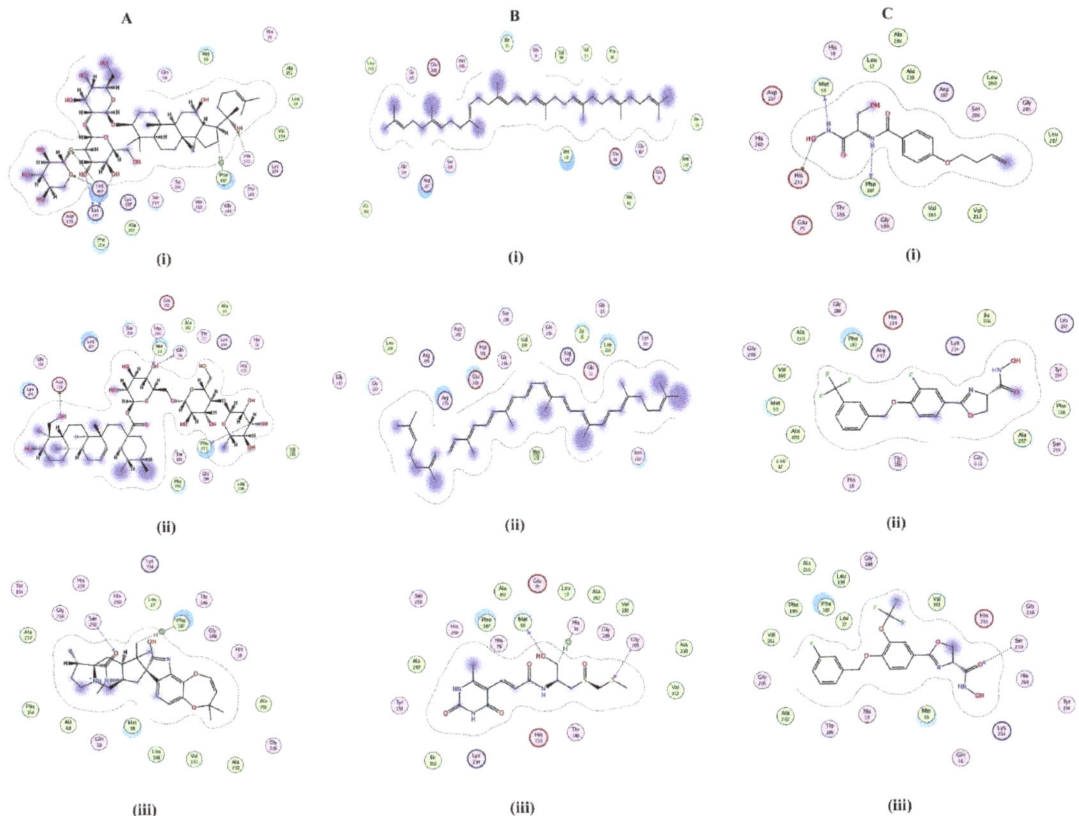

Figure 4. Two-dimensional depictions of shortlisted compounds from: (**A**) the postbiotics library, i.e., (**i**) Notoginsenoside St-4, (**ii**) Asiaticoside F, and (**iii**) Paraherquamide E; (**B**) the streptomycin library, i.e., (**i**) ZINC08219868 (Phytoene), (**ii**) ZINC08214943 (Lycopene), and (**iii**) ZINC04742519 (Sparsomycin); and (**C**) the ZINC LpxC library, i.e., (**i**) ZINC26844580, (**ii**) ZINC13474902, and (**iii**) ZINC13474878.

2.4. ADMET Profiling of Shortlisted Compounds

The majority of the shortlisted compounds were not found to inhibit the P450 class of enzymes, CYP1A2, CYP2C19, CYP2C9, CYP2D6, and CYP3A4. ZINC13474902 and ZINC13474878 inhibited several cytochrome P450s. Since these enzymes are involved in drug metabolism [50], deactivating or excreting them, their non-binding behavior indicates that the shortlisted drugs will not be rendered ineffective. Six compounds (ZINC26844580, ZINC13474902, ZINC13474878, Notoginsenoside St-4, Asiaticoside F, and Sparsomycin) were identified as non-permeable to the blood–brain barrier, while three (paraherquamide E, phytoene, and lycopene) were permeable. GI absorption was low for Notoginsenoside St-4, Asiaticoside F, and Sparsomycin. Skin sensitization was not observed, while significant permeability to Caco2 was seen. Caco2 being used as a model to study the intestinal displacement barrier, this means that the metabolites can cross the intestinal barrier. Two compounds (Notoginsenoside St-4 and Asiaticoside) were identified as P-glycoprotein inhibitors. This could have an impact on the tissue distribution of these compounds and may change the pharmacokinetics due to the drug–drug interaction properties of P-glycoprotein inhibitors [51]. Except for Notoginsenoside St-4 and Asiaticoside, all compounds followed Lipinski's rule of five. Compounds were non-toxic and AMES carcinogenicity was null (Table 3), except for ZINC13474878. These features make several of the screened inhibitors ideal candidates for future therapeutic development. For the compounds that do not fall in the ideal category of drug-likeliness or safety profile, their scaffolds could be used to make better compounds. Additionally, a conjugate could be made for the compounds that have low gastrointestinal tract absorption.

Table 3. ADMET properties analysis of the nine compounds shortlisted.

Name	Water Solubility	CaCo2 Permeability	HIA	Skin Permeability	Max. Tolerated Dose (Human)	Minnow Toxicity	T. Pyriformis Toxicity	Oral Rat Acute Toxicity (LD50)	Hepatotoxic
ZINC26844580	−2.348	−0.049	51.882	−2.898	1.177	2.262	0.269	2.238	Yes
ZINC13474902	−5.016	1.037	91.596	−3.066	0.193	2.117	0.422	2.983	Yes
ZINC13474878	−4.928	1.103	90.066	−2.993	0.272	2.587	0.356	2.96	Yes
Notoginsenoside St−4	−2.938	−1.241	0	−2.735	−1.755	10.516	0.285	2.738	No
Asiaticoside F	−2.922	−1.039	43.096	−2.735	−0.991	11.112	0.285	2.738	No
Paraherquamide E	−4.377	1.003	93.521	−3.226	−0.877	2.209	0.294	3.716	Yes
ZINC08219868 (Phytoene)	−6.345	1.28	91.213	−2.737	−0.33	−5.751	0.288	2.036	No
ZINC08214943 (Lycopene)	−6.514	1.317	93.238	−2.783	−0.447	−5.243	0.291	2.105	No
ZINC04742519 (Sparsomycin)	−2.348	−0.238	44.731	−3.08	1.279	2.726	0.146	2.11	Yes

2.5. MD Simulation Analysis

A classic MD analysis was performed to determine the free energies of complexes, i.e., MM/PBSAs were almost similar, with the lowest value for the LpxC–lycopene complex (Table 4). The atom-scale MD and the structural stabilities of the selected inhibitors and protein complex were determined in a time-dependent manner. RMSD, RMSF, hydrogen bonding, and Rg of the ligand-bound LpxC protein were studied to analyze the structural integrity during bonding. This was achieved through trajectory mapping graphs of the backbone carbon atoms.

Table 4. MM/PBSA values of the selected compounds.

Name	Ligand	Protein	Protein−Ligand Complex
ZINC1347878	0.04	−19.31	−18.95
ZINC13474902	0.03	−19.31	−18.95
ZINC26844580	0.14	−19.31	−19.03
Notoginsenoside St-4	−0.85	−19.31	−19.19
Asiaticoside F	−0.82	−19.31	−19.20
Paraherquamide E	−0.07	−19.31	−18.89
Phytoene	−0.96	−19.31	−19.46
Lycopene	−0.92	−19.31	−19.54
Sparsomycin	0.05	−19.31	−18.97

The RMSD analysis for ZINC LpxC database compounds (ZINC26844580, ZINC13474902, and ZINC13474878) showed the stability of compounds during the whole 50 ns simulation study, with a mild fluctuation found in ZINC26844580 at ~30 ns. The complete system was found to be at equilibrium within the range of 0.15–0.20 nm, as is evident from Figure 5A. The RMSD analysis for the postbiotic shortlisted compounds (Asiaticoside F and Paraherquamide E) showed considerable fluctuation during the initial 10 ns of simulation. However, after 15 ns, the system progressively stabilized and remained stable until the simulation was completed, with an average deviation of 0.15–0.20 nm, indicating the simulated system's convergence. Notoginsenoside St-4 showed stability throughout the simulation study, as is apparent from Figure 5B. Compounds from the streptomycin library (ZINC08219868 (Phytoene), ZINC08214943 (Lycopene), and ZINC04742519 (Sparsomycin)) exhibited mild to moderate fluctuations during the 50 ns simulation, showing variations in RMSDs values in an acceptable range of 0.15–0.25 nm (Figure 5C).

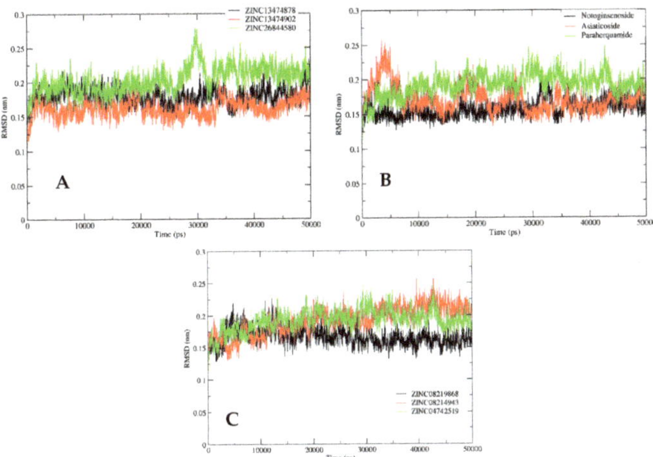

Figure 5. RMSD analysis for the shortlisted compounds: (**A**) ZINC26844580 (green), ZINC13474902 (red), and ZINC13474878 (black); (**B**) Notoginsenoside St-4 (black), Asiaticoside F (red), and Paraherquamide E (green); and (**C**) ZINC08219868 (Phytoene) (black), ZINC08214943 (Lycopene) (red), and ZINC04742519 (Sparsomycin) (green).

However, to evaluate the fluctuations found in amino acid residues of the ligand-bound protein during the simulation studies, RMSF data for these compounds were also plotted. Figure 6 depicts the RMSF plot for all nine shortlisted compounds, which displayed variations throughout the simulation. The average RMSF for the simulated system was determined to be 0.6 nm. RMSF was considerably stable for all of the amino acid residues of the ligand-bound protein in the simulated system.

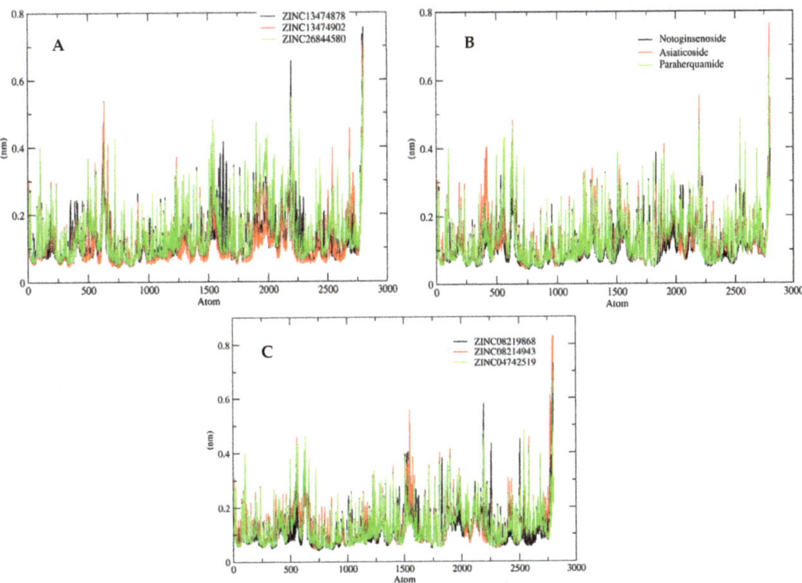

Figure 6. RMSF analysis for the shortlisted compounds: (**A**) ZINC26844580 (green), ZINC13474902 (red), and ZINC13474878 (black); (**B**) Notoginsenoside St-4 (black), Asiaticoside F (red), and Paraherquamide E (green); and (**C**) ZINC08219868 (Phytoene) (black), ZINC08214943 (Lycopene) (red), and ZINC04742519 (Sparsomycin) (green).

To evaluate the compactness of the structure complexes, Rg was plotted (Figure 7). The Rg plot for the catalytic site during the initial 10 ns of simulation showed large fluctuations. However, after 20 ns, the system gradually moved towards compaction, contributing to the overall stability of the simulated protein structure in the presence of all shortlisted compounds.

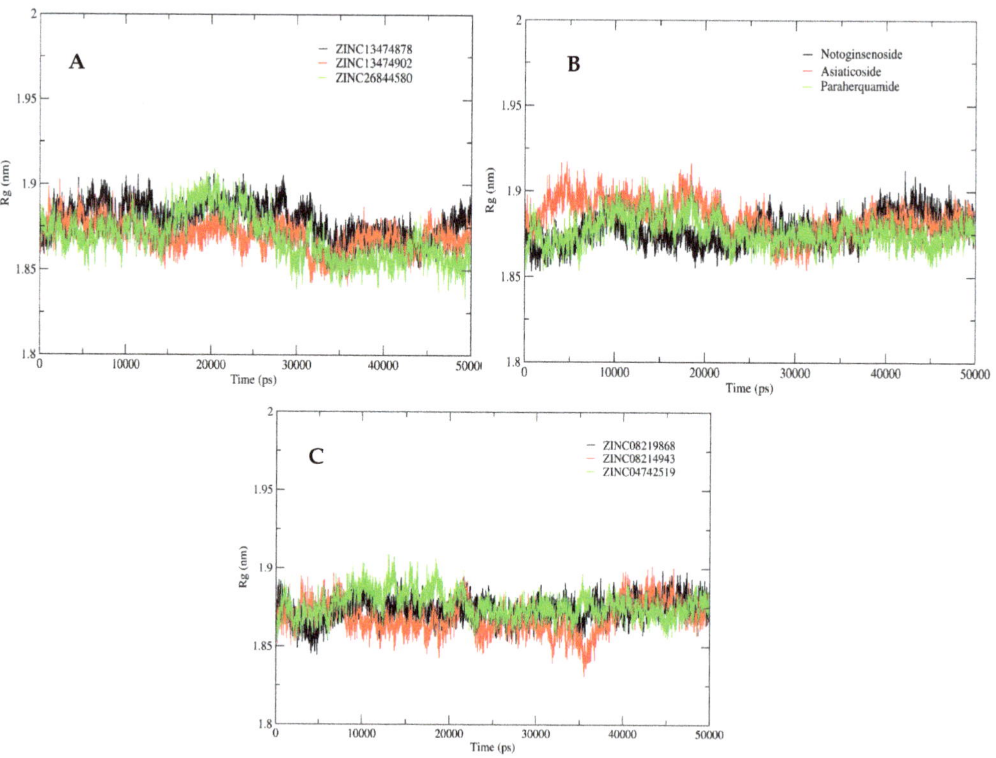

Figure 7. Rg analysis for the shortlisted compounds: (**A**) ZINC26844580 (green), ZINC13474902 (red), and ZINC13474878 (black); (**B**) Notoginsenoside St-4 (black), Asiaticoside F (red), and Paraherquamide E (green); and (**C**) ZINC08219868 (Phytoene) (black), ZINC08214943 (Lycopene) (red), and ZINC04742519 (Sparsomycin) (green).

The interactions between the protein and compounds were further determined through hydrogen bonding analysis. A robust interaction was observed between ZINC26844580, ZINC13474902, ZINC13474878 (ZINC LpxC inhibitor compounds), Asiaticoside F, Paraherquamide E, Notoginsenoside St-4 (Postbiotics), and LpxC protein in terms of hydrogen interaction, ranging from 8-10 bonds throughout the 50 ns simulation. The compound ZINC04742519 (Sparsomycin) from the streptomycin library mediated three hydrogen interactions, initially at 10 ns. However, after 15 ns of the simulation, five hydrogen bonds were observed constantly. ZINC08219868 (Phytoene) and ZINC08214943 (Lycopene) formed no hydrogen bonds, indicating hydrophobic interaction with the LpxC protein (as shown in Figure 8).

Eventually, the LpxC protein in combination with the nine compounds was found to be stable during the 50 ns simulation. This confirms that the docking interactions that the complexes formed were stable.

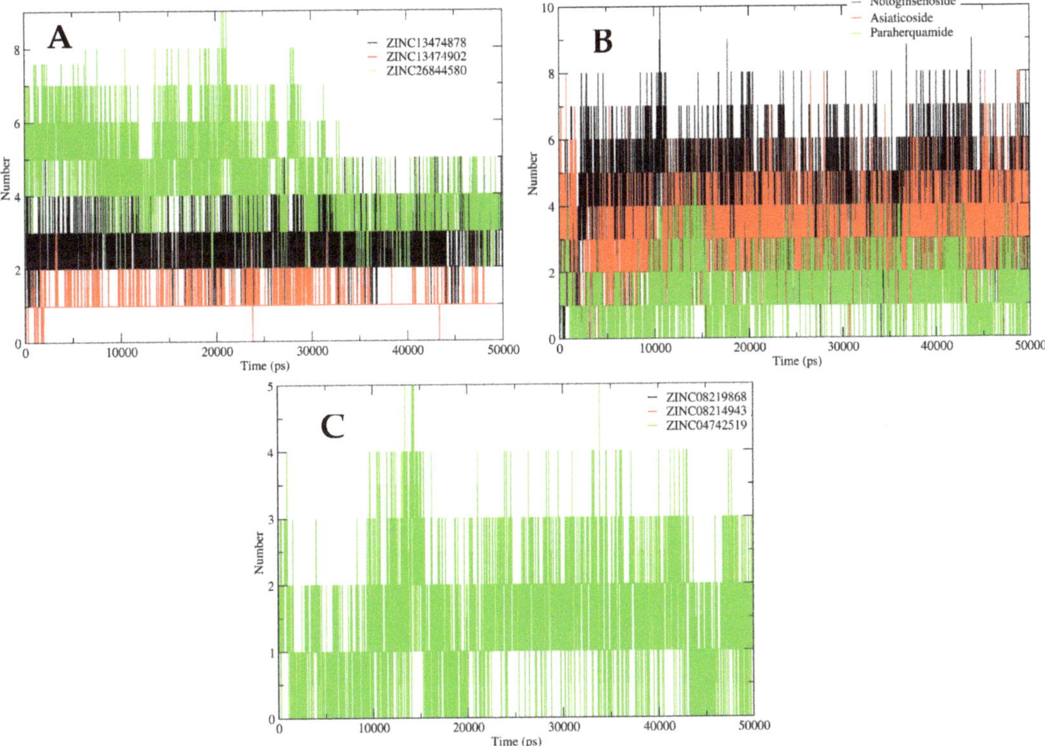

Figure 8. Hydrogen bond analysis for the shortlisted compounds: (**A**) ZINC26844580 (green), ZINC13474902 (red), and ZINC13474878 (black); (**B**) Notoginsenoside St-4 (black), Asiaticoside F (red), and Paraherquamide E (green); and (**C**) ZINC08219868 (Phytoene) (black), ZINC08214943 (Lycopene) (red), and ZINC04742519 (Sparsomycin) (green).

3. Discussion

Campylobacter infections are usually caused by the consumption of contaminated poultry products [9]. Although the frequency of human sickness caused by oral Campylobacter species is lower than that caused by zoonotic *Campylobacter* species, it is considered that non-*jejuni/coli Campylobacter* illness is underreported due to a lack of good culture-based detection methods [5]. Recently, *C. ureolyticus* has been reported to cause gastroenteritis in both developed and developing countries [3].

The current study intends to prioritize potential therapeutic targets, as well as lead drug candidates, against *C. ureolyticus* based on the results of a comprehensive pan-genome investigation. In this study, we used the pan-genome to detect and characterize the *C. ureolyticus* antimicrobial resistome as well. The pan-genome concept has previously been utilized to distinguish between commensal and pathogenic strains [52,53] and we were able to investigate the pan-resistome of the 13 strains as well. The pan-resistome analysis is useful for determining the ARG diversity of the species and revealing the occurrence of species-specific ARGs. Previously, Costa et al. [12] identified that some ARGs were conserved between *C. jejuni* and *C. lari*. In our analysis, we also found conservation of fluoroquinolone-resistant *gyrA* gene between *C. jejuni* and *C. ureolyticus* and its specific mutation T86I. TetM also exists in *C. jejuni* [54], but OXA-85 was unique to *C. ureolyticus* and has only been previously reported in *Fusobacterium nucleatum* subsp. *polymorphum* [21].

For drug target mining, the core genome of *C. ureolyticus* was subjected to subtractive genome analysis and 35 targets were obtained. Previously, 34 targets have been reported in *C. jejuni* [55]. Proteins with a role in metabolic pathways (especially lipopolysaccharide synthesis), secondary metabolite synthesis, two-component systems, and multi-drug resistance were identified as drug targets and some of them were similar to previously reported targets in *C. jejuni*. Eventually, UDP-3-O-acyl-N-acetylglucosamine deacetylase (LpxC) was selected as a potent drug target against *C. ureolyticus*. This protein previously has been identified as a potential drug target against *C. jejuni* [55], *Pseudomonas aeruginosa* [32], and several Gram-negative pathogens [56]. The 3D structure of this protein was modeled and a library of LpxC zinc inhibitors, probiotics, and streptomycins (957 total compounds) was screened against it. ADMET profiling was applied to identify associated adversities. This helps to identify decisive factors in relation to a molecule's potential to be further processed for use as a drug. Harmonizing toxicity and ADME helps to summarize the criteria for the ideal compound, and thus compounds can be narrowed down at the initial screening stage to decide whether to proceed further. Our studied compounds showed several good properties, and their scaffolds could be utilized further for designing and optimizing drugs against *C. ureolyticus* and possibly other bacteria harboring LpxC.

Despite the powerful potential of the screening and validation of binding properties with rigorous simulations, our study has its limitations. Computational predictions do not give 100% accurate results and failure is possible to some extent. Therefore, the results need to be interpreted with caution for use in clinical settings and pharmaceutical endeavors. Lab validations must be carried out before moving on to clinical trials of high-throughput screened molecules. Better algorithms and more computational power may be available in the future to resolve unstudied and necessary features of drug-like molecules but current investigations are rife with errors. This should not, however, discourage studies at the computational level; rather, it should stimulate further endeavors.

4. Materials and Methods

4.1. Data Retrieval

The complete genomes of *C. ureolyticus* strains ($n = 22$) were retrieved from the NCBI database [57]. Redundant species data were removed and 13 genomes were left (details in Supplementary File S1).

4.2. Pan-Genomic Analysis

The pan-genomic analysis was performed using the BPGA tool [58], the parameters for which were defined in our previous work [59]. The clustering of inputted FASTA files was performed using the USERACH algorithm [60], with defined cutoff values of 70% for homologous genes. The pan- and core genome dot plots were generated. Furthermore, the alignment of core, accessory, and unique genomes was performed using the MUSCLE tool [61] with default parameters. UPGMA based on maximum likelihood was used to infer the phylogenetic relationships between strains. Additionally, the resistome of the identified core, unique, and accessory genes was investigated via the Comprehensive Antibiotic Resistance Database (CARD) [62]. CARD employs the automated BLASTp algorithm and a cutoff of 70% for alignment. The homologous gene set was functionally annotated through the Clusters of Orthologous Groups of proteins (COG) database [63]. The genome fractions were annotated for gene functions.

4.3. Drug Target Prioritization

The obtained core genome was subjected to the standalone CD-HIT program [64] and homologous gene products with 60% similarity exclusion criteria were clustered. CD-HIT also reduces redundancy. Using an in-house script, all of the clusters were retrieved from the CD-HIT output file and further investigated for gene essentiality. The obtained data were BLAST-searched against the CEG [65] and DEG [66] databases (with an E-value of 10^{-10}) and a bit score of 100. The proteins found essential in both databases were used for further target identification. The significantly preferable drug targets must be non-homologous to the human genome in order to avoid the cross-reactivity of drug candidates.

For this purpose, the translated products of the obtained sets of genes were subjected to BLASTp against the human proteome (conditions: E-value: 10^{-3}; gap penalty: 11; gap extension penalty: 1).

Furthermore, a comparative analysis was performed between obtained non-homologous proteins and the proteomes of the 83 useful species of human microbial gut flora. The acquired set of proteins was processed against the DrugBank database. The identified druggable targets were also filtered through the virulence factor database (VFDB) [67] using an E-value of 10^{-3}.

4.4. Structural Modeling

The structure of the selected protein LpxC was modeled using SWISS-MODEL [68]. This tool uses a homology-based approach for deciphering protein structures. The final model was selected based on its percent identity and query coverage to its template LpxC protein from *Escherichia coli* (PDB ID: 4MDT). Furthermore, the validation of the modeled structure was achieved using a Ramachandran plot. Secondary structure evaluation, Z-score prediction, and stereochemical quality (via PROCHECK) were also assessed through the SWISS-MODEL server [69]. The secondary structure and the disordered regions were predicted using the Phyre2 server [70].

4.5. Virtual Screening

The interactions between the screened streptomycins ($n = 737$), postbiotics ($n = 78$), and LpxC zinc database inhibitors ($n = 142$) against the LpxC protein were appraised using a molecular docking-based screening approach [71]. Using MOE v2019 software, the pre-docking protonation of LpxC was carried out with the following parameters: flip: all atoms; atoms: all atoms; titrate: all atoms; solvent: 80, disconnected metal treatment: enabled; temperature = 300 K; salt = 0.1; pH = 7; van der Waals: 800R3, with cutoff (A): 10; dielectric: 1, 1; protection = none; electrostatics: GB/VI, with cutoff (A): 15. Correspondingly, the following parameters were set for energy minimization: forcefield: Amber99; gradient: 0.05; fix hydrogen and partial charges = yes. The 957 compounds from the mentioned libraries were taken and screened against LpxC. The screening parameters were: placement = Triangle Matcher; refinement = forcefield; rescoring 1 = London dG; rescoring 2 = affinity dG. The top docked compounds were chosen based on their S-values. The interactions of the docked protein–ligand complexes were examined to explore the hydrogen bonding and hydrophobic interactions among receptors and ligand atoms within a range of 5 Å [72].

4.6. ADMET Profiling of the Shortlisted Drug Candidates

After the molecular docking study, pharmacokinetic properties of the shortlisted lead drug compounds, such as absorption, distribution, metabolism, excretion (ADME), and toxicology, were determined using Swiss ADME [73]. Following this, the selected compounds were processed using the pkCSM tool (http://biosig.unimelb.edu.au/pkcsm/ (accessed on 28 December 2021)) to determine the optimal drug candidate with high penetration and the least number of side effects. SwissADME presented skin permeation values, and pkCSM provided drug tolerance values for a range of organisms, including humans. Drug safety evaluation is critical for new drugs. Early knowledge of medication toxicity and side effects is crucial for drug induction in the development pipeline [74,75]. With this in mind, the highest tolerated dosages, the impacts on various species, and the excretion of drug characteristics were all determined.

4.7. Molecular Dynamics (MD) Simulation

The stability and flexibility analyses of the identified compounds against the shortlisted drug targets were carried out through MD simulation studies. MM/PBSA values were calculated using MOE software [76], while the atom-scale MD simulations were carried out with the GROMACS 5.1.2 package24 using the gromos54a71 forcefield. The ligand topology parameters were generated using the Automated Topology Builder (ATB v3.0) server. Using periodic boundary conditions, the complex was placed at a distance of 1.0 from the box edge in a dodecahedron box, with the addition of solvent molecules from the SPC water model. The insertion of counter ions into the solvated system helps to neutralize the entire system. The neutralized system was minimized using the steepest descent method, with a maximum force of 1000 kJ mol^{-1}. The minimized systems were passed for 100 ps of NVT and NPT equilibration before the production dynamics to raise the system temperature by 300K and maintain a constant pressure of 1 bar for the system utilizing thermostat and Berendsen barostat algorithms. The Linear Constraint Solver (LINCS) algorithm was used to constrain all the bonds, and the Particle Mesh Ewald (PME) method was employed to compute the long-range electrostatics with a cutoff value of 1.0 nm. Finally, a 50 ns production run was carried out, with coordinates and energies being saved every 10 ps in the output trajectory file, as per the procedure. Root Mean Square Deviation (RMSD), Root Mean Square Fluctuation (RMSF), and Radius of Gyration (Rg) were plotted to evaluate the system's stability.

5. Conclusions

To the best of our knowledge, the current study is the first to map the resistome of *C. ureolyticus* as well as drug targets for this species. Resistance profiling is not only useful for the scientific community working in this domain but also for clinicians and health policymakers. Our computational analysis will aid the scientific community in further investigating the proposed therapeutic drug targets and inhibitors using experimental methodologies. The identified therapeutic proteins might be studied further in experimental investigations to prevent *C. ureolyticus*-mediated gastroenteritis, along with the compounds that we screened against the LpxC enzyme. The prioritized compounds ZINC26844580, ZINC13474902, ZINC13474878, Notoginsenoside St-4, Asiaticoside F, Paraherquamide E, Phytoene, Lycopene, and Sparsomycin need to be tested further in vivo and in vitro for their druggability.

Supplementary Materials: The following supporting information can be downloaded at: https://www.mdpi.com/article/10.3390/antibiotics11050680/s1, File S1: Details of genome data used in this study, File S2: Pan-genome statistics of the *C. ureolyticus* strains.

Author Contributions: Conceptualization, Z.B. and R.U.; methodology, Z.B. and R.U.; software, M.M.M., A.A. and S.A.; validation, Z.B., K.J. and K.K.; formal analysis, Z.B., K.K. and K.J.; resources, M.M.M., S.A. and A.A.; data curation, Z.B., K.J. and K.K.; writing—original draft preparation, K.J. and K.K.; writing—review and editing, Z.B.; visualization, M.M.M., S.A. and A.A.; supervision, R.U.; project administration, Z.B. All authors have read and agreed to the published version of the manuscript.

Funding: This research received no external funding.

Institutional Review Board Statement: Not applicable.

Informed Consent Statement: Not applicable.

Data Availability Statement: Publicly analyzed datasets used in this study can be found in NCBI GenBank by reference to the details provided in Supplementary File S1. Any other raw or generated data are available from the corresponding authors without reservation.

Acknowledgments: We would like to thank 'The SEARLE Company Ltd. (TSCL)', Karachi, Pakistan, for its capacity-building support of the Jamil-ur-Rahman Center for Genome Research, PCMD, ICCBS, University of Karachi, Pakistan.

Conflicts of Interest: The authors declare no conflict of interest.

References

1. Burgos-Portugal, J.A.; Kaakoush, N.O.; Raftery, M.J.; Mitchell, H.M. Pathogenic potential of Campylobacter ureolyticus. *Infect Immun.* **2012**, *80*, 883–890. [CrossRef] [PubMed]

2. Bullman, S.; O'Leary, J.; Lucey, B.; Byrne, D.; Sleator, R.D. Campylobacter ureolyticus: An emerging gastrointestinal pathogen? *FEMS Immunol. Med. Microbiol.* **2011**, *61*, 228–230. [CrossRef] [PubMed]
3. Fernández, H.; Vera, F.; Villanueva, M.P.; García, A. Occurrence of Campylobacter species in healthy well-nourished and malnourished children. *Braz. J. Microbiol.* **2008**, *39*, 56–58. [CrossRef]
4. Daisy, P.; Singh, S.K.; Vijayalakshmi, P.; Selvaraj, C.; Rajalakshmi, M.; Suveena, S. A database for the predicted pharmacophoric features of medicinal compounds. *Bioinformation* **2011**, *6*, 167. [CrossRef]
5. Serichantalergs, O.; Ruekit, S.; Pandey, P.; Anuras, S.; Mason, C.; Bodhidatta, L.; Swierczewski, B. Incidence of Campylobacter concisus and C. ureolyticus in traveler's diarrhea cases and asymptomatic controls in Nepal and Thailand. *Gut Pathog.* **2017**, *9*, 47. [CrossRef]
6. Boggess, K.A.; Trevett, T.N.; Madianos, P.N.; Rabe, L.; Hillier, S.L.; Beck, J.; Offenbacher, S. Use of DNA hybridization to detect vaginal pathogens associated with bacterial vaginosis among asymptomatic pregnant women. *Am. J. Obs. Gynecol.* **2005**, *193*, 752–756. [CrossRef]
7. O'Donovan, D.; Corcoran, G.D.; Lucey, B.; Sleator, R.D. Campylobacter ureolyticus: A portrait of the pathogen. *Virulence* **2014**, *5*, 498–506. [CrossRef]
8. Bennett, K.W.; Eley, A.; Woolley, P.D.; Duerden, B.I. Isolation of Bacteroides ureolyticus from the genital tract of men with and without non-gonococcal urethritis. *Eur. J. Clin. Microbiol. Infect. Dis.* **1990**, *9*, 825–826. [CrossRef]
9. Basic, A.; Enerbäck, H.; Waldenström, S.; Östgärd, E.; Suksuart, N.; Dahlen, G. Presence of Helicobacter pylori and Campylobacter ureolyticus in the oral cavity of a Northern Thailand population that experiences stomach pain. *J. Oral. Microbiol.* **2018**, *10*, 1527655. [CrossRef]
10. Fraczek, M.; Piasecka, M.; Gaczarzewicz, D.; Szumala-Kakol, A.; Kazienko, A.; Lenart, S.; Laszczynska, M.; Kurpisz, M. Membrane stability and mitochondrial activity of human-ejaculated spermatozoa during in vitro experimental infection with Escherichia coli, Staphylococcus haemolyticus and Bacteroides ureolyticus. *Andrologia* **2012**, *44*, 315–329. [CrossRef]
11. Mukhopadhya, I.; Thomson, J.M.; Hansen, R.; Berry, S.H.; El-Omar, E.M.; Hold, G.L. Detection of Campylobacter concisus and other Campylobacter species in colonic biopsies from adults with ulcerative colitis. *PLoS ONE* **2011**, *6*, e21490. [CrossRef] [PubMed]
12. Costa, D.; Iraola, G. Pathogenomics of emerging Campylobacter species. *Clin. Microbiol. Rev.* **2019**, *32*, e00072-18. [CrossRef] [PubMed]
13. Caputo, A.P.-E.F.; Raoult, D. Genome and pan-genome analysis to classify emerging bacteria. *Biol. Direct* **2019**, *14*, 5. [CrossRef]
14. Spreafico, R.; Soriaga, L.B.; Grosse, J.; Virgin, H.W.; Telenti, A. Advances in Genomics for Drug Development. *Genes* **2020**, *11*, 942. [CrossRef] [PubMed]
15. Kareem, S.M.; Al-Kadmy, I.M.S.; Kazaal, S.S.; Ali, A.N.M.; Aziz, S.N.; Makharita, R.R.; Algammal, A.M.; Al-Rejaie, S.; Behl, T.; Batiha, G.E.; et al. Detection of gyrA and parC Mutations and Prevalence of Plasmid-Mediated Quinolone Resistance Genes in Klebsiella pneumoniae. *Infect. Drug Resist.* **2021**, *14*, 555–563. [CrossRef] [PubMed]
16. Dahl, L.G.; Joensen, K.G.; Osterlund, M.T.; Kiil, K.; Nielsen, E.M. Prediction of antimicrobial resistance in clinical Campylobacter jejuni isolates from whole-genome sequencing data. *Eur. J. Clin. Microbiol. Infect. Dis.* **2021**, *40*, 673–682. [CrossRef]
17. Elhadidy, M.; Ali, M.M.; El-Shibiny, A.; Miller, W.G.; Elkhatib, W.F.; Botteldoorn, N.; Dierick, K. Antimicrobial resistance patterns and molecular resistance markers of Campylobacter jejuni isolates from human diarrheal cases. *PLoS ONE* **2020**, *15*, e0227833. [CrossRef]
18. Haldenby, S.; Bronowski, C.; Nelson, C.; Kenny, J.; Martinez-Rodriguez, C.; Chaudhuri, R.; Williams, N.J.; Forbes, K.; Strachan, N.J.; Pulman, J.; et al. Increasing prevalence of a fluoroquinolone resistance mutation amongst Campylobacter jejuni isolates from four human infectious intestinal disease studies in the United Kingdom. *PLoS ONE* **2020**, *15*, e0227535. [CrossRef]
19. Liao, Y.S.; Chen, B.H.; Teng, R.H.; Wang, Y.W.; Chang, J.H.; Liang, S.Y.; Tsao, C.S.; Hong, Y.P.; Sung, H.Y.; Chiou, C.S. Antimicrobial Resistance in Campylobacter coli and Campylobacter jejuni from Human Campylobacteriosis in Taiwan, 2016 to 2019. *Antimicrob. Agents Chemother.* **2022**, *66*, e0173621. [CrossRef]
20. Espinoza, N.; Rojas, J.; Pollett, S.; Meza, R.; Patino, L.; Leiva, M.; Camina, M.; Bernal, M.; Reynolds, N.D.; Maves, R.; et al. Validation of the T86I mutation in the gyrA gene as a highly reliable real time PCR target to detect Fluoroquinolone-resistant Campylobacter jejuni. *BMC Infect. Dis.* **2020**, *20*, 518. [CrossRef]
21. Voha, C.; Docquier, J.D.; Rossolini, G.M.; Fosse, T. Genetic and biochemical characterization of FUS-1 (OXA-85), a narrow-spectrum class D beta-lactamase from Fusobacterium nucleatum subsp. polymorphum. *Antimicrob. Agents Chemother.* **2006**, *50*, 2673–2679. [CrossRef] [PubMed]
22. Bidell, M.R.; Lodise, T.P. Use of oral tetracyclines in the treatment of adult outpatients with skin and skin structure infections: Focus on doxycycline, minocycline, and omadacycline. *Pharmacotherapy* **2021**, *41*, 915–931. [CrossRef] [PubMed]
23. Dhasmana, N.; Ram, G.; McAllister, K.N.; Chupalova, Y.; Lopez, P.; Ross, H.F.; Novick, R.P. Dynamics of Antibacterial Drone Establishment in Staphylococcus aureus: Unexpected Effects of Antibiotic Resistance Genes. *mBio* **2021**, *12*, e0208321. [CrossRef] [PubMed]
24. Dominguez-Perez, R.A.; de la Torre-Luna, R.; Ahumada-Cantillano, M.; Vázquez-Garcidueñas, M.S.; MarthaPérez-Serrano, R.; ElizabethMartínez-Martínez, R.; LauraGuillén-Nepita, A. Detection of the antimicrobial resistance genes blaTEM-1, cfxA, tetQ, tetM, tetW and ermC in endodontic infections of a Mexican population. *J. Glob. Antimicrob. Resist.* **2018**, *15*, 20–24. [CrossRef] [PubMed]

25. Haubert, L.; Cunha, C.; Lopes, G.V.; Silva, W.P.D. Food isolate Listeria monocytogenes harboring tetM gene plasmid-mediated exchangeable to Enterococcus faecalis on the surface of processed cheese. *Food Res. Int.* **2018**, *107*, 503–508. [CrossRef] [PubMed]

26. Roberts, M.C.; Kenny, G.E. Dissemination of the tetM tetracycline resistance determinant to Ureaplasma urealyticum. *Antimicrob. Agents Chemother.* **1986**, *29*, 350–352. [CrossRef] [PubMed]

27. Shaskolskiy, B.; Dementieva, E.; Leinsoo, A.; Petrova, N.; Chestkov, A.; Kubanov, A.; Deryabin, D.; Gryadunov, D. Tetracycline resistance of Neisseria gonorrhoeae in Russia, 2015-2017. *Infect. Genet. Evol.* **2018**, *63*, 236–242. [CrossRef]

28. Yamada, Y.; Takashima, H.; Walmsley, D.L.; Ushiyama, F.; Matsuda, Y.; Kanazawa, H.; Yamaguchi-Sasaki, T.; Tanaka-Yamamoto, N.; Yamagishi, J.; Kurimoto-Tsuruta, R. Fragment-based discovery of novel non-hydroxamate LpxC inhibitors with antibacterial activity. *J. Med. Chem.* **2020**, *63*, 14805–14820. [CrossRef]

29. Clayton, G.M.; Klein, D.J.; Rickert, K.W.; Patel, S.B.; Kornienko, M.; Zugay-Murphy, J.; Reid, J.C.; Tummala, S.; Sharma, S.; Singh, S.B.; et al. Structure of the bacterial deacetylase LpxC bound to the nucleotide reaction product reveals mechanisms of oxyanion stabilization and proton transfer. *J. Biol. Chem.* **2013**, *288*, 34073–34080. [CrossRef]

30. Aris, S.N.A.M.; Rahman, M.Z.A.; Rahman, R.N.Z.R.A.; Ali, M.S.M.; Salleh, A.B.; Teo, C.Y.; Leow, T.C. Identification of potential riboflavin synthase inhibitors by virtual screening and molecular dynamics simulation studies. *J. King Saud Univ.-Sci.* **2021**, *33*, 101270. [CrossRef]

31. Fujita, K.; Takata, I.; Yoshida, I.; Okumura, H.; Otake, K.; Takashima, H.; Sugiyama, H. TP0586532, a non-hydroxamate LpxC inhibitor, has in vitro and in vivo antibacterial activities against Enterobacteriaceae. *J. Antibiot.* **2022**, *75*, 98–107. [CrossRef] [PubMed]

32. Krause, K.M.; Haglund, C.M.; Hebner, C.; Serio, A.W.; Lee, G.; Nieto, V.; Cohen, F.; Kane, T.R.; Machajewski, T.D.; Hildebrandt, D. Potent LpxC inhibitors with in vitro activity against multidrug-resistant Pseudomonas aeruginosa. *Antimicrob. Agents Chemother.* **2019**, *63*, e00977-19. [CrossRef] [PubMed]

33. Xia, H.; Zhan, X.; Mao, X.M.; Li, Y.Q. The regulatory cascades of antibiotic production in Streptomyces. *World J. Microbiol. Biotechnol.* **2020**, *36*, 13. [CrossRef]

34. Zhang, Z.; Du, C.; de Barsy, F.; Liem, M.; Liakopoulos, A.; van Wezel, G.P.; Choi, Y.H.; Claessen, D.; Rozen, D.E. Antibiotic production in Streptomyces is organized by a division of labor through terminal genomic differentiation. *Sci. Adv.* **2020**, *6*, eaay5781. [CrossRef]

35. Zendeboodi, F.; Khorshidian, N.; Mortazavian, M.A.; Cruz, A.G. Probiotic: Conceptualization from a new approach. *Curr. Opin. Food Sci.* **2020**, *32*, 103–123. [CrossRef]

36. Rajaiah, A.; Ponraj, J.G.; Swasthikka, R.P.; Abirami, G.; Ragupathi, T.; Jayakumar, R.; Ravi, A.V. Anti-QS mediated anti-infection efficacy of probiotic culture-supernatant against Vibrio campbellii infection and the identification of active compounds through in vitro and in silico analyses. *Biocatal. Agric. Biotechnol.* **2021**, *35*, 102108.

37. Balmeh, N.; Mahmoudi, S.; Fard, N.A. Manipulated bio antimicrobial peptides from probiotic bacteria as proposed drugs for COVID-19 disease. *Inf. Med. Unlocked* **2021**, *23*, 100515. [CrossRef]

38. Pei, Y.; Du, Q.; Liao, P.Y.; Chen, Z.P.; Wang, D.; Yang, C.R.; Kitazato, K.; Wang, Y.F.; Zhang, Y.J. Notoginsenoside ST-4 inhibits virus penetration of herpes simplex virus in vitro. *J. Asian. Nat. Prod. Res.* **2011**, *13*, 498–504. [CrossRef]

39. Nhiem, N.X.; Tai, B.H.; Quang, T.H.; Kiem, P.V.; Minh, C.V.; Nam, N.H.; Kim, J.H.; Im, L.R.; Lee, Y.M.; Kim, Y.H. A new ursane-type triterpenoid glycoside from Centella asiatica leaves modulates the production of nitric oxide and secretion of TNF-alpha in activated RAW 264. 7 cells. *Bioorg. Med. Chem Lett.* **2011**, *21*, 1777–1781. [CrossRef]

40. Liesch, J.M.; Wichmann, C.F. Novel antinematodal and antiparasitic agents from Penicillium charlesii. II. Structure determination of paraherquamides B, C, D, E, F, and G. *J. Antibiot.* **1990**, *43*, 1380–1386.

41. Ceapa, C.D.; Vazquez-Hernandez, M.; Rodriguez-Luna, S.D.; Vazquez, A.P.C.; Suarez, V.J.; Rodriguez-Sanoja, R.; Alvarez-Buylla, E.R.; Sanchez, S. Genome mining of Streptomyces scabrisporus NF3 reveals symbiotic features including genes related to plant interactions. *PLoS ONE* **2018**, *13*, e0192618. [CrossRef] [PubMed]

42. Krügel, H.; Krubasik, P.; Weber, K.; Saluz, H.P. Functional analysis of genes from Streptomyces griseus involved in the synthesis of isorenieratene, a carotenoid with aromatic end groups, revealed a novel type of carotenoid desaturase. *Biochim. Et Biophys. Acta (BBA)–Mol. Cell Biol. Lipids* **1999**, *1439*, 57–64. [CrossRef]

43. Fuke, T.; Sato, T.; Jha, S.; Tansengco, M.L.; Atomi, H. Phytoene production utilizing the isoprenoid biosynthesis capacity of Thermococcus kodakarensis. *Extremophiles* **2018**, *22*, 301–313. [CrossRef] [PubMed]

44. Sui, Y.; Mazzucchi, L.; Acharya, P.; Xu, Y.; Morgan, G.; Harvey, P.J. A Comparison of beta-Carotene, Phytoene and Amino Acids Production in Dunaliella salina DF 15 (CCAP 19/41) and Dunaliella salina CCAP 19/30 Using Different Light Wavelengths. *Foods* **2021**, *10*, 2824. [CrossRef] [PubMed]

45. Pollmann, H.; Breitenbach, J.; Sandmann, G. Development of Xanthophyllomyces dendrorhous as a production system for the colorless carotene phytoene. *J. Biotechnol.* **2017**, *247*, 34–41. [CrossRef] [PubMed]

46. De Lourdes Moreno, M.; Sánchez-Porro, C.; García, M.T.; Mellado, E. Carotenoids' production from halophilic bacteria. *Methods Mol. Biol.* **2012**, *892*, 207–217. [PubMed]

47. Maia, E.H.B.; Assis, L.C.; de Oliveira, T.A.; da Silva, A.M.; Taranto, A.G. Structure-based virtual screening: From classical to artificial intelligence. *Front. Chem.* **2020**, *8*, 343. [CrossRef]

48. Miras-Moreno, B.; Pedreño, M.A.; Romero, L.A. Bioactivity and bioavailability of phytoene and strategies to improve its production. *Phytochem. Rev. Vol.* **2019**, *18*, 356–359. [CrossRef]

49. Zylicz, Z.; Hofs, H.P.; Wagener, D.J. Potentiation of cisplatin antitumor activity on L1210 leukemia s.c. by sparsomycin and three of its analogues. *Cancer Lett.* **1989**, *46*, 153–157.
50. Guengerich, F.P. Cytochrome p450 and chemical toxicology. *Chem. Res. Toxicol.* **2008**, *21*, 70–83. [CrossRef]
51. Lin, J.H.; Yamazaki, M. Role of P-glycoprotein in pharmacokinetics: Clinical implications. *Clin Pharm.* **2003**, *42*, 59–98. [CrossRef] [PubMed]
52. Conlan, S.; Mijares, L.A.; Program, N.C.S.; Becker, J.; Blakesley, R.W.; Bouffard, G.G.; Brooks, S.; Coleman, H.; Gupta, J.; Gurson, N.; et al. Staphylococcus epidermidis pan-genome sequence analysis reveals diversity of skin commensal and hospital infection-associated isolates. *Genome. Biol.* **2012**, *13*, R64. [CrossRef] [PubMed]
53. Rasko, D.A.; Rosovitz, M.J.; Myers, G.S.; Mongodin, E.F.; Fricke, W.F.; Gajer, P.; Crabtree, J.; Sebaihia, M.; Thomson, N.R.; Chaudhuri, R.; et al. The pangenome structure of *Escherichia coli*: Comparative genomic analysis of *E. coli* commensal and pathogenic isolates. *J. Bacteriol.* **2008**, *190*, 6881–6893. [CrossRef] [PubMed]
54. Hormeno, L.; Campos, M.J.; Vadillo, S.; Quesada, A. Occurrence of tet(O/M/O) Mosaic Gene in Tetracycline-Resistant Campylobacter. *Microorganisms* **2020**, *8*, 1710. [CrossRef]
55. Mehla, K.; Ramana, J. Novel Drug Targets for Food-Borne Pathogen Campylobacter jejuni: An. Integrated Subtractive Genomics and Compara-tive Metabolic Pathway Study. *OMICS* **2015**, *19*, 393–406.
56. Zhou, P.; Hong, J. Structure-and Ligand-Dynamics-Based Design of Novel Antibiotics Targeting Lipid A Enzymes LpxC and LpxH in Gram-Negative Bacteria. *Acc. Chem. Res.* **2021**, *54*, 1623–1634. [CrossRef]
57. Pruitt, K.D.; Tatusova, T.; Maglott, D.R. NCBI Reference Sequence (RefSeq): A curated non-redundant sequence database of genomes, transcripts and proteins. *Nucleic Acids Res.* **2005**, *33*, D501–D504. [CrossRef]
58. Chaudhari, N.M.; Gupta, V.K.; Dutta, C. BPGA-an ultra-fast pan-genome analysis pipeline. *Sci. Rep.* **2016**, *6*, 24373. [CrossRef]
59. Basharat, Z.; Jahanzaib, M.; Yasmin, A.; Khan, I.A. Pan-genomics, drug candidate mining and ADMET profiling of natural product inhibitors screened against Yersinia pseudotuberculosis. *Genomics* **2021**, *113*, 238–244. [CrossRef]
60. Edgar, R. *Usearch*; Lawrence Berkeley National Lab. (LBNL): Berkeley, CA, USA, 2010.
61. Edgar, R.C. MUSCLE: Multiple sequence alignment with high accuracy and high throughput. *Nucleic. Acids. Res.* **2004**, *32*, 1792–1797. [CrossRef]
62. Alcock, B.P.; Raphenya, A.R.; Lau, T.T.; Tsang, K.K.; Bouchard, M.; Edalatmand, A.; Huynh, W.; Nguyen, A.-L.V.; Cheng, A.A.; Liu, S. CARD 2020: Antibiotic resistome surveillance with the comprehensive antibiotic resistance database. *Nucleic. Acids Res.* **2020**, *48*, D517–D525. [CrossRef] [PubMed]
63. Tatusov, R.L.; Galperin, M.Y.; Natale, D.A.; Koonin, E.V. The COG database: A tool for genome-scale analysis of protein functions and evolution. *Nucleic. Acids. Res.* **2000**, *28*, 33–36. [CrossRef] [PubMed]
64. Huang, Y.; Niu, B.; Gao, Y.; Fu, L.; Li, W. CD-HIT Suite: A web server for clustering and comparing biological sequences. *Bioinformatics* **2010**, *26*, 680–682. [CrossRef] [PubMed]
65. Ye, Y.-N.; Hua, Z.-G.; Huang, J.; Rao, N.; Guo, F.-B. CEG: A database of essential gene clusters. *BMC Genom.* **2013**, *14*, 769. [CrossRef]
66. Luo, H.; Lin, Y.; Gao, F.; Zhang, C.-T.; Zhang, R. DEG 10, an update of the database of essential genes that includes both protein-coding genes and noncoding genomic elements. *Nucleic. Acids Res.* **2014**, *42*, D574–D580. [CrossRef]
67. Liu, B.; Zheng, D.; Jin, Q.; Chen, L.; Yang, J. VFDB 2019: A comparative pathogenomic platform with an interactive web interface. *Nucleic Acids Res.* **2019**, *47*, D687–D692. [CrossRef]
68. Pandit, S.B.; Zhang, Y.; Skolnick, J. TASSER-Lite: An automated tool for protein comparative modeling. *Biophys. J.* **2006**, *91*, 4180–4190. [CrossRef]
69. Laskowski, R.A.; MacArthur, M.W.; Moss, D.S.; Thornton, J.M. PROCHECK: A program to check the stereochemical quality of protein structures. *J. Appl. Crystallogr.* **1993**, *26*, 283–291. [CrossRef]
70. Kelley, L.A.; Mezulis, S.; Yates, C.M.; Wass, M.N.; Sternberg, M.J. The Phyre2 web portal for protein modeling, prediction and analysis. *Nat. Protoc.* **2015**, *10*, 845–858. [CrossRef]
71. Basharat, Z.; Akhtar, U.; Khan, k.; Alotaibi, G.; Jalal, K.; Abbas, M.N.; Hayat, A.; Ahmad, D.; Hassan, S.S. Differential analysis of Orientia tsutsugamushi genomes for therapeutic target identification and possible intervention through natural product inhibitor screening. *Comput. Biol. Med.* **2022**, *141*, 105165. [CrossRef]
72. Basharat, Z.; Jahanzaib, M.; Rahman, N. Therapeutic target identification via differential genome analysis of antibiotic resistant Shigella sonnei and inhibitor evaluation against a selected drug target. *Infect. Genet. Evol.* **2021**, *94*, 105004. [CrossRef] [PubMed]
73. Daina, A.; Michielin, O.; Zoete, V. SwissADME: A free web tool to evaluate pharmacokinetics, drug-likeness and medicinal chemistry friendliness of small molecules. *Sci. Rep.* **2017**, *7*, 42717. [CrossRef] [PubMed]
74. Chen, Y.; Ung, C. Prediction of potential toxicity and side effect protein targets of a small molecule by a ligand–protein inverse docking approach. *J. Mol. Graph. Model.* **2001**, *20*, 199–218. [CrossRef]
75. Pu, L.; Liu, T.; Wu, H.-C.; Mukhopadhyay, S.; Brylinski, M. eToxPred: A machine learning-based approach to estimate the toxicity of drug candidates. *BMC Pharmacol. Toxicol.* **2019**, *20*, 2. [CrossRef]
76. Basharat, Z.; Khan, K.; Jalal, K.; Ahmad, D.; Hayat, A.; Alotaibi, G.; al Mouslem, A.; Alkhayl, F.F.A.; Almatroudi, A. An in silico hierarchal approach for drug candidate mining and validation of natural product inhibitors against pyrimidine biosynthesis enzyme in the antibiotic-resistant Shigella flexneri. *Infect Genet. Evol.* **2022**, *98*, 105233. [CrossRef]

Review

Efficacy of Combination Therapies for the Treatment of Multi-Drug Resistant Gram-Negative Bacterial Infections Based on Meta-Analyses

Takumi Umemura [1], Hideo Kato [1], Mao Hagihara [1,2], Jun Hirai [1], Yuka Yamagishi [1] and Hiroshige Mikamo [1,*]

[1] Department of Clinical Infectious Diseases, Aichi Medical University, Nagakute 480-1195, Japan; umemuratakumi@gmail.com (T.U.); katou.hideo.233@mail.aichi-med-u.ac.jp (H.K.); hagimao@aichi-med-u.ac.jp (M.H.); hiraichimed@gmail.com (J.H.); y.yamagishi@mac.com (Y.Y.)

[2] Department of Molecular Epidemiology and Biomedical Sciences, Aichi Medical University, Nagakute 480-1195, Japan

* Correspondence: mikamo@aichi-med-u.ac.jp; Tel.: +81-561-61-1842

Abstract: There is increasing evidence regarding the optimal therapeutic strategies for multidrug-resistant (MDR) bacteria that cause common infections and are resistant to existing antibiotics. Combination therapies, such as β-lactam combined with β-lactamase inhibitors or combination antibiotics, is a therapeutic strategy to overcome MDR bacteria. In recent years, the therapeutic options have expanded as certain combination drugs have been approved in more countries. However, only a handful of guidelines support these options, and the recommendations are based on low-quality evidence. This review describes the significance and efficacy of combination therapy as a therapeutic strategy against Gram-negative MDR pathogens based on previously reported meta-analyses.

Keywords: antibiotics; combination therapy; multi-drug resistant infection; meta-analysis

check for **updates**

Citation: Umemura, T.; Kato, H.; Hagihara, M.; Hirai, J.; Yamagishi, Y.; Mikamo, H. Efficacy of Combination Therapies for the Treatment of Multi-Drug Resistant Gram-Negative Bacterial Infections Based on Meta-Analyses. *Antibiotics* **2022**, *11*, 524. https://doi.org/10.3390/antibiotics11040524

Academic Editor: Dóra Kovács

Received: 18 March 2022
Accepted: 12 April 2022
Published: 14 April 2022

Publisher's Note: MDPI stays neutral with regard to jurisdictional claims in published maps and institutional affiliations.

1. Introduction

Antibiotic resistance is one of the top ten global public health threats. The World Health Organization (WHO) has reported that there is mounting evidence that the spread of multidrug-resistant (MDR) bacteria that cause common infections and are resistant to treatment with existing antibiotics is increasing [1,2]. In recent years, only a few new antibiotics have been developed for the treatment of infections by MDR pathogens, and emergent resistance has been reported after clinical use [3]. Therefore, the Center for Disease Prevention and Control (CDC) identified MDR Gram-negative bacteria such as carbapenem-resistant Acinetobacter and carbapenem-resistant Enterobacterales (CRE) as urgent threats because the rapid spread of resistance and lack of appropriate treatment strategies are considered critical [4].

Combination therapy is a strategy for preventing infections caused by MDR Gram-negative pathogens [5–8]. New combinations are increasingly proposed as a therapeutic option with the increasing approval of novel drugs. The combinations include antibiotics plus drugs without antibiotic activity, or antibiotics plus other antibiotics. However, only a handful of guidelines support this option, and they are based on low-quality evidence [9], since there are few randomized controlled trials in the literature that provide high levels of evidence to support the clinical question.

Meta-analyses can contribute to the establishment of evidence-based strategies and resolve contradictory research outcomes. In fact, many researchers have compared combination therapy with monotherapy using the meta-analysis. However, the interpretation of combination therapy may be misleading since criteria such as infection type and pathogen are different among these studies. Therefore, we comprehensively summarize published meta-analyses and illustrate the significance and feature of combination therapy as a therapeutic option against MDR pathogens. Table 1 shows a list of studies that reported meta-analyses of antibiotic combinations [10–24].

Table 1. The reports of systematic review and meta-analysis on antibiotic combinations.

Study	Year Published (Research Duration from Databases)	Database	Study Design Included in Meta-Analysis	Antibiotic Regimens (Monotherapy vs. Combination Therapy)	Type of Infection	Pathogens	Main Outcomes (Monotherapy vs. Combination Therapy)
Sfeir et al [10]	2018 (up to 15 June 2017)	MEDLINE EMBASE Cochrane Library	pro or retrospective observational, cohort, and active surveillance	BL-BLI including piperacillin-tazobactam vs. carbapenem	BSI	ESBL-producing Enterobacterales	Mortality BL-BLI vs. carbapenem as definitive, OR 0.96, 95% CI 0.59–1.86, as empirical, OR 1.13, 95% CI 0.87–1.48 TAZ/PIPC vs. carbapenem as definitive, OR 0.97, 95% CI 0.59–1.6, as empirical, OR 1.27, 95% CI 0.96–1.66
Zhang et al. [11]	2021 (up to December 2020)	Cochrane Library PubMed EMBASE	RCT cohort	BL-BLI vs. carbapenem	cUTI APN	ESBL-producing Enterobacterales	mortality, RR = 0.63, 95% CI 0.30–1.32 clinical success, RR = 0.99, 95% CI 0.96–1.03 microbiological success, RR = 1.06, 95% CI 1.01–1.11
Sternbach et al. [12]	2018 (up to December 2017)	PubMed CENTRAL LILACS	RCT	CZA vs. comparator (mainly carbapenem)	cUTI cIAI NP	mostly Enterobacterales (~25% ESBL-carrying)	30-day mortality, RR 1.10, 95% CI 0.70–1.72 serious adverse events, RR 1.24, 95% CI 1.00–1.54
Che et al. [13]	2019 (up to December 2018)	Medline Embase Cochrane Library	RCT	CZA vs. carbapenem	cUTI APN	mostly Enterobacterales	clinical success, RD 0.00, 95% CI 0.06–0.06 microbiological success, RD 0.07, 95% CI 0.04–0.18 serious adverse events, RD 0.02, 95% CI 0.00–0.04
Isler et al. [25]	2020 (The dates of coverage were 27 January 2020 to 10 February 2020)	PubMed, CENTRAL, CINAHL, Scopus OvidMedline OvidEmbase Web of Science	RCT	CZA vs. carbapenem	cUTI cIAI HAP/VAP	ESBL and AmpC-producing Enterobacterales	clinical response for ESBL producers, RR 1.02, 95% CI, 0.97–1.08 for AmpC producers, RR, 0.91, 95% CI 0.76–1.10
Onorato et al. [15]	2019 (up to February 2019)	Medline Google Scholar Cochrane Library	cohort case-control case series	CZA vs. CZA plus other antibiotics	any	Carbapenem resistant Enterobacterales *P. aeruginosa*	mortality rate, RR 1.18, 95% CI 0.88–1.58 rate of microbiological cure, RR 1.04, 95% CI 0.85–1.28
Fiore et al. [16]	2020 (up to 2 February 2020)	Medline EMBASE CENTRAL	RCT cohort	CZA vs. CZA plus other antibiotics	any (mostly BSI)	Carbapenem resistant (mainly KPC producing) Enterobacterales	mortality rate, OR 0.96, 95% CI 0.65–1.41

Table 1. *Cont.*

Study	Year Published (Research Duration from Databases)	Database	Study Design Included in Meta-Analysis	Antibiotic Regimens (Monotherapy vs. Combination Therapy)	Type of Infection	Pathogens	Main Outcomes (Monotherapy vs. Combination Therapy)
Li et al. [17]	2021 (up to 31 March 2021)	PubMed EMBASE Web of Science CNKI Wanfang Data databases	cohort case series cross sectional	CZA vs. CZA plus other antibiotics	any	Any (carbapenem resistant)	overall mortality rates, OR 1.03, 95% CI 0.79–1.34 clinical success, OR 0.95, 95% CI 0.64–1.39 microbiologically negative, OR 0.99, 95% CI 0.54–1.81 posttreatment resistance of CZA, OR 0.65, 95% CI 0.34–1.26
Fiore et al. [18]	2021 (up to November 2020)	Medline EMBASE CENTRAL	retrospective cohort case-control	C/T vs. C/T plus other antibiotics	any	*P. aeruginosa* ESBL producing Enterobacterales	all-cause mortality, RR 0.31, 95% CI 0.10–0.97 clinical improvement, RR 0.97, 95%CI 0.54–1.74 microbiological cure, RR 0.83, 95%CI 0.12–5.70
Paul et al. [19]	2003 (up to March 2002)	Medline Embase Lilacs Cochrane Library	RCT	B-lactam vs. β-lactam-aminoglycoside combination	fever and neu-tropenia	any	all cause fatality, RR 0.85, 95% CI 0.72–1.02 treatment failure, RR 0.92, 95%CI 0.85–0.89 any adverse event, RR 0.85, 95%CI 0.73–1.00
Paul et al. [20]	2004 (up to March 2003)	Medline Embase Lilacs Cochrane Library	RCT	B-lactam vs. β-lactam-aminoglycoside combination	severe infections	any	all cause fatality, RR 0.90, 95% CI 0.77–1.06 clinical failure, RR 0.87, 95%CI 0.78–0.97 nephrotoxicity, RR 0.36, 95%CI 0.28–0.47
Zusman et al. [21]	2017 (up to 10 April 2016)	PubMed Cochrane Library	RCT retrospective observational	polymyxin monotherapy vs. polymyxin-based combination therapy	any	carbapenem-resistant or carbapenemase-producing Gram-negative bacteria	mortality, uOR 1.58, 95% CI 1.03–2.42 mortality compared with combination with TGC, AG and FOM, uOR 1.57, 95% CI 1.06–2.32 mortality for *K. pneumoniae* bacteremia, uOR 2.09, 95% CI 1.21–3.6
Samal et al. [22]	2021 (up to 31 December 2018)	PubMed Cochrane Library	RCT pro or retrospective observational	polymyxin monotherapy vs. polymyxin-based combination therapy	any	polymyxin-susceptible, carbapenem-resistant or carbapenemase-producing Gram-negative bacteria	mortality, RR 0.81, 95% CI 0.65–1.01 polymyxin-carbapenem combination in mortality, RR 0.64, 95% CI 0.40–1.03

Table 1. *Cont.*

Study	Year Published (Research Duration from Databases)	Database	Study Design Included in Meta-Analysis	Antibiotic Regimens (Monotherapy vs. Combination Therapy)	Type of Infection	Pathogens	Main Outcomes (Monotherapy vs. Combination Therapy)
Cheng IL et al. [23]	2018 (up to July 2018)	PubMed Embase Cochrane databases	RCT	colistin monotherapy vs. colistin-based combination therapy	any (mostly VAP)	carbapenem-resistant Gram-negative bacteria (mostly *A. baumanii*)	all-cause mortality, RR 1.03, 95% CI 0.89–1.20 infection-related mortality, RR 1.23, 95% CI 0.91–1.67 microbiologic response, RR 0.86, 95% CI 0.72–1.04
Vardakas K7 et al. [9]	2018 (up to November 2016)	PubMed Scopus	RCT	colistin monotherapy vs. colistin-based combination therapy	any (mostly VAP or BSI)	MDR or XDR Gram-negative bacteria (mainly *K. pneumoniae* or *A. baumanii*)	mortality, RR 0.91, 95% CI 0.81–1.02 mortality (in favor of combination with high-dose colistin), RR 0.80, 95% CI 0.69–0.93
Liu J. et al. [24]	2021 (up to March 2020)	PubMed Embase Cochrane Web of Science	RCT pro or retrospective observational	colistin monotherapy vs. colistin-based combination therapy	any (mostly VAP or BSI)	MDR or XDR *A. baumanii*	clinical improvement: RFP, RR 1.28, 95% CI 0.67–2.45; FOM, RR 1.08, 95% CI 0.76–1.53; sulbactam, RR 1.02, 95% CI 0.86–1.22 clinical cure: carbapenem, RR 1.34, 95% CI 0.92–1.95; sulbactam, RR 1.28, 95% CI 0.90–1.83 rate of microbiological eradication (in favor of combination with RFP or FOM), RFP, RR 1.31, 95% CI 1.01–1.69; FOM, RR 1.23, 95% CI 1.01–1.53

Abbreviations; BL-BLI, β-lactam and β-lactamase inhibitor combination; BSI, blood stream infection; ESBL, expended spectrum β-lactamase; OR, odds ratio; CI, confidence interval; RCT, randomized control trial; cUTI, complicated urinary tract infections; APN, acute pyelonephritis; CZA, ceftadizime avibactam; C/T, Ceftlozane-tazobactam; cIAI, complicated intra-abdominal infection; NP, nosocomial pneumonia; RR, risk ratio; RD, risk difference; uOR, TGC, tigecycline; AG, aminoglycoside; FOM, fosfomycin; unadjusted odds ratio;VAP, ventilator associated pneumonia; BSI, blood stream infection; MDR, multi drug resistant; XDR, extensively drug resistant; RFP, rifampicin.

2. Combinations of Antibiotics plus β-Lactamase Inhibitor

β-lactamase production is one of the main mechanisms of resistance against β-lactams. To date, many types of β-lactamases have been reported: Ambler class A, which includes TEM nova (TEM), sulfhydryl variable (SHV), cefotaxime (CTX), *Klebsiella pneumoniae* carbapenemases (KPC), and *Serratia marcescens* enzymes (SME); serine-based class B, which includes Verona integron-encoded metallo-β-lactamase (VIM), imipenemase metallo-β-lactamase (IMP), New Delhi metallo-β-lactamase (NDM), and Sao Paulo metallo-β-lactamase (SPM); zinc-based class C, which includes AmpC; and serine-based class D, which includes oxacillinase (OXA).

Recently, the Infectious Diseases Society of America (IDSA) published a guideline recommending carbapenems as the preferred treatment option for infections caused by resistant bacteria, without mentioning β-lactam and β-lactamase inhibitor combination therapies [26]. However, a recent study based on the MERINO trial reported that differences in mortality rates are less pronounced for β-lactams used in combination with β-lactamase inhibitors [25]. Moreover, several novel β-lactam and β-lactamase inhibitor combination therapies have been developed that target Gram-negative bacteria that produce β-lactamase. Therefore, β-lactam and β-lactamase inhibitor combination therapies may be a novel alternative option for the treatment of β-lactamase-producing pathogens.

2.1. Carbapenem versus β-Lactam and β-Lactamase Inhibitor Combinations

Extended-spectrum β-lactamases (ESBL) are class A β-lactamase enzymes that hydrolyze the β-lactam ring, conferring resistance to most β-lactam antibiotics, including expanded-spectrum cephalosporins, and are often resistant to other classes of antibiotics (e.g., fluoroquinolones, trimethoprim-sulfamethoxazole, aminoglycosides, and tetracyclines) [27–30]. Moreover, the overuse of carbapenems, which are generally recommended as first-line agents for infections caused by ESBL-producing pathogens [26], lead to an increase in selective pressures for carbapenem resistance [31]. Therefore, the interest in β-lactam and β-lactamase inhibitor combination therapies has increased due to the limited treatment options for infections caused by ESBL-producing pathogens.

Two meta-analyses have evaluated the efficacy of carbapenem versus β-lactam and β-lactamase inhibitor combination therapies for the treatment of infections caused by ESBL-producing pathogens (Table 2). Sfeir et al. conducted a meta-analysis to investigate the 30-day mortality of patients with blood stream infections caused by ESBL-producing pathogens [10]. A total of twenty-five cohort or case-control studies were included: eleven evaluated empiric treatment, eight evaluated definitive treatment, and six simultaneously evaluated empiric and definitive treatment. No statistically significant differences in mortality were found between β-lactam and β-lactamase inhibitor combination therapy and carbapenem administered as an empirical (odds ratio [OR] 1.13, 95% confidence interval [CI] 0.87–1.48) or definitive (OR 0.96, 95% CI 0.50–1.86) treatment. In subgroup analyses, there was no significant difference in mortality between patients treated with piperacillin-tazobactam (PTZ) and carbapenem when used as empirical (OR 1.26, 95% CI 0.96–1.66) or definitive (OR 0.97, 95% CI 0.59–1.6) treatment for ESBL-producing Enterobacterales bloodstream infections. Zhang et al. compared clinical outcomes, such as clinical success, microbiological success, and mortality for urinary tract infections (UTIs) caused by ESBL-producing Enterobacterales [11]. Three randomized controlled trials (RCTs) and seven cohort studies were included. There was no statistically significant difference between carbapenem and β-lactam and β-lactamase inhibitor combination therapy with regards to clinical success (risk rate [RR] 0.99, 95% CI 0.96–1.03) and mortality (RR 0.63, 95% CI 0.30–1.32). In contrast, a slightly higher rate of microbiological success was observed in patients treated with a β-lactam and β-lactamase inhibitor combination therapy (RR 1.06, 95% CI 1.01–1.11). However, this result was mainly attributed to treatment with ceftazidime-avibactam (CZA) based on a single RCT (RR 1.32, 95% CI 1.13–1.55).

Table 2. Meta-analyses evaluated the efficacy of carbapenem versus β-lactam and β-lactamase inhibitor combination therapies.

Study	No. of Studies Analyzed	Patients			Results with Significance
		Infections	Pathogens	Antibiotics	
Sfeir et al. [10]	25 (cohort or case-control)	BSI	ESBL-producing Enterobacterales	BL-BLI including PIPC/TAZ vs. carbapenem	None
Zhang et al. [11]	10 (3 RCTs and 7 cohort)	cUTI APN	ESBL-producing Enterobacterales	BL-BLI vs. carbapenem	None

Abbreviations; BL-BLI, β-lactam and β-lactamase inhibitor combination; PIPC/TAZ, piperacillin-tazobactam; BSI, blood stream infection; ESBL, expended spectrum β-lactamase; RCT, randomized control trial; cUTI, complicated urinary tract infections; APN, acute pyelonephritis.

2.2. Carbapenem versus Ceftazidime-Avibactam

CZA is a combination of the third-generation cephalosporin ceftazidime and the novel non-β-lactam β-lactamase inhibitor avibactam. CZA shows in vitro activity against Amber class A, class C, and some class D β-lactamase-producing bacteria, including Enterobacterales and *Pseudomonas aeruginosa*, whereas it has no activity against metallo-β-lactamase-producing organisms [32–34]. CZA has been approved by the United States Food and Drug Administration (FDA) and the European Medicines Agency (EMA) for the treatment of infections caused by Gram-negative bacteria for which limited therapeutic options exist [35,36].

There have been three meta-analyses on the efficacy and safety of carbapenem versus CZA for infections caused by Enterobacterales (Table 3). Sternbach et al. assessed the efficacy and safety of this treatment for complicated infections [12]. Eligible studies included RCTs for the treatment of complicated UTIs, complicated intra-abdominal infections, and nosocomial pneumonia among adult patients. Seven RCTs were included, of which less than 25% of cases were ESBL-producing Enterobacterales at the initiation of treatment. All-cause 30-day mortality was reported in six trials, which included one UTI, four intra-abdominal infection, and one pneumonia trial, that showed no significant difference in response between carbapenem and CZA (RR 1.10, 95% CI 0.70–1.72). Moreover, a microbiological response was reported in five trials, including three UTI, one intra-abdominal infection, and one pneumonia trial, that showed no significant difference between carbapenem and CZA (RR 1.04, 95% CI 0.93–1.17). In contrast, serious adverse events were reported in six trials including three UTI, two intra-abdominal infections, and one pneumonia trial, and a significantly higher rate of discontinuation was demonstrated for patients treated with CZA (RR 1.24, 95% CI 1.00–1.54). Che et al. evaluated the feasibility of treating Enterobacterales infections with CZA instead of carbapenem [13]. In this meta-analysis, eligible studies included RCTs that treated adult patients for Enterobacterales infections or mixed infections with Enterobacterales bacteria accounting for more than 90% of infections in the population, and three RCTs were included. The meta-analysis of the three trials showed that there were no significant differences between CZA and carbapenems in the rate of clinical success [risk difference (RD) 0.00, 95% CI −0.06–0.06] and microbiological success (RD 0.07, 95% CI −0.04–0.18). In contrast, serious adverse events were reported in two studies and tended to occur more frequently in patients treated with CZA than in those treated with carbapenems (RD 0.02, 95% CI −0.00–0.04). Isler et al. conducted a meta-analysis of adult patients who were suspected to have or were diagnosed with infection with ESBL- or AmpC-producing Enterobacterales in complicated UTI or complicated intra-abdominal infection and pneumonia [14]. Five RCTs were available to evaluate clinical and microbiological responses; four studies reported outcome data for ESBL-producing Enterobacterales; three studies reported outcome data for AmpC-producing Enterobacterales; and four studies reported outcome data for ceftazidime non-susceptible Enterobacterales. The clinical response for ESBL- or AmpC-producing Enterobacterales showed no significant differences between patients treated with carbapenem and CZA (clinical response for ESBL producers, RR 1.02, 95% CI 0.97–1.08; for AmpC producers, RR 0.91, 95% CI 0.76–1.10). CZA showed a better microbiologic response than carbapenem for ceftazidime non-susceptible Enterobacterales in patients with complicated UTI and pneumonia (RR 1.21, 95% CI 1.07–1.37).

Table 3. Meta-analyses evaluated the efficacy of carbapenem versus ceftazidime avibactam.

Study	No. of Studies Analyzed	Patients			Results with Significance
		Infections	Pathogens	Antibiotics	
Sternbach et al. [12]	7 RCTs	cUTI cIAI NP	mostly Enterobacterales (~25% ESBL-carrying)	CZA vs. comparator (mainly carbapenem)	Significantly higher rate treated with CZA (RR 1.24, 95% CI 1.00–1.54)
Che et al. [13]	3 RCTs	cUTI APN	mostly Enterobacterales	CZA vs. carbapenem	SAEs with CZA were numerically higher (RD = 0.02, 95% CI 0.00 to 0.04; p = 0.06).
Isler et al. [14]	5 RCTs	cUTI cIAI HAP/VAP	ESBL and AmpC-producing Enterobacterales	CZA vs. carbapenem	CZA showed a better microbiologic response for ceftazidime non-susceptible Enterobacterales (RR 1.21, 95% CI 1.07–1.37)

Abbreviations; CI, confidence interval; RCT, randomized control trial; cUTI, complicated urinary tract infections; APN, acute pyelonephritis; CZA, ceftazidime avibactam; cIAI, complicated intra-abdominal infection; NP, nosocomial pneumonia; ESBL, expended spectrum β-lactamase; SAE, severe adverse effect; RR, risk ratio.

2.3. CZA versus CZA Combination Therapy

CZA monotherapy led to the emergence of resistance-conferring mutations in blaKPC-3 in *K. pneumoniae* [37]. Moreover, a recent study reported that *P. aeruginosa* that produce metallo-β-lactamases (VIM, IMP, and NDM) have a high resistance rate against CZA [38]. Therefore, CZA monotherapy may not be sufficient to treat all MDR bacteria. However, combination therapy with CZA is expected to be a therapeutic strategy in patients infected with MDR bacteria, since in vitro studies have shown positive effects with CZA plus other antibiotics [39,40].

Three meta-analyses compared CZA monotherapy with CZA combination therapy in patients with carbapenem-resistant Gram-negative pathogens (Table 4). Onorato et al. conducted a meta-analysis of infections caused by CRE and carbapenem-resistant *P. aeruginosa* [15]. Eleven retrospective studies were included, and three were case series. Seven studies included only patients with CRE, one included only patients with carbapenem-resistant *P. aeruginosa*, and three reported infections caused by both pathogens. The meta-analysis included various types of infections, including pneumonia, bacteremia, intra-abdominal infection, UTI, skin and soft-tissue infection, bone and joint infection, and multiple infections. Six studies included patients with a KPC-producing strain, five studies included patients with an OXA-48-producing strain, and the strain of resistance was unknown in some patients. The mortality rate was similar in patients treated with monotherapy and combination therapy (RR 1.18, 95% CI 0.88–1.58). Similarly, the mortality rate was not significantly different between the two groups in five studies evaluating in-hospital mortality (RR 1.37, 95% CI 0.80–2.34), four studies evaluating 30-day mortality (RR 1.07, 95% CI 0.75–1.53), and two studies evaluating 90-day mortality (RR 1.42, 95% CI 0.44–4.60). Regarding microbiological outcome, no difference was observed between the two groups in seven studies (RR 1.04, 95% CI 0.85–1.28). In three studies microbiological outcome was defined as the presence of at least one negative culture during therapy (RR 1.02, 95% CI 0.67–1.56), in two studies microbiological outcome was defined as negative cultures after more than 7 days of treatment (RR 0.92, 95% CI 0.87–1.97), and in six studies patients infected with carbapenem-resistant *P. aeruginosa* (RR 1.05, 95% CI 0.84–1.31) were excluded. Fiore et al. conducted a network meta-analysis of patients with CRE infections [16]. Six retrospective studies were included, and mortality was reported at different time points. The network meta-analysis showed no significant difference in the mortality rate between patients who received CZA monotherapy and those who received CAZ combination therapy (OR 0.96, 95% CI: 0.65–1.41). Li et al. conducted a meta-analysis to compare mortality rate, microbiological response, clinical success, and development of resistance [17]. Seventeen retrospective observational studies were included: eleven cohort studies, one case-cohort study, two case series, and three cross-sectional studies. The pathogen reported by most of these studies was CRE, with carbapenem-resistant *K. pneumoniae* as the most frequently reported pathogen, while one study included both carbapenem-resistant *K. pneumoniae* and *P. aeruginosa*, and one study included CRE, carbapenem-resistant *P. aeruginosa*, and *A. baumannii*. Regarding infection types, most studies reported multiple infections, two studies reported only bacteremia, and one study reported pneumonia. Antibiotics that were used in combination with CZA therapy included aminoglycosides, polymyxin, tigecycline, colistin, amikacin, imipenem, gentamicin, ciprofloxacin, meropenem, fosfomycin, carbapenems, fluoroquinolones, minocycline, and sulfamethoxazole-trimethoprim. There was no statistically significant difference in mortality rate at any time point (overall, 14, 30, and 90 days, in-hospital) between CZA therapy alone and CZA-based combination therapy (overall, OR 1.03, 95% CI 0.79–1.34; 14-day, OR 0.90, 95% CI 0.32–2.50; 30-day, OR 0.96, 95% CI 0.69–1.33; 90-day, OR 1.74, 95% CI 0.79–3.82; in-hospital, OR 1.01, 95% CI 0.55–1.86). No statistically significant difference was found between groups in the microbiological response in five studies (OR 0.99, 95% CI 0.54–1.81), in the clinical success in ten studies (OR 0.95, 95% CI 0.64–1.39), or in the post-treatment resistance to CZA in six studies (OR 0.65, 95% CI 0.34–1.26). However, the combination therapy was more strongly associated with lower resistance to CZA in the three pooled studies (OR 0.18, 95% CI 0.04–0.78).

Table 4. Meta-analyses evaluated the efficacy of ceftazidime avibactam (CZA) versus CZA combination therapy.

Study	No. of Studies Analyzed	Patients			Results with Significance
		Infections	Pathogens	Antibiotics	
Onorato et al. [15]	11 (cohort, case-control, case series)	any	CRE P. aeruginosa	CZA vs. CZA plus other antibiotics	None
Fiore et al. [16]	13 (7 RCTs, 6 cohorts)	any (mostly BSI)	CRE (mainly KPC producing)	CZA vs. CZA plus other antibiotics	None
Li et al. [17]	17 (11 cohort, 1 case series, 2 case-control, 3 cross sectional)	any	any (carbapenem resistant)	CZA vs. CZA plus other antibiotics	A trend of post-treatment resistance occurred more likely in CZA monotherapy (according to the pooled three studies, OR 0.18, 95% CI 0.04–0.78).

Abbreviations; CRE, carbapenem resistant Enterobacterales; RCT, randomized control trial; CZA, ceftazidime avibactam; OR, odds ratio; CI, confidence interval.

2.4. Ceftlozane-Tazobactam (C/T) versus C/T Combination Therapy

Tazobactam targets the active site of serine-based β-lactamases, which are mainly class A β-lactamases [18]. Ceftlozane, an oxyamino-aminothiazolyl cephalosporin, is structurally similar to ceftazidime and is active against *P. aeruginosa* [18]. A combination with ceftlozane and tazobactam has been approved for the treatment of complicated UTI, complicated intra-abdominal infections, and hospital-acquired and ventilator-associated pneumonia (HAP/VAP) [41–43]. In vitro studies evaluating combination regimens containing C/T plus other antibiotics showed a decline in the bacterial burden of MDR pathogens [44–47]. In contrast, clinical studies have shown discrepancies with the results of preclinical studies.

One meta-analysis reported a comparison between C/T alone and C/T in association with other antibiotics for the treatment of adult patients with microbiologically confirmed bacterial infections in any setting [48]. The study included seven retrospective cohort studies (two multicenter and five single-center) and one single-center case-control study. Seven of the eight studies evaluated infections caused by *P. aeruginosa* and the other evaluated infections caused by ESBL-producing Enterobacterales. Patients developed sepsis in two of the eight studies, lower respiratory tract infection in one study, and osteomyelitis in another study. Four studies that evaluated all-cause mortality enrolled 148 patients (C/T, 87 patients; C/T combination therapy, 61 patients), and the mortality rate was significantly decreased with C/T combination therapy compared to that with C/T monotherapy (OR 0.31, 95% CI 0.10–0.97, *p* = 0.045). Seven studies that enrolled 391 patients evaluated clinical improvement outcomes (C/T, 261 patients; C/T combination therapy,

130 patients), and the clinical outcome did not improve using C/T combination therapy (OR 0.97, 95% CI 0.54–1.74, p = 0.909). Two studies that enrolled 33 patients evaluated microbiological cure outcomes (C/T, 13 patients; C/T combination therapy, 20 patients), and there was no significant difference in microbiological cure between C/T combination therapy and C/T monotherapy (OR 0.83, 95% CI 0.12–5.70, p value was not reported). The discrepancy between the overall estimate of the effect between the mortality and clinical outcome (clinical cure and microbiologic cure) is difficult to interpret clinically, but heterogeneities in sample size and patient backgrounds may explain the difference.

3. Antibiotics Combinations

The IDSA guidelines do not recommend routine antibiotic combination therapy for infections caused by carbapenem-resistant Enterobacterales and *P. aeruginosa* [10] since the combination therapy increases the incidence of nephrotoxicity [19,20,49,50]. However, few reports have shown an increase in other adverse events caused by the combination therapy [6]. Currently, new β-lactam and β-lactamase inhibitor combination therapies, such as CZA, C/T, imipenem-cilastatin-relebactam, and meropenem-vaborbactam, have been developed to treat infections caused by MDR pathogens. Tazobactam targets the active site of serine-based β-lactamases, mainly class A β-lactamases such as SHV [33], and REL and VAB inhibit class A and C β-lactamases, but not class B and D [51,52]. Avibactam inhibits class A, C, and D serine-based β-lactamase inhibitors [33]. Notably, these β-lactam and β-lactamase inhibitor combinations are unable to treat all carbapenemase-producing pathogens. Therefore, attention has been focused on antibiotic combinations as treatment options for these pathogens. Here, we summarize antibiotic combinations for the treatment of infections caused by MDR Gram-negative bacteria based on previous meta-analyses.

3.1. β-Lactam versus β-Lactam plus Aminoglycoside

Antibiotic synergy has traditionally been demonstrated with β-lactam–aminoglycoside combinations for the treatment of infections with Gram-negative pathogens. The combination of β-lactams and aminoglycosides provides different mechanisms by which bacteria are eliminated [6]. However, clinical studies have shown results that are contrary to those of in vitro and in vivo studies.

Two meta-analyses have evaluated the efficacy of β-lactam versus β-lactam plus aminoglycoside (Table 5). Paul et al. performed a meta-analysis of RCTs on the treatment of patients with fever and neutropenia [19]. Forty-seven RCTs were included in this meta-analysis. Ceftazidime, PTZ, imipenem, and cefoperazone were administered as β-lactam regimens, while amikacin, gentamicin, and tobramycin were administered as aminoglycoside regimens. Although combination therapy did not improve all-cause mortality compared to monotherapy (RR 0.85, 95% CI 0.72–1.02), there was a significantly higher treatment success rate using combination therapy for the treatment of severe neutropenia (<100/mm3; RR 1.49, 95% CI 1.13–1.97) in both adults >16 years old (RR 1.21, 95% CI 1.07–1.37) and children (RR 2.74, 95% CI 1.08–6.98). In addition, Paul et al. performed a meta-analysis of RCTs for severe infections in patients without neutropenia [20]. Sixty-four RCTs were included in this meta-analysis. Infection types included severe sepsis, pneumonia, Gram-negative infections, abdominal infections, UTIs, and Gram-positive infections. Twelve RCTs reported all-cause mortality, and there was no difference between monotherapy and combination therapy (RR 1.02, 95% CI 0.76–1.38). Clinical and bacteriological failures were evaluated in data from twenty and fourteen RCTs, respectively, and combination therapy showed a tendency to improve clinical failure compared with same β-lactam monotherapy (RR 1.09, 95% CI 0.94–1.27), while there was no significant difference in bacteriological failure between the two groups (RR 1.08, 95% CI 0.71–1.64). Furthermore, nephrotoxicity was significantly more common with combination therapy (RR 0.36, 95% CI 0.28–0.47).

Table 5. Meta-analyses evaluated the efficacy of β-lactam versus β-lactam-AG combination.

Study	No. of Studies Analyzed	Patients			Results with Significance
		Infections	Pathogens	Antibiotics	
Paul et al. [19]	47 RCTs	fever and neutropenia	any	β-lactam vs. β-lactam-AG combination	Higher treatment success rate using combination therapy for the treatment of severe neutropenia (<100/mm3; RR 1.49, 95% CI 1.13–1.97) in both adults > 16 years old (RR 1.21, 95% CI 1.07–1.37) and children (RR 2.74, 95% CI 1.08–6.98).
Paul et al. [20]	64 RCTs	severe infections	any	β-lactam vs. β-lactam-AG combination	Clinical failure was more common with combination treatment overall (RR 0.87, 95% CI 0.78–0.97) Nephrotoxicity was significantly more common with combination therapy (RR 0.36, 95% CI 0.28–0.47).

Abbreviations; RCT, randomized control trial; AG, aminoglycoside; RR, risk ratio; CI, confidence interval.

3.2. Carbapenem plus Carbapenem (Double Carbapenems) versus Other Antibiotic Regimens

Bulik et al. reported the efficacy of double carbapenem therapy (DCT) against carbapenemase-producing *K. pneumoniae* in a mouse thigh infection model [53]. Moreover, it was reported that DCT had a synergistic effect in isolates with a high minimum inhibitory concentration for meropenem (up to 128 mg/L) [54]. In 2013, DCT successfully cured three patients with KPC-producing *K. pneumoniae* [55]. Over the past few years, DCT has emerged as a promising treatment strategy for carbapenem-resistant *K. pneumoniae* [56].

One meta-analysis evaluated the efficacy and safety of DCT and other antibiotic regimens in patients with infections caused by MDR Gram-negative pathogens [57]. The study included three cohort or case-control studies comprising 235 patients with CRE infection, and the infection types mainly included pneumonia, bloodstream infections, and UTIs. The DCT regimens were combinations of ertapenem (1–2 g daily) and meropenem (2 g every 8 h daily) or doripenem (2 g every 8 h daily), and other antibiotic regimens were colistin, tigecycline, aminoglycoside monotherapies, or combination regimens. There were no obvious advantages of DCT in clinical response (OR 1.74, 95% CI 0.99–3.06, $p = 0.05$) or microbiological response (OR 1.90, 95% CI 0.95–3.80, $p = 0.07$), but the mortality rate in the DCT group was significantly lower than that in the control group (OR 0.44, 95% CI 0.24–0.82, $p = 0.009$). No adverse events resulted in treatment interruption. A critical limitation of this study was the low grade of evidence, because the analysis was based on only three retrospective cohort or case-control studies. However, current data suggest that DCT may be an effective and safe strategy for treating carbapenem-resistant pathogens.

3.3. Polymyxin versus Polymyxin Combination Therapy

The alarming increase in MDR Gram-negative bacteria has resulted in the resurgence of polymyxin use, although its use has been limited for the past decade [58]. Polymyxin is at times the only therapeutic option because of its variable susceptibility to other antibiotics [59]. However, the use of polymyxin has several disadvantages, including poor efficacy compared to β-lactam [59], nephrotoxicity induced by high doses [60], and the emergence of resistance during therapy [61]. Therefore, in clinical settings, polymyxin is used in combination with other antibiotics to improve clinical outcomes.

Two meta-analyses have evaluated the efficacy of polymyxin versus polymyxin combination therapy (Table 6). Zusman et al. performed a meta-analysis examining the effectiveness of polymyxin monotherapy versus polymyxin-based combination therapy by antibiotic type and bacterial species [21]. Eligible studies included retrospective studies, prospective studies, or RCTs in adult patients infected with polymyxin-susceptible, carbapenem-resistant, or carbapenemase-producing Gram-negative bacteria, and twenty-two studies were included: nineteen retrospective observational studies and three RCTs. Overall, nine studies assessed tigecycline, seven studies assessed carbapenems, three studies assessed rifampicin, three studies assessed aminoglycosides, three studies assessed sulbactam, two studies assessed vancomycin, one study assessed PTZ, and one study assessed intravenous fosfomycin. Mortality rates were significantly higher with polymyxin monotherapy compared to those with carbapenem combination therapy in seven studies (OR 1.58, 95% CI 1.03–2.42), and those with tigecycline, aminoglycoside, or fosfomycin in eleven studies (OR 1.57, 95% CI 1.06–2.32). In particular, seven studies on the treatment of *K. pneumoniae* bacteremia found that combination therapy with tigecycline or aminoglycoside resulted in significantly lower mortality rates than those with polymyxin monotherapy (OR 2.09, 95% CI 1.21–3.60). When monotherapy was compared to any combination therapy for the type of infection, any combination therapy was associated with improved mortality rates in the treatment of bacteremia (OR 2.23, 95% CI 1.51–3.30), while there was no significant difference in mortality rates between monotherapy and combination therapy for the treatment of ventilator-associated or hospital-acquired pneumonia in five studies (OR 0.69, 95% CI 0.39–1.24).

Table 6. Meta-analyses evaluated the efficacy of polymyxin versus polymyxin combination therapy.

Study	No. of Studies Analyzed	Patients			Results with Significance
		Infections	Pathogens	Antibiotics	
Zusman et al. [21]	22 (RCT, retrospective observational)	any	CR or CP-GNB	polymyxin monotherapy vs. polymyxin-based combination therapy	Mortality rates were significantly higher with polymyxin monotherapy (OR 1.58, 95% CI 1.03–2.42)
Samal et al. [22]	39 (6 RCTs, 11 prospective and 22 retrospective observational)	any	polymyxin-susceptible, CR or CP GNB	polymyxin monotherapy vs. polymyxin-based combination therapy	Mortality rates were significantly lower with combination (OR 0.81, 95% CI 0.65–1.01)

Abbreviations; RCT, randomized control trial; CR, carbapenem resistant; GNB, Gram-negative bacteria; CP, carbapenemase producing; OR, odds ratio; CI, confidence interval.

Samal et al. [22] performed a meta-analysis with similar eligibility criteria as the previously mentioned study [21]. This study included seventeen prospective studies (6 RCTs) and twenty-two retrospective studies. A meta-analysis of all-cause mortality in all 39 studies yielded an OR of 0.81 with a 95% CI of 0.65–1.01. Nine studies that used only carbapenems yielded an OR of 0.64 with a 95% CI of 0.40–1.03. Moreover, a separate meta-analysis of only RCTs yielded an OR of 0.82 with a 95% CI of 0.58–1.16.

3.4. Colistin versus Colistin Combination Therapy

Colistin and polymyxin B differ by a single amino acid in the peptide ring, with phenylalanine in polymyxin B and leucine in colistin [62]. Colistin is a mixture of colistin A and colistin B, with colistin B accounting for the majority of the total dose [63], and is administered as the inactive pro-drug colistimethate sodium [64]. The different degrees of protein binding between colistin A and B cause inter- and intra-individual variability in the plasma concentrations of colistin [65,66]. Several approaches have been proposed to overcome these problems. Among the implemented strategies, combination regimens have been identified as the most promising [67]. In vitro studies have demonstrated synergistic and additive effects when combination regimens including colistin were evaluated [68–71]. In particular, the combination of colistin and tigecycline has been found to be effective against strains that form biofilms, although this may be dependent on the concentration of each drug [72].

Recently, 3 meta-analyses have evaluated the efficacy of colistin versus colistin combination therapy (Table 7). Cheng et al. conducted a meta-analysis of five RCTs on the treatment of carbapenem-resistant Gram-negative bacterial infections [23]. All *A. baumannii* isolates were carbapenem-resistant. The colistin-based combination regimens included rifampicin (two studies), fosfomycin (one study), meropenem (one study), and ampicillin-sulbactam (one study). Compared to colistin combination therapy, colistin monotherapy had no association with higher mortality in five studies (RR 1.03, 95% CI 0.89–1.20), higher infection-related mortality in four studies (RR 1.23, 95% CI 0.91–1.67), and lower microbiologic response in five studies (RR 0.86, 95% CI 0.72–1.04). In addition, compared to colistin combination therapy, colistin monotherapy was not associated with lower nephrotoxicity in three studies (RR 0.98, 95% CI 0.84–1.21). Vardakas et al. conducted a meta-analysis of all study types, except case reports and case series, on MDR or extensively drug-resistant (XDR) Gram-negative infections [9]. Thirty-two studies were included: twenty-two retrospective studies, six prospective studies, one study with both prospective and retrospective aspects, and three RCTs. The studies focused mainly on infections caused by *A. baumannii* and *K. pneumoniae*; infections caused by *P. aeruginosa* and other Enterobacterales were also included. Carbapenem, tigecycline, gentamicin, rifampin, sulbactam, fosfomycin, non-carbapenem β-lactams, and ciprofloxacin were the main antibiotics used in the combination regimens in various proportions. The rates of resistance to individual antibiotics varied. Compared to colistin monotherapy, colistin-based combination therapy was not associated with lower mortality (RR 0.91, 95% CI 0.81–1.02). In sub-analyses of high-dose treatments (>6 million international units; RR 0.80, 95% CI 0.69–0.93), combination therapy was found to be significantly more effective in patients with bacteremia (RR 0.75, 95% CI 0.57–0.98), and in patients with *A. baumannii* infections (RR 0.88, 95% CI 0.78–1.00). Liu et al. performed a network meta-analysis of the treatment of MDR *A. baumannii* infections [24]. Eighteen studies were included: seven RCTs and eleven retrospective studies. Eleven studies focused on pneumonia, whereas the other studies included patients with mixed infections. Rifampicin, fosfomycin, sulbactam, and carbapenem were used in combination regimens with colistin. There was no significant difference in clinical improvement and cure between monotherapy and any combination therapy (clinical improvement: rifampicin, RR 1.28, 95% CI 0.67–2.45; fosfomycin, RR 1.08, 95% CI 0.76–1.53; sulbactam, RR 1.02, 5% CI 0.86–1.22; clinical cure: carbapenem, RR 1.34, 95% CI 0.92–1.95; sulbactam, RR 1.28, 95% CI 0.90–1.83). The combination of rifampicin and fosfomycin was associated with a significantly higher rate of microbiological eradication than colistin monotherapy (rifampicin, RR 1.31, 95% CI 1.01–1.69; fosfomycin, RR 1.23, 95% CI 1.01–1.53). There were no statistically significant differences between colistin monotherapy and any combination therapy (sulbactam, RR 0.74, 95% CI 0.41–1.34; carbapenem, RR 0.75, 95% CI 0.43–1.30; fosfomycin, RR 0.89, 95% CI 0.62–1.28).

Table 7. Meta-analyses evaluated the efficacy of colistin versus colistin combination therapy.

Study	No. of Studies Analyzed	Patients			Results with Significance
		Infections	Pathogens	Antibiotics	
Cheng IL et al. [23]	5 RCTs	any (mostly VAP)	CR-GNB (mostly *A. baumanii*)	colistin vs. colistin-based combination	None
Vardakas KZ et al. [9]	32 (3 RCTs, 6 prospective, 22 retrospective and one in both observational)	any (mostly VAP or BSI)	MDR or XDR-GNB (mainly *K. pneumoniae* or *A. baumanii*)	colistin vs. colistin-based combination	High-dose treatments (>6 million international units; RR 0.80, 95% CI 0.69–0.93), com-bination therapy was found to be significantly more effective in patients with bacteremia (RR 0.75, 95% CI 0.57–0.98)
Liu J et al. [24]	18 (7 RCTs, 11 retrospective)	any (mostly VAP or BSI)	MDR or XDR *A. baumanii*	colistin vs. colistin-based combination	Combination of RFP and FOM was associated with a significantly higher rate of microbiological eradication (RFP, RR 1.31, 95% CI 1.01–1.69; FOM, RR 1.23, 95% CI 1.01–1.53)

Abbreviations; RCT, randomized control trial; VAP, ventilator associated pneumonia; CR, carbapenem resistant; GNB, Gram-negative bacteria; RR, risk ratio; CI, confidence interval; BSI, blood stream infection; MDR, multi drug resistant; XDR, extensively drug resistant; OR, odds ratio FOM, fosfomycin; RFP, rifampicin.

4. Conclusions

This review provides a comprehensive and critical evaluation of the current evidence from meta-analyses on antibiotic combinations for the treatment of MDR Gram-negative bacteria. The relatively low number of patients included in these meta-analyses suggests that further appropriately designed studies should be conducted to evaluate the efficacy of combination therapy versus monotherapy. Several clinical studies have been conducted for the approval of β-lactamase inhibitor combination therapies, but fewer trials have targeted patients infected solely with MDR pathogens. Real-world data must be accumulated and analyzed to determine the efficacy of treatments for patients with MDR bacterial infections. Numerous in vitro and in vivo studies have evaluated the efficacy of antibiotics and antibiotic combinations in the treatment of MDR Gram-negative pathogens. However, clinical studies are limited, except those on colistin combination therapies. It is therefore necessary to collect further evidence on this topic. Combination therapies may prevent the emergence of resistance, achieve higher rates of success, and allow lower doses or shorter treatment periods, albeit with higher costs, potential side effects, and a potential for the emergence of a higher level of resistance than that predicted by in vitro studies.

Author Contributions: Conceptualization, T.U., H.K. and M.H.; writing—original draft preparation, T.U. and H.K.; writing—review and editing, H.K., M.H., J.H., Y.Y. and H.M. All authors have read and agreed to the published version of the manuscript.

Funding: This research received no external funding.

Institutional Review Board Statement: Not applicable.

Informed Consent Statement: Not applicable.

Data Availability Statement: The data presented in this review are available upon reasonable request from the corresponding author.

Acknowledgments: The authors thank all the co-authors who assisted in this project.

Conflicts of Interest: The authors declare no conflict of interest.

References

1. World Health Organization. Global Antimicrobial Resistance and Use Surveillance System (GLASS) Report: 2021. Available online: https://apps.who.int/iris/rest/bitstreams/1350455/retrieve (accessed on 30 November 2021).
2. European Centre for Disease Prevention and Control. Surveillance of Antimicrobial Resistance in Europe, 2020 Data. Available online: https://www.ecdc.europa.eu/sites/default/files/documents/Surveillance-antimicrobial-resistance-in-Europe-2020.pdf (accessed on 30 November 2021).
3. Petty, L.A.; Henig, O.; Patel, T.S.; Pogue, J.M.; Kaye, K.S. Overview of meropenem-vaborbactam and newer antimicrobial agents for the treatment of carbapenem-resistant Enterobacteriaceae. *Infect. Drug Resist.* **2018**, *11*, 1461–1472. [CrossRef] [PubMed]
4. Centre for Disease Prevention and Control. 2019 AR Threats Report. Available online: https://www.cdc.gov/drugresistance/pdf/threats-report/2019-ar-threats-report-508.pdf (accessed on 30 November 2021).
5. Hagihara, M.; Crandon, J.L.; Nicolau, D.P. The efficacy and safety of antibiotic combination therapy for infections caused by Gram-positive and Gram-negative organisms. *Expert Opin. Drug Saf.* **2012**, *11*, 221–233. [CrossRef] [PubMed]
6. Tamma, P.D.; Cosgrove, S.E.; Maragakis, L.L. Combination therapy for treatment of infections with gram-negative bacteria. *Clin. Microbiol. Rev.* **2012**, *25*, 450–470. [CrossRef] [PubMed]
7. Tängdén, T. Combination antibiotic therapy for multidrug-resistant Gram-negative bacteria. *Upsala J. Med. Sci.* **2014**, *119*, 149–153. [CrossRef]
8. Jean, S.S.; Chang, Y.C.; Lin, W.C.; Lee, W.S.; Hsueh, P.R.; Hsu, C.W. Epidemiology, Treatment, and Prevention of Nosocomial Bacterial Pneumonia. *J. Clin. Med.* **2020**, *9*, 275. [CrossRef]
9. Vardakas, K.Z.; Mavroudis, A.D.; Georgiou, M.; Falagas, M.E. Intravenous colistin combination antimicrobial treatment vs. monotherapy: A systematic review and meta-analysis. *Int. J. Antimicrob. Agents* **2018**, *51*, 535–547. [CrossRef]
10. Sfeir, M.M.; Askin, G.; Christos, P. Beta-lactam/beta-lactamase inhibitors versus carbapenem for bloodstream infections due to extended-spectrum beta-lactamase-producing Enterobacteriaceae: Systematic review and meta-analysis. *Int. J. Antimicrob. Agents* **2018**, *52*, 554–570. [CrossRef]
11. Zhang, H.; Liang, B.; Wang, J.; Cai, Y. Non-carbapenem β-lactam/β-lactamase inhibitors versus carbapenems for urinary tract infections caused by extended-spectrum β-lactamase-producing Enterobacteriaceae: A systematic review. *Int. J. Antimicrob. Agents* **2021**, *58*, 106410. [CrossRef]
12. Sternbach, N.; Leibovici Weissman, Y.; Avni, T.; Yahav, D. Efficacy and safety of ceftazidime/avibactam: A systematic review and meta-analysis. *J. Antimicrob. Chemother.* **2018**, *73*, 2021–2029. [CrossRef]
13. Che, H.; Wang, R.; Wang, J.; Cai, Y. Ceftazidime/avibactam versus carbapenems for the treatment of infections caused by Enterobacteriaceae: A meta-analysis of randomised controlled trials. *Int. J. Antimicrob. Agents* **2019**, *54*, 809–813. [CrossRef]

14. Isler, B.; Ezure, Y.; Romero, J.L.G.; Harris, P.; Stewart, A.G.; Paterson, D.L. Is Ceftazidime/Avibactam an Option for Serious Infections Due to Extended-Spectrum-β-Lactamase- and AmpC-Producing Enterobacterales?: A Systematic Review and Meta-analysis. *Antimicrob. Agents Chemother.* **2020**, *65*, e01052-20. [CrossRef] [PubMed]

15. Onorato, L.; Di Caprio, G.; Signoriello, S.; Coppola, N. Efficacy of ceftazidime/avibactam in monotherapy or combination therapy against carbapenem-resistant Gram-negative bacteria: A meta-analysis. *Int. J. Antimicrob. Agents* **2019**, *54*, 735–740. [CrossRef] [PubMed]

16. Fiore, M.; Alfieri, A.; Di Franco, S.; Pace, M.C.; Simeon, V.; Ingoglia, G.; Cortegiani, A. Ceftazidime-Avibactam Combination Therapy Compared to Ceftazidime-Avibactam Monotherapy for the Treatment of Severe Infections Due to Carbapenem-Resistant Pathogens: A Systematic Review and Network Meta-Analysis. *Antibiotics* **2020**, *9*, 388. [CrossRef] [PubMed]

17. Li, D.; Fei, F.; Yu, H.; Huang, X.; Long, S.; Zhou, H.; Zhang, J. Ceftazidime-Avibactam Therapy Versus Ceftazidime-Avibactam-Based Combination Therapy in Patients With Carbapenem-Resistant Gram-Negative Pathogens: A Meta-Analysis. *Front. Pharmacol.* **2021**, *12*, 707499. [CrossRef] [PubMed]

18. Van Duin, D.; Bonomo, R.A. Ceftazidime/Avibactam and Ceftolozane/Tazobactam: Second-generation β-Lactam/β-Lactamase Inhibitor Combinations. *Clin. Infect. Dis.* **2016**, *63*, 234–241. [CrossRef]

19. Paul, M.; Soares-Weiser, K.; Leibovici, L. Beta lactam monotherapy versus beta lactam-aminoglycoside combination therapy for fever with neutropenia: Systematic review and meta-analysis. *BMJ* **2003**, *326*, 1111. [CrossRef]

20. Paul, M.; Benuri-Silbiger, I.; Soares-Weiser, K.; Leibovici, L. Beta lactam monotherapy versus beta lactam-aminoglycoside combination therapy for sepsis in immunocompetent patients: Systematic review and meta-analysis of randomised trials. *BMJ* **2004**, *328*, 668. [CrossRef]

21. Zusman, O.; Altunin, S.; Koppel, F.; Dishon Benattar, Y.; Gedik, H.; Paul, M. Polymyxin monotherapy or in combination against carbapenem-resistant bacteria: Systematic review and meta-analysis. *J. Antimicrob. Chemother.* **2017**, *72*, 29–39. [CrossRef]

22. Samal, S.; Samir, S.B.; Patra, S.K.; Rath, A.; Dash, A.; Nayak, B.; Mohanty, D. Polymyxin Monotherapy vs. Combination Therapy for the Treatment of Multidrug-resistant Infections: A Systematic Review and Meta-analysis. *Indian J. Crit. Care Med.* **2021**, *25*, 199–206.

23. Cheng, I.L.; Chen, Y.H.; Lai, C.C.; Tang, H.J. Intravenous Colistin Monotherapy versus Combination Therapy against Carbapenem-Resistant Gram-Negative Bacteria Infections: Meta-Analysis of Randomized Controlled Trials. *J. Clin. Med.* **2018**, *7*, 208. [CrossRef]

24. Liu, J.; Shu, Y.; Zhu, F.; Feng, B.; Zhang, Z.; Liu, L.; Wang, G. Comparative efficacy and safety of combination therapy with high-dose sulbactam or colistin with additional antibacterial agents for multiple drug-resistant and extensively drug-resistant *Acinetobacter baumannii* infections: A systematic review and network meta-analysis. *J. Glob. Antimicrob. Resist.* **2021**, *24*, 136–147. [PubMed]

25. Henderson, A.; Paterson, D.L.; Chatfield, M.D.; Tambyah, P.A.; Lye, D.C.; De, P.P.; Lin, R.T.P.; Chew, K.L.; Yin, M.; Lee, T.H.; et al. MERINO Trial Investigators and the Australasian Society for Infectious Disease Clinical Research Network (ASID-CRN). Association Between Minimum Inhibitory Concentration, Beta-lactamase Genes and Mortality for Patients Treated with Piperacillin/Tazobactam or Meropenem from the MERINO Study. *Clin. Infect. Dis.* **2021**, *73*, e3842–e3850. [PubMed]

26. Tamma, P.D.; Aitken, S.L.; Bonomo, R.A.; Mathers, A.J.; van Duin, D.; Clancy, C.J. Infectious Diseases Society of America Guidance on the Treatment of Extended-Spectrum β-lactamase Producing Enterobacterales (ESBL-E), Carbapenem-Resistant Enterobacterales (CRE), and *Pseudomonas aeruginosa* with Difficult-to-Treat Resistance (DTR-*P. aeruginosa*). *Clin. Infect. Dis.* **2021**, *72*, 1109–1116. [PubMed]

27. Ben-Ami, R.; Rodríguez-Baño, J.; Arslan, H.; Pitout, J.D.; Quentin, C.; Calbo, E.S.; Azap, O.K.; Arpin, C.; Pascual, A.; Livermore, D.M.; et al. A multinational survey of risk factors for infection with extended-spectrum beta-lactamase-producing enterobacteriaceae in nonhospitalized patients. *Clin. Infect. Dis.* **2009**, *49*, 682–690. [CrossRef]

28. Pitout, J.D.; Laupland, K.B. Extended-spectrum beta-lactamase-producing Enterobacteriaceae: An emerging public-health concern. *Lancet Infect. Dis.* **2008**, *8*, 159–166. [CrossRef]

29. Prakash, V.; Lewis, J.S., 2nd; Herrera, M.L.; Wickes, B.L.; Jorgensen, J.H. Oral and parenteral therapeutic options for outpatient urinary infections caused by enterobacteriaceae producing CTX-M extended-spectrum beta-lactamases. *Antimicrob. Agents Chemother.* **2009**, *53*, 1278–1280. [CrossRef]

30. Pitout, J.D. Enterobacteriaceae that produce extended-spectrum β-lactamases and AmpC β-lactamases in the community: The tip of the iceberg? *Curr. Pharm. Des.* **2013**, *19*, 257–263. [CrossRef]

31. Karaiskos, I.; Giamarellou, H. Carbapenem-Sparing Strategies for ESBL Producers: When and How. *Antibiotics* **2020**, *9*, 61. [CrossRef]

32. Livermore, D.M.; Meunier, D.; Hopkins, K.L.; Doumith, M.; Hill, R.; Pike, R.; Staves, P.; Woodford, N. Activity of ceftazidime/avibactam against problem Enterobacteriaceae and *Pseudomonas aeruginosa* in the UK, 2015–2016. *J. Antimicrob. Chemother.* **2018**, *73*, 648–657. [CrossRef]

33. Tuon, F.F.; Rocha, J.L.; Formigoni-Pinto, M.R. Pharmacological aspects and spectrum of action of ceftazidime-avibactam: A systematic review. *Infection* **2018**, *46*, 165–181. [CrossRef]

34. Wright, H.; Bonomo, R.A.; Paterson, D.L. New agents for the treatment of infections with Gram-negative bacteria: Restoring the miracle or false dawn? *Clin. Microbiol. Infect.* **2017**, *23*, 704–712. [CrossRef] [PubMed]

35. Karaiskos, I.; Lagou, S.; Pontikis, K.; Rapti, V.; Poulakou, G. The "Old" and the "New" Antibiotics for MDR Gram-Negative Pathogens: For Whom, When, and How. *Front. Public Health* **2019**, *7*, 151. [CrossRef] [PubMed]

36. Rodríguez-Baño, J.; Gutiérrez-Gutiérrez, B.; Machuca, I.; Pascual, A. Treatment of Infections Caused by Extended-Spectrum-Beta-Lactamase-, AmpC-, and Carbapenemase-Producing Enterobacteriaceae. *Clin. Microbiol. Rev.* **2018**, *31*, e00079-17. [CrossRef] [PubMed]

37. Shields, R.K.; Chen, L.; Cheng, S.; Chavda, K.D.; Press, E.G.; Snyder, A.; Pandey, R.; Doi, Y.; Kreiswirth, B.N.; Nguyen, M.H.; et al. Emergence of Ceftazidime-Avibactam Resistance Due to Plasmid-Borne blaKPC-3 Mutations during Treatment of Carbapenem-Resistant *Klebsiella pneumoniae* Infections. *Antimicrob. Agents Chemother.* **2017**, *61*, e02097-16. [CrossRef] [PubMed]

38. Gill, C.M.; Aktaþ, E.; Alfouzan, W.; Bourassa, L.; Brink, A.; Burnham, C.D.; Canton, R.; Carmeli, Y.; Falcone, M.; Kiffer, C.; et al. The ERACE-PA Global Surveillance Program: Ceftolozane/tazobactam and Ceftazidime/avibactam in vitro Activity against a Global Collection of Carbapenem-resistant *Pseudomonas aeruginosa*. *Eur. J. Clin. Microbiol. Infect. Dis.* **2021**, *40*, 2533–2541. [CrossRef]

39. Gaibani, P.; Lewis, R.E.; Volpe, S.L.; Giannella, M.; Campoli, C.; Landini, M.P.; Viale, P.; Re, M.C.; Ambretti, S. In vitro interaction of ceftazidime-avibactam in combination with different antimicrobials against KPC-producing Klebsiella pneumoniae clinical isolates. *Int. J. Infect. Dis.* **2017**, *65*, 1–3. [CrossRef]

40. Romanelli, F.; De Robertis, A.; Carone, G.; Dalfino, L.; Stufano, M.; Del Prete, R.; Mosca, A. In Vitro Activity of Ceftazidime/Avibactam Alone and in Combination with Fosfomycin and Carbapenems against KPC-producing Klebsiella Pneumoniae. *New Microbiol.* **2020**, *43*, 136–138.

41. Wagenlehner, F.M.; Umeh, O.; Steenbergen, J.; Yuan, G.; Darouiche, R.O. Ceftolozane-tazobactam compared with levofloxacin in the treatment of complicated urinary-tract infections, including pyelonephritis: A randomised, double-blind, phase 3 trial (ASPECT-cUTI). *Lancet* **2015**, *385*, 1949–1956. [CrossRef]

42. Solomkin, J.; Hershberger, E.; Miller, B.; Popejoy, M.; Friedland, I.; Steenbergen, J.; Yoon, M.; Collins, S.; Yuan, G.; Barie, P.S.; et al. Ceftolozane/Tazobactam Plus Metronidazole for Complicated Intra-abdominal Infections in an Era of Multidrug Resistance: Results From a Randomized, Double-Blind, Phase 3 Trial (ASPECT-cIAI). *Clin. Infect. Dis.* **2015**, *60*, 1462–1471. [CrossRef]

43. Kollef, M.H.; Nováček, M.; Kivistik, Ü.; Réa-Neto, Á.; Shime, N.; Martin-Loeches, I.; Timsit, J.F.; Wunderink, R.G.; Bruno, C.J.; Huntington, J.A.; et al. Ceftolozane-tazobactam versus meropenem for treatment of nosocomial pneumonia (ASPECT-NP): A randomised, controlled, double-blind, phase 3, non-inferiority trial. *Lancet Infect. Dis.* **2019**, *19*, 1299–1311. [CrossRef]

44. Rico Caballero, V.; Almarzoky Abuhussain, S.; Kuti, J.L.; Nicolau, D.P. Efficacy of Human-Simulated Exposures of Ceftolozane-Tazobactam Alone and in Combination with Amikacin or Colistin against Multidrug-Resistant *Pseudomonas aeruginosa* in an In Vitro Pharmacodynamic Model. *Antimicrob. Agents Chemother.* **2018**, *62*, e02384-17. [CrossRef] [PubMed]

45. Galani, I.; Papoutsaki, V.; Karantani, I.; Karaiskos, I.; Galani, L.; Adamou, P.; Deliolanis, I.; Kodonaki, A.; Papadogeorgaki, E.; Markopoulou, M.; et al. In vitro activity of ceftolozane/tazobactam alone and in combination with amikacin against MDR/XDR *Pseudomonas aeruginosa* isolates from Greece. *J. Antimicrob. Chemother.* **2020**, *75*, 2164–2172. [CrossRef] [PubMed]

46. Grohs, P.; Taieb, G.; Morand, P.; Kaibi, I.; Podglajen, I.; Lavollay, M.; Mainardi, J.L.; Compain, F. In Vitro Activity of Ceftolozane-Tazobactam against Multidrug-Resistant Nonfermenting Gram-Negative Bacilli Isolated from Patients with Cystic Fibrosis. *Antimicrob. Agents Chemother.* **2017**, *61*, e02688-16. [CrossRef] [PubMed]

47. Cuba, G.T.; Rocha-Santos, G.; Cayô, R.; Streling, A.P.; Nodari, C.S.; Gales, A.C.; Pignatari, A.C.C.; Nicolau, D.P.; Kiffer, C.R.V. In vitro synergy of ceftolozane/tazobactam in combination with fosfomycin or aztreonam against MDR *Pseudomonas aeruginosa*. *J. Antimicrob. Chemother.* **2020**, *75*, 1874–1878. [CrossRef] [PubMed]

48. Fiore, M.; Corrente, A.; Pace, M.C.; Alfieri, A.; Simeon, V.; Ippolito, M.; Giarratano, A.; Cortegiani, A. Ceftolozane-Tazobactam Combination Therapy Compared to Ceftolozane-Tazobactam Monotherapy for the Treatment of Severe Infections: A Systematic Review and Meta-Analysis. *Antibiotics* **2021**, *10*, 79. [CrossRef]

49. Paul, M.; Soares-Weiser, K.; Grozinsky, S.; Leibovici, L. Beta-lactam versus beta-lactam-aminoglycoside combination therapy in cancer patients with neutropenia. *Cochrane Database Syst. Rev.* **2003**, *3*, CD003038.

50. Paul, M.; Silbiger, I.; Grozinsky, S.; Soares-Weiser, K.; Leibovici, L. Beta lactam antibiotic monotherapy versus beta lactam-aminoglycoside antibiotic combination therapy for sepsis. *Cochrane Database Syst. Rev.* **2006**, *1*, CD003344.

51. Heo, Y.A. Imipenem/Cilastatin/Relebactam: A Review in Gram-Negative Bacterial Infections. *Drugs* **2021**, *81*, 377–388. [CrossRef]

52. Hecker, S.J.; Reddy, K.R.; Totrov, M.; Hirst, G.C.; Lomovskaya, O.; Griffith, D.C.; King, P.; Tsivkovski, R.; Sun, D.; Sabet, M.; et al. Discovery of a Cyclic Boronic Acid β-Lactamase Inhibitor (RPX7009) with Utility vs. Class A Serine Carbapenemases. *J. Med. Chem.* **2015**, *58*, 3682–3692. [CrossRef]

53. Bulik, C.C.; Nicolau, D.P. Double-carbapenem therapy for carbapenemase-producing *Klebsiella pneumoniae*. *Antimicrob. Agents Chemother.* **2011**, *55*, 3002–3004. [CrossRef]

54. Oliva, A.; Scorzolini, L.; Cipolla, A.; Mascellino, M.T.; Cancelli, F.; Castaldi, D.; D'Abramo, A.; D'Agostino, C.; Russo, G.; Ciardi, M.R.; et al. In vitro evaluation of different antimicrobial combinations against carbapenemase-producing *Klebsiella pneumoniae*: The activity of the double-carbapenem regimen is related to meropenem MIC value. *J. Antimicrob. Chemother.* **2017**, *72*, 1981–1984. [CrossRef] [PubMed]

55. Giamarellou, H.; Galani, L.; Baziaka, F.; Karaiskos, I. Effectiveness of a double-carbapenem regimen for infections in humans due to carbapenemase-producing pandrug-resistant *Klebsiella pneumoniae*. *Antimicrob. Agents Chemother.* **2013**, *57*, 2388–2390. [CrossRef] [PubMed]

56. Mashni, O.; Nazer, L.; Le, J. Critical Review of Double-Carbapenem Therapy for the Treatment of Carbapenemase-Producing *Klebsiella pneumoniae*. *Ann. Pharmacother.* **2019**, *53*, 70–81. [CrossRef] [PubMed]

57. Li, Y.Y.; Wang, J.; Wang, R.; Cai, Y. Double-carbapenem therapy in the treatment of multidrug resistant Gram-negative bacterial infections: A systematic review and meta-analysis. *BMC Infect. Dis.* **2020**, *20*, 408. [CrossRef] [PubMed]
58. Falagas, M.E.; Kasiakou, S.K.; Saravolatz, L.D. Colistin: The revival of polymyxins for the management of multidrug-resistant gram-negative bacterial infections. *Clin. Infect. Dis.* **2005**, *40*, 1333–1341. [CrossRef] [PubMed]
59. Paul, M.; Bishara, J.; Levcovich, A.; Chowers, M.; Goldberg, E.; Singer, P.; Lev, S.; Leon, P.; Raskin, M.; Yahav, D.; et al. Effectiveness and safety of colistin: Prospective comparative cohort study. *J. Antimicrob. Chemother.* **2010**, *65*, 1019–1027. [CrossRef]
60. Hartzell, J.D.; Neff, R.; Ake, J.; Howard, R.; Olson, S.; Paolino, K.; Vishnepolsky, M.; Weintrob, A.; Wortmann, G. Nephrotoxicity associated with intravenous colistin (colistimethate sodium) treatment at a tertiary care medical center. *Clin. Infect. Dis.* **2009**, *48*, 1724–1728. [CrossRef]
61. Tan, C.H.; Li, J.; Nation, R.L. Activity of colistinagainst heteroresistant *Acinetobacter baumannii* and emergence of resistance in an in vitro pharmacokinetic/pharmacodynamic model. *Antimicrob. Agents Chemother.* **2007**, *51*, 3413–3415.
62. Nation, R.L.; Velkov, T.; Li, J. Colistin and polymyxin B: Peas in a pod, or chalk and cheese? *Clin. Infect. Dis.* **2014**, *59*, 88–94. [CrossRef]
63. Brink, A.J.; Richards, G.A.; Colombo, G.; Bortolotti, F.; Colombo, P.; Jehl, F. Multicomponent antibiotic substances produced by fermentation: Implications for regulatory authorities, critically ill patients and generics. *Int. J. Antimicrob. Agents* **2014**, *43*, 1–6. [CrossRef]
64. Nation, R.L.; Li, J.; Cars, O.; Couet, W.; Dudley, M.N.; Kaye, K.S.; Mouton, J.W.; Paterson, D.L.; Tam, V.H.; Theuretzbacher, U.; et al. Framework for optimisation of the clinical use of colistin and polymyxin B: The Prato polymyxin consensus. *Lancet Infect. Dis.* **2015**, *15*, 225–234. [CrossRef]
65. Li, J.; Milne, R.W.; Nation, R.L.; Turnidge, J.D.; Smeaton, T.C.; Coulthard, K. Use of high-performance liquid chromatography to study the pharmacokinetics of colistin sulfate in rats following intravenous administration. *Antimicrob. Agents Chemother.* **2003**, *47*, 1766–1770. [CrossRef] [PubMed]
66. Mohamed, A.F.; Karaiskos, I.; Plachouras, D.; Karvanen, M.; Pontikis, K.; Jansson, B.; Papadomichelakis, E.; Antoniadou, A.; Giamarellou, H.; Armaganidis, A.; et al. Application of a loading dose of colistin methanesulfonate in critically ill patients: Population pharmacokinetics, protein binding, and prediction of bacterial kill. *Antimicrob. Agents Chemother.* **2012**, *56*, 4241–4249. [CrossRef] [PubMed]
67. Tumbarello, M.; Trecarichi, E.M.; De Rosa, F.G.; Giannella, M.; Giacobbe, D.R.; Bassetti, M.; Losito, A.R.; Bartoletti, M.; Del Bono, V.; Corcione, S.; et al. Infections caused by KPC-producing *Klebsiella pneumoniae*: Differences in therapy and mortality in a multicentre study. *J. Antimicrob. Chemother.* **2015**, *70*, 2133–2143. [CrossRef] [PubMed]
68. Ku, Y.H.; Chen, C.C.; Lee, M.F.; Chuang, Y.C.; Tang, H.J.; Yu, W.L. Comparison of synergism between colistin, fosfomycin and tigecycline against extended-spectrum β-lactamase-producing *Klebsiella pneumoniae* isolates or with carbapenem resistance. *J. Microbiol. Immunol. Infect.* **2017**, *50*, 931–939. [CrossRef] [PubMed]
69. Leelasupasri, S.; Santimaleeworagun, W.; Jitwasinkul, T. Antimicrobial Susceptibility among Colistin, Sulbactam, and Fosfomycin and a Synergism Study of Colistin in Combination with Sulbactam or Fosfomycin against Clinical Isolates of Carbapenem-Resistant *Acinetobacter baumannii*. *J. Pathog.* **2018**, *2018*, 3893492. [CrossRef] [PubMed]
70. Petersen, P.J.; Labthavikul, P.; Jones, C.H.; Bradford, P.A. In vitro antibacterial activities of tigecycline in combination with other antimicrobial agents determined by chequerboard and time-kill kinetic analysis. *J. Antimicrob. Chemother.* **2006**, *57*, 573–576. [CrossRef]
71. Arroyo, L.A.; Mateos, I.; González, V.; Aznar, J. In vitro activities of tigecycline, minocycline, and colistin-tigecycline combination against multi- and pandrug-resistant clinical isolates of *Acinetobacter baumannii* group. *Antimicrob. Agents Chemother.* **2009**, *53*, 1295–1296. [CrossRef]
72. Sato, Y.; Ubagai, T.; Tansho-Nagakawa, S.; Yoshino, Y.; Ono, Y. Effects of colistin and tigecycline on multidrug-resistant *Acinetobacter baumannii* biofilms: Advantages and disadvantages of their combination. *Sci. Rep.* **2021**, *11*, 11700. [CrossRef]

Article

Optimizing Doses of Ceftolozane/Tazobactam as Monotherapy or in Combination with Amikacin to Treat Carbapenem-Resistant *Pseudomonas aeruginosa*

Worapong Nasomsong [1], Parnrada Nulsopapon [2,3], Dhitiwat Changpradub [1], Supanun Pungcharoenkijkul [4], Patomroek Hanyanunt [5], Tassanawan Chatreewattanakul [5] and Wichai Santimaleeworagun [2,3,*]

[1] Division of Infectious Diseases, Department of Internal Medicine, Phramongkutklao Hospital and College of Medicine, Bangkok 10400, Thailand; nasomsong.w@gmail.com (W.N.); dhitiwat@yahoo.com (D.C.)

[2] Department of Pharmacy, Faculty of Pharmacy, Silpakorn University, Nakhon Pathom 73000, Thailand; aom.olivegreen1812@gmail.com

[3] Pharmaceutical Initiative for Resistant Bacteria and Infectious Diseases Working Group [PIRBIG], Faculty of Pharmacy, Silpakorn University, Nakhon Pathom 73000, Thailand

[4] Pharmacy Unit, Nopparat Rajathanee Hospital, Bangkok 10230, Thailand; supanun.pung@gmail.com

[5] Division of Microbiology, Department of Clinical Pathology, Phramongkutklao Hospital, Bangkok 10400, Thailand; patomroek.han@pcm.ac.th (P.H.); joymicro14@gmail.com (T.C.)

* Correspondence: swichai1234@gmail.com; Tel.: +66-34-255-800; Fax: +66-34-255-801

Citation: Nasomsong, W.; Nulsopapon, P.; Changpradub, D.; Pungcharoenkijkul, S.; Hanyanunt, P.; Chatreewattanakul, T.; Santimaleeworagun, W. Optimizing Doses of Ceftolozane/Tazobactam as Monotherapy or in Combination with Amikacin to Treat Carbapenem-Resistant *Pseudomonas aeruginosa*. *Antibiotics* **2022**, *11*, 517. https://doi.org/10.3390/antibiotics11040517

Academic Editor: Dóra Kovács

Received: 16 February 2022
Accepted: 7 April 2022
Published: 13 April 2022

Publisher's Note: MDPI stays neutral with regard to jurisdictional claims in published maps and institutional affiliations.

Abstract: Carbapenem-resistant *Pseudomonas aeruginosa* (CRPA) is a hospital-acquired pathogen with a high mortality rate and limited treatment options. We investigated the activity of ceftolozane/tazobactam (C/T) and its synergistic effects with amikacin to extend the range of optimal therapeutic choices with appropriate doses. The E-test method is used to determine in vitro activity. The optimal dosing regimens to achieve a probability of target attainment (PTA) and a cumulative fraction of response (CFR) of $\geq$90% were simulated using the Monte Carlo method. Of the 66 CRPA isolates, the rate of susceptibility to C/T was 86.36%, with an MIC_{50} and an MIC_{90} of 0.75 and 24 µg/mL, respectively. Synergistic and additive effects between C/T and amikacin were observed in 24 (40%) and 18 (30%) of 60 CRPA isolates, respectively. The extended infusion of C/T regimens achieved a $\geq$90% PTA of 75% and a 100% $fT >$ MIC at C/T MICs of 4 and 2 µg/mL, respectively. Only the combination of either a short or prolonged C/T infusion with a loading dose of amikacin of 20–25 mg/kg, followed by 15–20 mg/kg/day amikacin dosage, achieved $\geq$90% CFR. The C/T infusion, combined with currently recommended amikacin dose regimens, should be considered to manage CRPA infections.

Keywords: antibiotic combination; minimum inhibitory concentration; Monte Carlo; synergistic effect

1. Introduction

Pseudomonas aeruginosa is a significant cause of hospital-acquired infections and is frequently multidrug-resistant (MDR) [1]. MDR *P. aeruginosa* was reported in 14.7% and 22.0% of cases of bloodstream infections and pneumonia, respectively. The high mortality rate of infections with MDR *P. aeruginosa* makes treating serious infections more challenging. Furthermore, MDR *P. aeruginosa* is frequently resistant to carbapenems and other β-lactams. β-lactam resistance is mediated via multiple mechanisms, including the acquisition of metallo-β-lactamases, the increased production of chromosomal AmpC, increased efflux, and changes in membrane permeability [2,3]. The most consistently active drugs against MDR *P. aeruginosa* are aminoglycosides and polymyxins, but pharmacokinetic limitations and their association with worse outcomes when given as monotherapy make their use suboptimal [4,5].

Ceftolozane is a novel cephalosporin with enhanced activity against *P. aeruginosa*. When combined with the well-described β-lactamase inhibitor tazobactam, ceftolozane/tazobactam (C/T) is a safe and effective treatment for complicated urinary tract infections and complicated intraabdominal infections (in combination with metronidazole) caused by Gram-negative organisms, including *P. aeruginosa*. C/T also has good in vitro activity against MDR *P. aeruginosa* [6]. Relative to most β-lactams, ceftolozane has improved activity against *P. aeruginosa* because it is stable against AmpC enzymes produced by *P. aeruginosa*, is unaffected by active efflux, and is not appreciably affected by porin channel changes. Tazobactam protects ceftolozane from destruction by most extended-spectrum β-lactamases, but does not add to its activity against *P. aeruginosa* [7,8].

Shortridge et al. described the excellent activity of C/T against 3851 isolates of *P. aeruginosa*, including MDR or extensively drug-resistant (XDR) isolates. As the most active β-lactam agent tested against *P. aeruginosa*, C/T may be an important agent to treat severe bacterial infections [9]. Additionally, antipseudomonal antibiotics, such as aztreonam, amikacin, and meropenem, produce a synergistic effect with C/T [10–12]. However, among the few reports of the synergistic effect of C/T with other antibiotics, studies involving isolates of *P. aeruginosa* are limited.

Treatment failure may occur in critically ill patients infected with MDR or XDR *P. aeruginosa*. A minimum inhibitory concentration (MIC) of >2 μg/mL for C/T was recently reported as being associated with 30-day mortality in patients infected with MDR *P. aeruginosa* treated with the approved recommended dosage regimens [13]. Furthermore, the pharmacokinetic/pharmacodynamic (PK/PD) properties of antibiotics become altered due to pathophysiological changes in critically ill patients. Thus, PK/PD analysis using Monte Carlo simulation is used to design optimal antibiotic dosing regimens, thereby maximizing antibiotic activity, increasing the probability of clinical success, and reducing the rate of antibiotic resistance [14,15].

As described above, C/T seems an attractive agent to treat infections with carbapenem-resistant *P. aeruginosa* (CRPA). However, in vitro susceptibility data of C/T against CRPA in Thailand has never been reported. Thus, this study investigated the in vitro activity of C/T and the synergistic effect of C/T and amikacin against CRPA isolates from clinical specimens of patients at Phramongkutklao Hospital, Thailand, to assess the optimal dosing regimens of C/T as a monotherapy or in combination with amikacin in critically ill patients.

2. Results

2.1. In Vitro Susceptibility of C/T and Comparator Agents

Of the 66 CRPA strains isolated from unique patients, 55 (83.33%) were MDR, 9 (13.64%) were XDR, and 2 (3.03%) were PDR. The susceptibility test results of C/T and other comparator agents are presented in Tables 1 and 2.

Table 1. In vitro susceptibility and percentage of susceptibility among ceftolozane/tazobactam (C/T) and comparator agents against clinical isolates (*n* = 66) of carbapenem-resistant *Pseudomonas aeruginosa* (CRPA) from the broth micro dilution method.

Agents	MIC Range (μg/mL)	MIC$_{50}$ (μg/mL)	MIC$_{90}$ (μg/mL)	Percentage of Susceptible Strains [a]
Ceftazidime	1–>32	8	>32	63.64
Cefepime	1–>32	8	>32	63.64
Piperacillin/tazobactam	8/4–>64/4	32/4	>64/4	45.45
Imipenem	≤0.5–>8	>8	>8	4.55
Meropenem	0.5–>8	8	>8	13.64
Ciprofloxacin	0.06–>2	0.25	>2	66.67
Levofloxacin	0.06–>8	2	>8	51.52
Gentamicin	2–>8	2	>8	83.33
Amikacin	8–>32	8	32	89.39
Colistin	1–>4	1	2	96.97

Abbreviations: MIC, minimum inhibitory concentration; MIC$_{50}$, minimum inhibitory concentration required to inhibit the growth of 50% of organisms; MIC$_{90}$, minimum inhibitory concentration required to inhibit the growth of 90% of organisms. [a] *Pseudomonas aeruginosa* strains are defined susceptible to the studied antibiotics and intermediate to colistin following the Clinical & Laboratory Standards Institute (CLSI) 2021. The cut-off for the susceptible breakpoint were ≤8 μg/mL for ceftazidime and cefepime, ≤16/4 μg/mL for piperacillin/tazobactam, ≤2 μg/mL for imipenem and meropenem, ≤0.5 μg/mL for ciprofloxacin, ≤1 μg/mL for levofloxacin, ≤4 μg/mL for gentamicin, ≤16 μg/mL for amikacin. The cut-off for the intermediate MIC breakpoint was ≤2 μg/mL for colistin.

Table 2. E-test results of antibiotic susceptibility to ceftolozane/tazobactam (C/T) (*n* = 66) and synergy testing of C/T with amikacin (*n* = 60) for all carbapenem-resistant *Pseudomonas aeruginosa* (CRPA) isolates.

No.	MIC [a] (µg/mL)				Synergistic Testing Results	
	C/T [b]	AMK	C/T [b] Combined with AMK	AMK Combined with C/T [b]	ΣFICI	Interpretation
1	0.50	4.00	0.38	0.75	0.94	ADD
2	1.50	3.00	0.75	0.5	0.66	ADD
3	1.00	0.50	0.38	0.094	0.56	ADD
4	0.38	3.00	0.38	1	1.33	IND
5	0.75	1.50	0.25	0.19	0.46	SYN
6	0.38	1.50	0.5	0.38	1.56	IND
7	0.25	2.00	0.25	0.75	1.37	IND
8	0.50	2.00	0.38	0.5	1.01	IND
9	1.00	3.00	0.25	0.19	0.31	SYN
10	1.50	2.00	0.75	0.38	0.69	ADD
11	0.50	4.00	0.5	4	2.00	IND
12	2.00	2.00	1	0.38	0.69	ADD
13	1.00	3.00	1	0.75	1.25	IND
14	0.75	8.00	0.75	4	1.50	IND
15	0.50	2.00	0.5	0.75	1.37	IND
16	0.50	2.00	0.5	0.75	1.37	IND
17	0.50	0.75	0.25	0.125	0.66	ADD
18	1.00	4.00	0.75	1	1.00	ADD
19	1.00	1.00	0.25	0.094	0.34	SYN
20	0.50	2.00	0.5	0.5	1.25	IND
21	0.75	1.00	0.25	0.19	0.52	ADD
22	2.00	4.00	0.75	0.5	0.50	SYN
23	1.00	2.00	1	0.75	1.37	IND
24	1.50	1.00	1	0.25	0.91	ADD
25	0.50	8.00	0.5	2	1.25	IND
26	1.50	8.00	0.5	1	0.45	SYN
27	0.75	1.50	0.5	0.25	0.83	ADD
28	1.50	6.00	0.5	1	0.50	SYN
29	2.00	4.00	0.38	0.25	0.25	SYN
30	2.00	2.00	0.75	0.25	0.50	SYN
31	64.00	1.50	32	0.25	0.66	ADD
32	3.00	1.50	1	0.19	0.46	SYN
33	0.19	2.00	0.19	0.5	1.25	IND
34	0.50	1.50	0.5	0.38	1.25	IND
35	0.38	2.00	0.25	0.38	0.84	ADD
36	0.19	0.75	0.19	0.75	2.00	IND
37	0.50	6.00	0.5	1.5	1.25	IND
38	0.75	4.00	0.25	0.5	0.45	SYN
39	0.75	2.00	0.19	0.25	0.37	SYN
40	0.25	1.00	0.125	0.19	0.69	ADD
41	0.75	1.50	0.19	0.125	0.33	SYN
42	0.75	6.00	0.25	1	0.50	SYN
43	0.38	1.50	0.19	0.25	0.66	ADD
44	0.75	3.00	0.19	0.38	0.38	SYN
45	0.50	1.00	0.5	0.25	1.25	IND
46	1.50	6.00	0.5	0.75	0.45	SYN
47	24.00	32.00	24	8	1.25	IND
48	0.75	3.00	0.094	0.19	0.18	SYN
49	0.50	3.00	0.094	0.19	0.25	SYN
50	4.00	12.00	1.5	1	0.45	SYN
51	4.00	6.00	2	1	0.66	ADD
52	0.75	1.50	0.5	0.38	0.92	ADD
53	0.75	1.50	0.25	0.19	0.46	SYN
54	1.00	3.00	0.25	0.19	0.31	SYN

Table 2. *Cont.*

No.	MIC ^{a} (µg/mL)				Synergistic Testing Results	
	C/T ^{b}	AMK	C/T ^{b} Combined with AMK	AMK Combined with C/T ^{b}	ΣFICI	Interpretation
55	1.50	6.00	0.38	0.5	0.33	SYN
56	8.00	6.00	2	0.5	0.33	SYN
57	4.00	3.00	1.5	0.5	0.54	ADD
58	1.50	6.00	0.5	0.75	0.45	SYN
59	0.75	3.00	0.5	0.75	0.91	ADD
60	1.00	6.00	0.38	0.75	0.50	SYN
61	>256	96.00	N/A	N/A	N/A	N/A
62	>256	12.00	N/A	N/A	N/A	N/A
63	>256	3.00	N/A	N/A	N/A	N/A
64	>256	>256	N/A	N/A	N/A	N/A
65	>256	3.00	N/A	N/A	N/A	N/A
66	>256	12.00	N/A	N/A	N/A	N/A
MIC range (µg/mL)	0.19–>256	0.5–>256	0.094–32	0.094–8	-	-
MIC$_{50}$ (µg/mL)	0.75	3	0.5	0.5	-	-
MIC$_{90}$ (µg/mL)	24	8	1	1	-	-
%S ^{c}	86.36	95.45	96.67	100	-	-

Abbreviations: MIC, minimum inhibitory concentration; MIC$_{50}$, MIC required to inhibit the growth of 50% of organisms; MIC$_{90}$, MIC required to inhibit the growth of 90% of organisms; %S, percentage of susceptible strains; AMK, amikacin; C/T, ceftolozane/tazobactam; FICI, fractional inhibitory concentration index; SYN, synergistic effect (FICI ≤ 0.5); ADD; additive effect (FICI > 0.5–≤ 1); IND, indifference (FICI > 1–≤ 4); N/A, not-available. ^{a} The susceptibility testing was performed using the E-test method. ^{b} The MIC values of ceftolozane were combined with fixed tazobactam concentration (4 µg/mL). ^{c} The cut-off for the susceptible breakpoint for *Pseudomonas aeruginosa* strains following the Clinical & Laboratory Standards Institute (CLSI) 2021 were ≤4/4 µg/mL for ceftolozane/tazobactam and ≤16 µg/mL for amikacin.

According to broth microdilution (BMD), the CRPA strains were most susceptible to colistin (96.97%), amikacin (89.39%), and gentamicin (83.33%), and only 4.55% and 13.64% were susceptible to imipenem and meropenem, respectively. Regarding other antipseudomonal β-lactams, the susceptibility rates of CRPA were 63.64%, 63.64%, and 45.45% to ceftazidime, cefepime, and piperacillin/tazobactam, respectively. Among non-β-lactams, amikacin and gentamicin showed good activity against CRPA.

C/T was the most active β-lactam agent against the 66 strains of CRPA from the E-test method, of which 86.36% showed susceptibility at an MIC$_{50}$ of 0.75 µg/mL and an MIC$_{90}$ of 24 µg/mL.

2.2. Synergistic Activities

Synergistic, additive, and indifference effects of C/T and amikacin were observed in 24 of 60 (40%), 18 of 60 (30%), and 18 of 60 (30%) CRPA isolates (Tables 2 and 3). No antagonistic activity was observed between the combination of C/T and amikacin.

Table 3. In vitro synergistic testing of ceftolozane/tazobactam (C/T) with amikacin against CRPA isolates ($n = 60$).

Antibiotic Combination	No (%)			
	Synergism	Additive Effect	Indifference	Antagonism
C/T + AMK	24 (40%)	18 (30%)	18 (30%)	0 (0%)

Abbreviations: AMK, amikacin; C/T, ceftolozane/tazobactam.

2.3. PTA and CFR

The PTA, including 75% and 100% fT > MIC for ceftolozane and 20% fT ≥ 1 µg/mL for tazobactam, for each dosing regimen at specific MICs is shown in Tables 4 and 5. All antibiotic dosing regimens (detailed in Section 4.4) met the criteria of ≥90% PTA of 20%

$f\mathrm{T} \geq 1$ µg/mL for tazobactam. The PTA of achieving 75% $f\mathrm{T} >$ MIC following ceftolozane administration by a 0.5-h infusion, a prolonged 4-h infusion, and a loading dose followed by continuous infusion was 93.92%, 93.61%, and 93.61% at MICs of 2, 4, and 8 µg/mL, respectively. Furthermore, the PTA of achieving 100% $f\mathrm{T} >$ MIC following ceftolozane administration by a 0.5-h infusion, a prolonged 4-h infusion, and a loading dose followed by continuous infusion was 95.50%, 94.75%, and 93.60% at MICs of 1, 2, and 8 µg/mL, respectively. None of the C/T dosing regimens achieved the PTA target when the MIC was $\geq$16 µg/mL.

Table 4. The probability of target attainment (PTA), including 75% $f\mathrm{T} >$ MIC for ceftolozane and 20% $f\mathrm{T} \geq 1$ µg/mL for tazobactam, at specific MICs for each dosing regimen.

| Dosage Regimens of C/T | | | PTA (%) | | | | | | | | | | | Tazo-Bactam |
| | | | Ceftolozane MICs (µg/mL) | | | | | | | | | | | |
LD	MD	IT	0.0625	0.125	0.25	0.5	1	2	4	8	16	32	64	128	20% $f\mathrm{T} \geq 1$ µg/mL
-	1.5 g q 8 h	0.5 h	100.00	99.99	99.85	99.59	98.34	93.92	80.64	50.97	14.09	1.14	0.01	0.00	97.50
-	1.5 g q 8 h	4 h	100.00	100.00	100.00	100.00	99.87	98.97	93.61	69.46	23.10	1.53	0.00	0.00	97.71
1.5 g	4.5 g	CI	100.00	100.00	100.00	100.00	100.00	100.00	99.95	93.61	44.75	3.83	0.05	0.00	99.97

Abbreviations: C/T, ceftolozane/tazobactam; PTA, the probability of target attainment; MIC, minimum inhibitory concentration; LD, loading dose; MD, maintenance dose; IT, infusion time; g, gram; h, hours; q, every; CI, continuous infusion over 24 h.

Table 5. The probability of target attainment (PTA), including 100% $f\mathrm{T} >$ MIC for ceftolozane and 20% $f\mathrm{T} \geq 1$ µg/mL for tazobactam, at specific MICs for each dosing regimen.

| Dosage Regimens of C/T | | | PTA (%) | | | | | | | | | | | Tazo-Bactam |
| | | | Ceftolozane MICs (µg/mL) | | | | | | | | | | | |
LD	MD	IT	0.0625	0.125	0.25	0.5	1	2	4	8	16	32	64	128	20% $f\mathrm{T} \geq 1$ µg/mL
-	1.5 g q 8 h	0.5 h	99.93	99.81	99.46	98.39	95.50	87.36	69.12	38.63	8.89	0.57	0.00	0.00	97.50
-	1.5 g q 8 h	4 h	100.00	100.00	99.95	99.68	98.58	94.75	82.35	52.07	14.56	0.81	0.00	0.00	97.71
1.5 g	4.5 g	CI	100.00	100.00	100.00	100.00	100.00	100.00	99.95	93.60	44.71	3.82	0.05	0.00	99.97

Abbreviations: C/T, ceftolozane/tazobactam; PTA, the probability of target attainment; MIC, minimum inhibitory concentration; LD, loading dose; MD, maintenance dose; IT, infusion time; g, gram; h, hours; q, every; CI, continuous infusion over 24 h.

For C/T, only combination regimens achieved the target of $\geq$90% CFR at 75% and 100% $f\mathrm{T} >$ MIC. None of the amikacin dosing regimens achieved the CFR target when administered as monotherapy. Fortunately, when amikacin was combined with C/T, the amikacin dosage, a loading dose of 20–25 mg/kg, followed by 15–20 mg/kg every 24 h, met the CFR target. The CFR for each dosing regimen of C/T, amikacin, and its combination are presented in Tables 6 and 7.

Table 6. The cumulative fraction of response (CFR) for each dosing regimen of ceftolozane/tazobactam (C/T) and ceftolozane/tazobactam (C/T), combined with amikacin (AMK).

Dosage Regimens of C/T			CFR (%)			
			75% fT > MIC		100% fT > MIC	
LD	MD	IT	C/T	C/T Combined with AMK	C/T	C/T Combined with AMK
-	1.5 g q 8 h	0.5 h	84.24	95.79	80.93	94.24
-	1.5 g q 8 h	4 h	86.82	96.62	84.61	95.94
1.5 g	4.5 g	CI	87.84	96.79	87.84	96.79

Abbreviations: C/T, ceftolozane/tazobactam; AMK, amikacin; CFR, cumulative fraction of response; MIC, minimum inhibitory concentration; LD, loading dose; MD, maintenance dose; IT, infusion time; g, gram; h, hours; q, every; CI, continuous infusion over 24 h.

Table 7. The cumulative fraction of response (CFR) for each dosing regimen of amikacin (AMK) and amikacin (AMK) combined with ceftolozane/tazobactam (C/T).

Dosage Regimens of AMK			CFR (%)	
			C_{max}/MIC $\geq$ 8	
LD	MD	IT	AMK	AMK Combined with C/T
20 mg/kg	15 mg/kg q 24 h	0.5 h	62.52	97.30
25 mg/kg	15 mg/kg q 24 h	0.5 h	62.79	97.36
25 mg/kg	20 mg/kg q 24 h	0.5 h	67.08	98.03

Abbreviations: C/T, ceftolozane/tazobactam; AMK, amikacin; CFR, cumulative fraction of response; MIC, minimum inhibitory concentration; LD, loading dose; MD, maintenance dose; IT, infusion time; mg, milligram; kg, kilogram; h, hours; q, every.

3. Discussion

There are limited treatment options for CRPA infections. The agents to which CRPA is susceptible include colistin or polymyxin B, which exhibit a suboptimal treatment outcome and a high rate of adverse reactions, especially nephrotoxicity. Our study in a Thai university hospital demonstrated the attractive in vitro activity of C/T against CRPA, with a susceptibility rate of 86.36%, which correlated with other studies, where the susceptibility rate of MDR *P. aeruginosa* or CRPA ranged from 67.2–85.9% [9,16]. Although C/T is considered the most active β-lactam against both susceptible and resistant strains of *P. aeruginosa*, a study from Singapore revealed a susceptibility rate of 37.9% for CRPA to C/T. This discrepancy was mainly attributed to the presence of carbapenemase-producing isolates, particularly metallo-β-lactamases [17].

Due to the high rate of susceptibility of CRPA to C/T, C/T may be useful as an empirical or definite antibiotic therapy for the treatment of infections suspected or known to be caused by CRPA. A clinical study showed that C/T was successful in treating 71% of patients with MDR *P. aeruginosa* infections [18]. Therefore, a polymyxin-sparing strategy may reduce antibiotic-related adverse events, as well as minimize the over-prescription of polymyxin.

Among the few reports of the synergistic effect of C/T with other antibiotics, studies involving isolates of *P. aeruginosa* are limited. Antipseudomonal antibiotics that produce a synergistic effect with C/T include aztreonam, amikacin, and meropenem [10–12]. In our study, we observed 40% synergism between C/T and amikacin against CRPA isolates using E-test methods. A recent study from Greece performed synergistic testing between C/T and amikacin against MDR *P. aeruginosa* using a time-kill assay, and a synergistic effect was observed in 85% [19]. These differences and discordant results may be due to several factors. First, the agreement between the time-kill assay and E-test crossing method ranged from 3–71% in MDR Gram-negative bacilli, including *P. aeruginosa* [20]. Second, differences in the genotypic resistance of bacterial strains may affect the synergistic result [19]. The good bactericidal and synergistic activity observed with the combination of C/T and amikacin may be attributed to β-lactam-mediated membrane damage, leading to increased aminoglycoside uptake [21]. Clinical data regarding the treatment outcome of C/T combination therapy against MDR Gram-negative bacilli, mostly MDR *P. aeruginosa*, revealed a significant decrease in mortality. A systematic review and meta-analysis included 8 non-randomized studies of C/T for treatment as monotherapy and combination therapy. The results showed that C/T in combination was associated with clinical improvement (OR, 0.97; 95% CI, 0.54 to 1.74; $p = 0.954$) and statistically lower mortality at 30 days (OR, 0.31; 95% CI, 0.10 to 0.97; $p = 0.045$) than the patient receiving C/T monotherapy [22]. Furthermore, a successful treatment outcome and rapid microbiological clearance using combination therapy of C/T with tobramycin against C/T-resistant *P. aeruginosa* in a severely neutropenic patient were also reported [23]. Thus, C/T combination therapy is potentially beneficial when combating refractory infections of MDR *P. aeruginosa*. Aminoglycosides, particularly amikacin, should be considered as the combination agent with C/T to achieve a good synergistic or additive effect.

The magnitude of the PK/PD target for cephalosporins that provided a maximal bactericidal effect was reported to range from 60–70% $fT > MIC$ in preclinical studies, whereas the magnitude of the cephalosporin PK/PD target to achieve a clinical cure and a microbiological cure was reported as 100% and 60–100% $fT > MIC$, respectively, in clinical studies [24]. When using ceftolozane against *P. aeruginosa*, the % $fT > MIC$ to achieve a 1- or 2-log reduction ranged from 30–40% [25,26]. Therefore, using 75% and 100% $fT > MIC$ as the PK/PD targets of ceftolozane in the simulated regimens may be adequate to predict the maximum bactericidal effect and microbiological cure.

The simulation studies revealed that antibiotic dosing regimens by prolonged infusion or continuous infusion had greater potential than intermittent infusion. Prolonged infusion or continuous infusion regimens achieved a target of $\geq$90% PTA of 75% and 100% $fT > MIC$ with C/T MICs $\leq$ 4 µg/mL (CLSI susceptible breakpoint) and $\leq$2 µg/mL, respectively, whereas a 1.5 g intermittent infusion every 8 h (an approved C/T dosing regimen for complicated urinary tract and intraabdominal infections) achieved $\geq$90% PTA at C/T MICs of $\leq$2 and $\leq$1 µg/mL, respectively, which agreed with previous studies [27,28]. There are concerns about the stability of C/T when using prolonged infusion or continuous infusion regimens. However, it was recently reported that C/T is stable for at least 24 h in 0.9% normal saline and 5% glucose solution in real-world conditions when stored in polypropylene tubes at room temperature (22 °C) without light exposure [29]. Therefore, the extended infusion of C/T is feasible, and because it increases the probability of treatment success, it may be a recommended regimen for the treatment of infections.

The synergistic effects of C/T plus amikacin contributed to achieving a target of $\geq$90% CFR. None of the prolonged or continuous infusions of C/T monotherapy achieved the target CFR. A previous study showed that only a continuous infusion of C/T monotherapy met the target of $\geq$85% CFR at 40%, 60%, and 100% $fT > MIC$ [28]. However, when C/T combined with amikacin, all C/T dosing regimens except intermittent regimens and all amikacin dosing regimens (loading dose 20–25 mg/kg, followed by 15–20 mg/kg/day) reached the CFR target of $\geq$90%. In 2017, Kato et al. recommended that the amikacin initial dose required to achieve $C_{max}/MIC \geq 8$ was 15 mg/kg/day, and the amikacin maintenance dosage was 15 mg/kg/day at amikacin MICs $\leq$ 4 µg/mL [30]. Fortunately, all clinical CRPA isolates except no.47 had amikacin MICs $\leq$ 4 µg/mL when combined with C/T. If synergism occurs, the MIC values of each antibiotic were reduced by at least 1-fold dilution. The decrease in MIC affects an increase $fT > MIC$ for C/T and C_{max}/MIC for amikacin, respectively, resulting in a greater achievement of the probability of CFR target in each antibiotic. Thus, it may be advantageous to consider C/T plus amikacin as an empirical therapy.

To our knowledge, this is the first study to determine the in vitro susceptibility and synergistic effect of C/T against CRPA isolates in Thailand. However, several limitations were encountered. First, the BMD method, which is the gold standard of antimicrobial susceptibility testing, was not performed for C/T. However, the E-test method for C/T is a simple and acceptable method for susceptibility testing [31]. Second, the E-test crossing method was selected as the synergistic testing method in this study. It is widely used in clinical practice because it is easy to perform. However, the other methods for synergistic study, especially time-kill assay as the gold standard, can be evaluated. Third, the genotypic resistance characteristics of the CRPA isolates were not investigated. Thus, the molecular basis of the characteristics of these CRPA strains, which might contribute to a better understanding the results, were not explored. Fourth, this was a limited single- center study, which may affect the generalizability of the CFR results. Thus, our findings should be appraised and compared with other cohorts. Furthermore, we recommend that a nationwide multicenter study using standardized antimicrobial susceptibility testing methods based on various types of CRPA should be undertaken. Fifth, the antibiotic dosing regimens were simulated using typical pharmacokinetic parameters [28]. Thus, antibiotic dosing regimens for C/T and for amikacin based on creatinine clearance should be determined and monitored by the therapeutic drug monitoring (TDM) for amikacin in order to be more effective and less nephrotoxic. Finally, our clinical CRPA isolates were mostly susceptible to amikacin; thus, the optimal amikacin dosing recommendation in the combination therapy should be used if any CRPA isolates are susceptible to amikacin. Despite these limitations, this study provides essential information for treating *P. aeruginosa* in clinical practice, particularly CRPA, where treatment options are extremely limited. In summary, C/T had a fair synergistic effect with amikacin and may be considered as a combination therapy in CRPA infection.

4. Materials and Methods

4.1. Bacterial Identification and Antimicrobial Susceptibility Test

From January–December 2020, CRPA isolates were collected from patients by the microbiology laboratory at Phramongkutklao Hospital, a 1200-bed teaching hospital of the Phramongkutklao College of Medicine, Royal Thai Army, Bangkok, Thailand. All studied CRPA isolates were identified as *P. aeruginosa* using matrix-assisted laser desorption ionization time-of-flight mass spectrometry.

Antimicrobial susceptibility tests were determined by the standard BMD method using frozen 96-well plates (THAN1F, Sensititre^{TM}, Thermo Fisher Scientific, Waltham, MA, USA). The colonies were picked up, suspended in distilled water, and adjusted the turbidity to 0.5 McFarland. Next, 10 µL of the prepared bacterial suspension (~10^8 cfu/mL) was diluted in cation-adjusted Mueller Hinton Broth (Sensititre^{TM} cation-adjusted Mueller Hinton Broth with TES; TREK Diagnostic Systems Ltd., East Grinstead, West Sussex, UK) at ~1:1000 dilution. The final bacterial suspension for testing contained an inoculum density of ~10^5 cfu/mL. The antimicrobial concentrations tested were as follows: amikacin, 8–32 µg/mL; cefepime, 1–32 µg/mL; ceftazidime, 1–32 µg/mL; ciprofloxacin, 0.06–2 µg/mL; colistin, 1–4 µg/mL; gentamicin, 2–8 µg/mL; imipenem, 0.5–8 µg/mL; meropenem, 0.5–8 µg/mL; piperacillin/tazobactam, 8/4–64/4 µg/mL; levofloxacin 0.06–8 µg/mL. The clinical isolates of *P. aeruginosa* were considered carbapenem-non-susceptible if their MIC values for meropenem or imipenem were $\geq$4 µg/mL [32].

The clinical isolates of CRPA were selected and underwent antimicrobial susceptibility testing against C/T and amikacin. The MIC values of each studied antibiotic were determined using E-test strips (Liofilchem, Inc., Waltham, MA, USA). A purified colony of isolated strains was picked up and suspended to 0.9% in normal saline (Univar$^®$, Ajax Finechem Pty Ltd., Taren Point, Australia) as 0.5 McFarland standard. The susceptibility data of C/T and amikacin against the CRPA isolates were collected and analyzed. The MICs of C/T and amikacin ranged from 0.008/4–128/4 µg/mL to 0.016–256 µg/mL, respectively. The results were interpreted according to the criteria of the Clinical and Laboratory Standards Institute (CLSI) [32]. All CRPA strains were stored at $-80\ °C$ until analysis. *P. aeruginosa* ATCC 27853 was used as a reference strain for quality control.

MIC values, MIC$_{50}$, and MIC$_{90}$ were measured for a tested population. MIC$_{50}$ and MIC$_{90}$ are defined as the MIC values inhibiting 50% and 90% of the tested isolates, respectively.

4.2. Synergy Test of C/T against CRPA

We designed a synergy test for C/T with amikacin against CRPA. The MIC values of each studied antibiotic were initially determined in order to further perform synergistic testing. For the synergy test, two E-test strips of studied antibiotics crossing formation in each MIC value with 90° angle were placed on an inoculated Mueller–Hinton Agar plate with bacteria spread. The resulting ellipse of inhibition was checked after 16–18 h at $35 \pm 2\,°C$.

The fractional inhibitory concentration index (FICI) was calculated for each antibiotic in each combination using the following formula: FICA + FICB = $\sum$FICI, where FICA equals the MIC of drug A in combination divided by the MIC of drug A alone, and FICB equals the MIC of drug B in combination divided by the MIC of drug B alone. The $\sum$FICI were interpreted as follows: synergy, FICI $\leq$ 0.5; additive effect, FICI > 0.5–$\leq$ 1; no interaction (indifference), FICI > 1–$\leq$ 4; or antagonism, FICI > 4 [33,34].

4.3. Phenotypic Classification

The CRPA isolates were classified as MDR (resistant to at least one agent in three or more antimicrobial categories), XDR (resistant to at least one agent in all but two or fewer antimicrobial categories), or PDR (resistant to all agents in all antimicrobial categories) based on the CLSI criteria described by Magiorakos et al. [35].

4.4. Antibiotic Dosing Regimen Simulations

We used two-compartment pharmacokinetic models of C/T and amikacin with linear elimination to simulate the plasma concentration time [28,36]. Pharmacokinetic parameters, the PK/PD targets, and indices for the simulation are described in Table 8.

Table 8. Pharmacokinetic parameters, the PK/PD targets, and indices of ceftolozane/tazobactam (C/T) and amikacin used for the simulation.

Antibiotics	Parameters	Mean	SD	%RSE	PK/PD Targets and Indices
Ceftolozane	V (L)	20.4	3.7	-	
	Kcp (h^{-1})	0.46	0.74	-	$75\%\,fT > MIC$,
	Kpc (h^{-1})	0.39	0.37	-	$100\%\,fT > MIC$
	CL (L/h)	7.2	3.2	-	
Tazobactam	V (L)	32.4	10	-	
	Kcp (h^{-1})	2.96	8.69	-	
	Kpc (h^{-1})	26.5	8.4	-	$20\%\,fT \geq 1\ \mu g/mL$
	CL (L/h)	25.4	9.4	-	
Amikacin	CL (L/h)	0.77	-	28.4	
	V (L)	19.2	-	5.31	
	Q (L/h)	4.38	-	18.3	$C_{max}/MIC \geq 8$
	V$_p$ (L)	9.38	-	7.15	

Abbreviations: V, typical volume of distribution of the central compartment; Vp, peripheral volume of distribution; Kcp, rate constant for distribution of unbound ceftolozane or tazobactam from central to peripheral compartment; Kpc, rate constant for distribution of unbound ceftolozane or tazobactam from peripheral to central compartment; CL, clearance; SD, standard deviation; PK/PD, pharmacokinetics/pharmacodynamics; MIC, minimum inhibitory concentration; RSE, relative standard error.

Simulated dosing regimens of C/T with log-normal distributions included intermittent infusion (1.5 g infusion over 0.5 h every 8 h) and extended infusion (1.5 g infusion over 4 h every 8 h and 1.5 g loading dose over 0.5 h, followed by 4.5 g continuous infusion over 24 h). The simulated dosing regimens of amikacin with log-normal distributions included a loading dose of 20–25 mg/kg, followed by maintenance doses of 15–20 mg/kg every 24 h.

The optimal antibiotic dosing regimens were simulated using a 10,000-subject Monte Carlo simulation (Oracle Crystal Ball Classroom Faculty Edition-Oracle 1-Click Crystal Ball 201, Thailand). PK/PD targets were set for each studied antimicrobial. The percentages of targets for the duration of time (fT) that the free drug concentration had to remain above the MIC ($fT > MIC$) were 75% and 100% for ceftolozane, and the target for tazobactam was $20\%\,fT \geq 1\ \mu g/mL$. The target ratio between the maximum drug concentration obtained after a single dose and the MIC (C_{max}/MIC) was ≥ 8 for amikacin.

The probability of target attainment (PTA) was determined as the percentage of probability at which the pharmacodynamic indices with specific MICs were achieved. The cumulative fraction of response (CFR) was calculated as the proportion of % PTA for each MIC of each of the pharmacodynamic indices according to the MIC distribution [37]. At least 90% PTA at a steady state for documented therapy and 90% CFR were considered the achievement targets for empirical therapy.

5. Conclusions

The CRPA clinical isolates from Thailand in our study showed high in vitro susceptibility (86.36%) to C/T. Furthermore, a 40% synergistic effect was observed in combination with amikacin. C/T extended infusion regimens may be considered empirical or definite antibiotic therapies when CRPA is suspected or detected. Amikacin with a loading dose of 20 mg/kg, followed by 15 mg/kg/day, seems an attractive combination agent with C/T when combination therapy is necessary.

Author Contributions: Conceptualization, W.N., D.C. and W.S.; methodology, W.N., D.C. and W.S.; software, W.S.; validation, P.N., S.P. and W.S.; formal analysis, P.N. and W.S.; investigation, P.N., S.P. and W.S.; resources, P.H., T.C.; data curation, P.N., S.P. and W.S.; writing—original draft preparation, W.N. and P.N.; writing—review and editing, W.N., D.C., P.N., S.P., P.H., T.C. and W.S.; visualization, W.N., D.C. and W.S.; supervision, W.N., D.C. and W.S.; project administration, W.N. and W.S.; funding acquisition, W.N. and W.S. All authors have read and agreed to the published version of the manuscript.

Funding: This research was funded by the Research and Creativity Fund, Faculty of Pharmacy, Silpakorn University (RAF 009/2564).

Institutional Review Board Statement: The study was conducted according to the guidelines of the Declaration of Helsinki and approved by the Royal Thai Army Medical Department and Phramongkutklao Hospital, Bangkok, Thailand (the Ethics number: R206b/62_Exp).

Informed Consent Statement: Patient consent was waived due to in vitro research being of minimal risk to subjects.

Data Availability Statement: Data available on request due to restrictions.

Acknowledgments: This research was financially supported through RAF 009/2564 from Faculty of Pharmacy, Silpakorn University.

Conflicts of Interest: The authors declare no conflict of interest.

References

1. Zilberberg, M.D.; Shorr, A.F. Prevalence of multidrug-resistant pseudomonas aeruginosa and carbapenem-resistant enterobacteriaceae among specimens from hospitalized patients with pneumonia and bloodstream infections in the United States from 2000 to 2009. *J. Hosp. Med.* **2013**, *8*, 559–563. [CrossRef] [PubMed]
2. Castanheira, M.; Deshpande, L.M.; Costello, A.; Davies, T.A.; Jones, R.N. Epidemiology and carbapenem resistance mechanisms of carbapenem-non-susceptible Pseudomonas aeruginosa collected during 2009–2011 in 14 European and Mediterranean countries. *J. Antimicrob. Chemother.* **2014**, *69*, 1804–1814. [CrossRef] [PubMed]
3. Lister, P.D.; Wolter, D.J.; Hanson, N.D. Antibacterial-Resistant Pseudomonas aeruginosa: Clinical Impact and Complex Regulation of Chromosomally Encoded Resistance Mechanisms. *Clin. Microbiol. Rev.* **2009**, *22*, 582–610. [CrossRef] [PubMed]
4. Furtado, G.H.C.; D'Azevedo, P.A.; Santos, A.F.; Gales, A.; Pignatari, A.C.C.; Medeiros, E.A.S. Intravenous polymyxin B for the treatment of nosocomial pneumonia caused by multidrug-resistant Pseudomonas aeruginosa. *Int. J. Antimicrob. Agents* **2007**, *30*, 315–319. [CrossRef]
5. Vidal, L.; Gafter-Gvili, A.; Borok, S.; Fraser, A.; Leibovici, L.; Paul, M. Efficacy and safety of aminoglycoside monotherapy: Systematic review and meta-analysis of randomized controlled trials. *J. Antimicrob. Chemother.* **2007**, *60*, 247–257. [CrossRef]
6. Farrell, D.J.; Flamm, R.K.; Sader, H.S.; Jones, R.N. Antimicrobial Activity of Ceftolozane-Tazobactam Tested against Enterobacteriaceae and Pseudomonas aeruginosa with Various Resistance Patterns Isolated in U.S. Hospitals (2011–2012). *Antimicrob. Agents Chemother.* **2013**, *57*, 6305–6310. [CrossRef]
7. Toussaint, K.A.; Gallagher, J.C. β-lactam/β-lactamase inhibitor combinations: From then to now. *Ann. Pharmacother.* **2015**, *49*, 86–98. [CrossRef]

8. Zhanel, G.G.; Chung, P.; Adam, H.; Zelenitsky, S.; Denisuik, A.; Schweizer, F.; Lagacé-Wiens, P.; Rubinstein, E.; Gin, A.S.; Walkty, A.; et al. Ceftolozane/Tazobactam: A Novel Cephalosporin/β-Lactamase Inhibitor Combination with Activity Against Multidrug-Resistant Gram-Negative Bacilli. *Drugs* **2013**, *74*, 31–51. [CrossRef]

9. Shortridge, D.; Pfaller, M.A.; Castanheira, M.; Flamm, R.K. Antimicrobial Activity of ceftolozane-tazobactam tested against Enterobacteriaceae and Pseudomonas aeruginosa with various resistance patterns isolated in U.S. Hospitals (2013–2016) as part of the surveillance program: Program to assess ceftolozane-tazobactam susceptibility. *Microb. Drug Resist.* **2018**, *24*, 563–577.

10. Dassner, A.M.; Sutherland, C.; Girotto, J.; Nicolau, D.P. In vitro Activity of Ceftolozane/Tazobactam Alone or with an Aminoglycoside Against Multi-Drug-Resistant Pseudomonas aeruginosa from Pediatric Cystic Fibrosis Patients. *Infect. Dis. Ther.* **2016**, *6*, 129–136. [CrossRef]

11. Jacqueline, C.; Howland, K.; Chesnel, L. In vitro activity of ceftolozane/tazobactam in combination with other classes of antibacterial agents. *J. Glob. Antimicrob. Resist.* **2017**, *10*, 326–329. [CrossRef] [PubMed]

12. Montero, M.; VanScoy, B.D.; López-Causapé, C.; Conde, H.; Adams, J.; Segura, C.; Zamorano, L.; Oliver, A.; Horcajada, J.P.; Ambrose, P.G. Evaluation of Ceftolozane-Tazobactam in Combination with Meropenem against Pseudomonas aeruginosa Sequence Type 175 in a Hollow-Fiber Infection Model. *Antimicrob. Agents Chemother.* **2018**, *62*, e00026-18. [CrossRef] [PubMed]

13. Rodríguez-Núñez, O.; Perianez-Parraga, L.; Oliver, A.; Munita, J.M.; Boté, A.; Gasch, O.; Nuvials, X.; Dinh, A.; Shaw, R.; Lomas, J.M.; et al. Higher MICs (>2 mg/L) Predict 30-Day Mortality in Patients With Lower Respiratory Tract Infections Caused by Multidrug- and Extensively Drug-Resistant Pseudomonas aeruginosa Treated With Ceftolozane/Tazobactam. *Open Forum Infect. Dis.* **2019**, *6*, ofz416. [CrossRef] [PubMed]

14. Tängdén, T.; Martín, V.R.; Felton, T.W.; Nielsen, E.I.; Marchand, S.; Brüggemann, R.J.; Bulitta, J.; Bassetti, M.; Theuretzbacher, U.; Tsuji, B.T.; et al. The role of infection models and PK/PD modelling for optimising care of critically ill patients with severe infections. *Intensiv. Care Med.* **2017**, *43*, 1021–1032. [CrossRef]

15. Asín-Prieto, E.; Rodríguez-Gascón, A.; Isla, A. Applications of the pharmacokinetic/pharmacodynamic (PK/PD) analysis of antimicrobial agents. *J. Infect. Chemother.* **2015**, *21*, 319–329. [CrossRef]

16. Pfaller, M.; Shortridge, D.; Sader, H.; Castanheira, M.; Flamm, R. Ceftolozane/tazobactam activity against drug-resistant Enterobacteriaceae and Pseudomonas aeruginosa causing healthcare-associated infections in the Asia-Pacific region (minus China, Australia and New Zealand): Report from an Antimicrobial Surveillance Programme (2013–2015). *Int. J. Antimicrob. Agents* **2018**, *51*, 181–189. [CrossRef]

17. Teo, J.Q.-M.; Lim, J.C.; Tang, C.Y.; Lee, S.J.-Y.; Tan, S.H.; Sim, J.H.-C.; Ong, R.T.-H.; Kwa, A.L.-H. Ceftolozane/Tazobactam Resistance and Mechanisms in Carbapenem-Nonsusceptible Pseudomonas aeruginosa. *mSphere* **2021**, *6*, 01026-20. [CrossRef]

18. Haidar, G.; Philips, N.J.; Shields, R.K.; Snyder, D.; Cheng, S.; Potoski, B.A.; Doi, Y.; Hao, B.; Press, E.G.; Cooper, V.; et al. Ceftolozane-Tazobactam for the Treatment of Multidrug-Resistant Pseudomonas aeruginosa Infections: Clinical Effectiveness and Evolution of Resistance. *Clin. Infect. Dis.* **2017**, *65*, 110–120. [CrossRef]

19. Galani, I.; Papoutsaki, V.; Karantani, I.; Karaiskos, I.; Galani, L.; Adamou, P.; Deliolanis, I.; Kodonaki, A.; Papadogeorgaki, E.; Markopoulou, M.; et al. In vitro activity of ceftolozane/tazobactam alone and in combination with amikacin against MDR/XDR Pseudomonas aeruginosa isolates from Greece. *J. Antimicrob. Chemother.* **2020**, *75*, 2164–2172. [CrossRef]

20. Gaudereto, J.J.; Neto, L.V.P.; Leite, G.C.; Espinoza, E.P.S.; Martins, R.C.R.; Villas Boa Prado, G.; Rossi, F.; Guimarães, T.; Levin, A.S.; Costa, S.F. Comparison of methods for the detection of in vitro synergy in multidrug-resistant gram-negative bacteria. *BMC Microbiol.* **2020**, *20*, 97. [CrossRef]

21. Kohanski, M.A.; Dwyer, D.J.; Collins, J.J. How antibiotics kill bacteria: From targets to networks. *Nat. Rev. Microbiol.* **2010**, *8*, 423–435. [CrossRef] [PubMed]

22. Fiore, M.; Corrente, A.; Pace, M.C.; Alfieri, A.; Simeon, V.; Ippolito, M.; Giarratano, A.; Cortegiani, A. Ceftolozane-Tazobactam Combination Therapy Compared to Ceftolozane-Tazobactam Monotherapy for the Treatment of Severe Infections: A Systematic Review and Meta-Analysis. *Antibiotics* **2021**, *10*, 79. [CrossRef] [PubMed]

23. So, W.; Shurko, J.; Galega, R.; Quilitz, R.; Greene, J.N.; Lee, G.C. Mechanisms of high-level ceftolozane/tazobactam resistance in Pseudomonas aeruginosa from a severely neutropenic patient and treatment success from synergy with tobramycin. *J. Antimicrob. Chemother.* **2018**, *74*, 269–271. [CrossRef] [PubMed]

24. Roberts, J.A.; Abdul-Aziz, M.-H.; Lipman, J.; Mouton, J.W.; Vinks, A.A.; Felton, T.W.; Hope, W.W.; Farkas, A.; Neely, M.N.; Schentag, J.J.; et al. Individualised antibiotic dosing for patients who are critically ill: Challenges and potential solutions. *Lancet Infect. Dis.* **2014**, *14*, 498–509. [CrossRef]

25. Craig, W.A.; Andes, D.R. In Vivo Activities of Ceftolozane, a New Cephalosporin, with and without Tazobactam against Pseudomonas aeruginosa and Enterobacteriaceae, Including Strains with Extended-Spectrum β-Lactamases, in the Thighs of Neutropenic Mice. *Antimicrob. Agents Chemother.* **2013**, *57*, 1577–1582. [CrossRef]

26. Lepak, A.J.; Reda, A.; Marchillo, K.; Van Hecker, J.; Craig, W.A.; Andes, D. Impact of MIC Range for Pseudomonas aeruginosa and Streptococcus pneumoniae on the Ceftolozane In Vivo Pharmacokinetic/Pharmacodynamic Target. *Antimicrob. Agents Chemother.* **2014**, *58*, 6311–6314. [CrossRef]

27. Natesan, S.; Pai, M.P.; Lodise, T.P. Determination of alternative ceftolozane/tazobactam dosing regimens for patients with infections due to Pseudomonas aeruginosa with MIC values between 4 and 32 mg/L. *J. Antimicrob. Chemother.* **2017**, *72*, 2813–2816. [CrossRef]

28. Sime, F.B.; Lassig-Smith, M.; Starr, T.; Stuart, J.; Pandey, S.; Parker, S.L.; Wallis, S.C.; Lipman, J.; Roberts, J.A. Population Pharmacokinetics of Unbound Ceftolozane and Tazobactam in Critically Ill Patients without Renal Dysfunction. *Antimicrob. Agents Chemother.* **2019**, *63*, e01265-19. [CrossRef]
29. Kratzer, A.; Rothe, U.; Dorn, C. NP-008 Stability of ceftolozane/tazobactam in solution as infusion for prolonged or continuous application. *Eur. J. Hosp. Pharm.* **2019**, *26*, A294. [CrossRef]
30. Kato, H.; Hagihara, M.; Hirai, J.; Sakanashi, D.; Suematsu, H.; Nishiyama, N.; Koizumi, Y.; Yamagishi, Y.; Matsuura, K.; Mikamo, H. Evaluation of Amikacin Pharmacokinetics and Pharmacodynamics for Optimal Initial Dosing Regimen. *Drugs R&D* **2017**, *17*, 177–187. [CrossRef]
31. Shields, R.K.; Clancy, C.J.; Pasculle, A.W.; Press, E.G.; Haidar, G.; Hao, B.; Chen, L.; Kreiswirth, B.N.; Nguyen, M.H. Verification of Ceftazidime-Avibactam and Ceftolozane-Tazobactam Susceptibility Testing Methods against Carbapenem-Resistant Enterobacteriaceae and Pseudomonas aeruginosa. *J. Clin. Microbiol.* **2018**, *56*, e01093-17. [CrossRef] [PubMed]
32. Clinical and Laboratory Standard Institute (CLSI). Performance Standards for Antimicrobial Susceptibility Testing; 31st Informational Supplement. CLSI Document M100-MS20. Available online: https://clsi.org/ (accessed on 18 May 2021).
33. Doern, C.D. When Does 2 Plus 2 Equal 5? A Review of Antimicrobial Synergy Testing. *J. Clin. Microbiol.* **2014**, *52*, 4124–4128. [CrossRef] [PubMed]
34. Odds, F.C. Synergy, antagonism, and what the chequerboard puts between them. *J. Antimicrob. Chemother.* **2003**, *52*, 1. [CrossRef] [PubMed]
35. Magiorakos, A.-P.; Srinivasan, A.; Carey, R.B.; Carmeli, Y.; Falagas, M.E.; Giske, C.G.; Harbarth, S.; Hindler, J.F.; Kahlmeter, G.; Olsson-Liljequist, B.; et al. Multidrug-resistant, extensively drugresistant and pandrug-resistant bacteria: An international expert proposal for interim standard definitions for acquired resistance. *Clin. Microbiol. Infect.* **2012**, *18*, 268–281. [CrossRef] [PubMed]
36. Delattre, I.K.; Musuamba, F.T.; Nyberg, J.; Taccone, F.S.; Laterre, P.-F.; Verbeeck, R.K.; Jacobs, F.; Wallemacq, P.E. Population Pharmacokinetic Modeling and Optimal Sampling Strategy for Bayesian Estimation of Amikacin Exposure in Critically Ill Septic Patients. *Ther. Drug Monit.* **2010**, *32*, 749–756. [CrossRef] [PubMed]
37. Mouton, J.W.; Dudley, M.N.; Cars, O.; Derendorf, H.; Drusano, G.L. Standardization of pharmacokinetic/pharmacodynamic (PK/PD) terminology for anti-infective drugs: An update. *J. Antimicrob. Chemother.* **2005**, *55*, 601–607. [CrossRef] [PubMed]

MDPI
St. Alban-Anlage 66
4052 Basel
Switzerland
www.mdpi.com

Antibiotics Editorial Office
E-mail: antibiotics@mdpi.com
www.mdpi.com/journal/antibiotics

9 783036 591094